Insight into Optics

Oliver S. Heavens

Oliver S. Heavens is professor of physics in the University of York. His BSc and DSc degrees are from the University of London and his PhD from the University of Reading. After several years in industrial research he joined the University of Reading as a demonstrator in 1947, then lecturer. In 1957 he moved to a readership at Royal Holloway College, with Sam Tolansky, pursuing long-standing interests in thin film and surface physics. 1959–60 was spent working with Charles Townes in Columbia University, New York, trying (unsuccessfully) to make the world's first laser work. He became the first Head of Department of Physics at the University of York in 1964, establishing research groups in lasers, X-ray crystallography, surface physics, geomagnetism and others. His current interests include, in addition to thin film optical studies, the development of internal reflectance fluoresence methods for protein adsorption studies. He has lectured in France, North America, the USSR, Australia, China and Eastern Europe.

Robert W. Ditchburn

Robert W. Ditchburn entered the University of Liverpool at the age of 16, graduating at 19, and obtained a major entrance scholarship to Trinity College Cambridge where after two years he passed the Tripos examination. He worked with J. J. Thomson, obtaining his PhD in 1928, the same year that he became a fellow of Trinity College Dublin. Apart from wartime secondment to the Admiralty Research Laboratory, Teddington, from 1942–45, he remained at Dublin until appointment to the Chair of Physics at the University of Reading in 1946. He established active groups in spectroscopy and physiological optics and presided over the department during a period of immense expansion. He retired in 1968, but continued to write until his death in 1987.

Insight into Optics

O. S. Heavens

Department of Physics, University of York

and the late
R. W. Ditchburn

JOHN WILEY & SONS

Chichester · New York · Brisbane · Toronto · Singapore

Other Wiley Editorial Offices

John Wiley & Sons, Inc., 605 Third Avenue,
New York, NY 10158-0012, USA

Jacaranda Wiley Ltd, G.P.O. Box 859, Brisbane,
Queensland 4001, Australia

John Wiley & Sons (Canada) Ltd, 22 Worcester Road,
Rexdale, Ontario M9W 1L1, Canada

John Wiley & Sons (SEA) Pte Ltd, 37 Jalan Pemimpin 05-04,
Block B, Union Industrial Building, Singapore 2057

Library of Congress Cataloging-in-Publication Data:
Heavens, O. S.
　　Insight into optics / O. S. Heavens and the late R. W. Ditchburn.
　　　　p.　　cm.
　　Includes bibliographical references and index.
　　ISBN 0 471 92769 4—ISBN 0 471 92901 8 (pbk.)
　　1. Optics.　I. Ditchburn, R. W.　II. Title.
　　QC358.H43　　1991
　　535—dc20　　　　　　　　　　　　　　　90-43729
　　　　　　　　　　　　　　　　　　　　　　CIP

British Library Cataloguing in Publication Data:
Heavens, O. S.
　Insight into optics.
　1. Optics
　I. Title　II. Ditchburn, R. W. (Robert), *d. 1987*
　535

　ISBN 0 471 92769 4 (cloth)
　ISBN 0 471 92901 8 (paper)

Typeset in 10/11pt Palatino by Mathematical Composition Setters Ltd, Salisbury, Wiltshire
Printed and bound in Great Britain by Courier International, Tiptree, Essex

Contents

Preface

The subject of optics continues to develop at a pace that requires frequent updating of textbooks suitable for the undergraduate. This book started as a joint venture but, sadly, Robert Ditchburn died in April 1987. Although much had been written by that time, there was no opportunity for a joint review of the material before the book finally went to press.

The changes taking place in the educational scene are such that departments of physics will in due course be dealing with incoming students with a more limited experience of their subject than has been the case in the past. This text therefore starts at an elementary level and should be suitable for students whose previous grounding in physics is limited to that likely to be covered in an integrated science course. It is, however, assumed that the student will be following, concurrently with the optics course, an appropriate course of mathematics. Thus familiarity with complex numbers and vectors will be assumed.

Much of the material in the first nine chapters may be thought of as laying the foundations for what follows. Chapters 10 to 13 cover topics whose importance has increased enormously since the arrival of the laser. Chapter 14 heralds the impending takeover of the communication field from electrons to photons, and Chapter 18 deals with the classical ideas of non-linear optics and introduces the exciting developments resulting, again, from the availability of the laser. If the student does not feel a sense of wonder at the subject of phase-conjugation the writer has failed! Three chapters on various aspects of optical measurements follow, illustrating the very wide range of types of measurement that are possible using optical means. How do we know how well optical instruments perform? Chapter 19 reveals how we go beyond the levels of subjective judgement in a scientific approach to the question of assessment. The second of the chapters dealing with lasers takes a closer look at the details of operation of this remarkable device and examines a number of important applications, highlighting the dramatic advances in spectroscopy which have been made possible.

In an elementary introduction to any subject we generally need to make some simplifying assumptions. Chapter 21 brings us closer to reality by tackling the problem of the real, as opposed to the idealised, light source. We see there that this is no mere curiosity, pursued only for the sake of completeness, but that it leads to new and exciting possibilities of immense practical importance.

We may need reminding that the present, phenomenally accurate, systems of navigation which we take for granted need to take account of the effects of relativity and would be seriously in error if these were ignored. Such matters form the subject of Chapter 22. In Chapter 23 we see that in the same way that the behaviour of matter can be explained only by recourse to quantum theory, the same situation exists for radiation.

At this stage we shall be aware of the definitive contribution that optics has made to other parts of physics and to other disciplines. Is there no limit to the power of optical experiments to reveal more and finer detail of what is being examined? There are limits and we look at these in the concluding chapter—and may be surprised!

There are some optical phenomena that really need the use of colour for a full appreciation. The decision to exclude colour (on ground of cost) was made on the assumption that those for whom the book is intended will have access to laboratories and will, as a necessary part of any optics course, be exposed to experiments and demonstrations.

A modest number of examples is included, often with a view to expanding on or clarifying the text. Instructor's attitudes to examples are sometimes like those towards other people's lecture notes—an overwhelming preference to produce one's own.

O. S. Heavens
York, May 1990

General Background

1.1 Why study optics?

It might be argued that at least the part of optics concerned with visible light is worth studying on the grounds that many optical manifestations are aesthetically pleasing. One has only to stick a piece of drawn-out glass rod in a laser beam to produce a beautiful swirl of coloured fringes on the wall. In fact, as we shall see throughout this book, the subject of optics not only fits elegantly and logically into the general physical scene, but has enabled immense strides to be made in our understanding of many areas of physics. It is inconceivable that we could have reached our present deep and detailed understanding of atomic structure without the contributions that optical spectroscopy has made. The rapid verification of the predictions of the theory of relativity was made with the help of optics and the development of biology without the assistance of the microscope would have been severely retarded. However, far from being merely (boring and) useful the study of optics has an inherent fascination: its aesthetic aspects are a bonus.

1.2 Historical background

We can afford no more space than to paint a general picture of the history of the development of optical ideas, which we do with the help of Tables 1.1 and 1.2. Table 1.1 takes us up to around the middle of the present century whilst the latter highlights the more recent progress as exhibited by the Nobel Prizes earned by the optics fraternity. There was in mid-century perhaps a feeling that the field of optics would be unlikely to provide any significant new excitements. Maxwell had given a comprehensive theory underpinning the behaviour of *all* kinds of radiation. Light, or the optical region, formed a miniscule slice of a spectrum of enormous width, from the γ-rays with wavelengths much less than the sizes of atoms to the waves that sail around the Earth and have wavelengths of tens of thousands of kilometres. From our point of view the visible region of this vast spectrum *is* of some importance: it is, after all, the only bit that we can see.

The order in Table 1.1 is by date of birth, not by the date at which the contributions appeared. Also the list of topics is not comprehensive: many of the people quoted made contributions over many areas of optics.

The Nobel prizewinners in the period from 1950 are given in Table 1.2.

Table 1.1

Date	Where or who	What
BC 1900	Ancient Egypt	Mirrors
424	Aristophanes	Burning glass
300	Euclid	Rectilinear propagation
AD 60	Seneca	Refraction
70	Pliny	
130	Ptolemy	Angles of reflection, refraction
1000	Alhazen	Spherical, parabolic mirrors
1215–1294	Bacon	Spectacles
1452–1519	Leonardo da Vinci	Cameras obscura
1535–1615	Della Porta	
1564–1642	Galilei	Refracting telescope
1587–1619	Lippershez	
1571–1630	Kepler	Total internal reflection
1580–1656	Fontana	Eyepiece lens
1588–1632	Janssen(?)	Microscope
1591–1626	Snell	Refraction law
1596–1650	Descartes	
1601–1665	Fermat	Principle of least time
1618–1663	Grimaldi	Diffraction
1629–1695	Huygens	Early ideas of wavefronts
1635–1703	Hooke	Thin film colours
1642–1727	Newton	Nature of light
		Dispersion
1644–1710	Römer	Speed of light
1693–1762	Bradley	Aberration of starlight
1698–1765	Klingenstjerna	Achromatic lens
1706–1761	Dollond	
1707–1783	Euler	
1766–1828	Wollaston	Solar absorption lines
1773–1829	Young	Wave theory
1775–1812	Malus	Polarisation
1786–1853	Arago	Solar absorption lines
1787–1826	Fraunhofer	Absorption lines, diffraction
1788–1827	Fresnel	Interference, wave theory
1791–1867	Faraday	Light/electromagnetism
1801–1892	Airy	Aberration, diffraction
1819–1896	Fizeau	Speed of light
1819–1868	Foucault	
1831–1879	Maxwell	Electromagnetic theory
1838–1923	Morley	Speed of light, aether
1852–1931	Michelson	
1853–1928	Lorentz	Aether theory
1854–1912	Poincaré	Aether
1857–1894	Hertz	Electromagnetic waves
1858–1947	Planck	Quanta
1879–1955	Einstein	Relativity
1885–1962	Bohr	H atom spectrum, model

Table 1.2 Nobel prizes in optics

When	Who	For what
1955	Lamb	Fine structure of hydrogen spectrum
1958	Cherenkov, Frank and Tamm	Cherenkov radiation
1964	Townes, Prokhorov and Basov	Laser
1966	Kastler	Resonance in atoms by optical methods
1971	Gabor	Holography
1981	Bloembergen and Schawlow	Laser spectroscopy

1.3 The scope of the subject

The choice of title of this book—'optics', rather than 'light'—arises essentially from the fact that the spectral range covered is much broader than the narrow slice of the spectrum perceived as 'light'. Many of the experimental methods used for the visible spectrum are equally useful for a considerable range on either side of the visible band. Yet we must not stray too far. For all the universal nature of the electromagnetic theory, the methods used for the study of γ-rays and of very long wavelength radio waves are significantly different from those most successfully employed for the visible. The methods used to produce microscope lenses of quite staggering performance are not available to us for the X-ray region, for example, where for practical purposes the refractive indices of all materials are unity. Similarly, a Fabry–Perot interferometer, requiring high-reflecting, parallel plates whose diameters are an enormous number of wavelengths, would be somewhat bulky for radio waves of kilometre wavelengths. Yet we shall need to consider matters that would have been absent from an optics textbook of a few decades past. In some experiments we stray into the technology of the γ-ray physicist, but counting photons rather than detecting waves. Elsewhere, we shunt optical light pulses around rather in the way that the electronics worker has shunted electrical pulses. It is not entirely fanciful to speculate that the use of electrons for the digital computer may

prove to have been a temporary phase in its development. There appear distinct advantages of using *photons* rather than electrons for this purpose. Not only do they mind their own business and so ignore one another if they cross paths, but the process of producing and fiddling with two-dimensional images represents a simple but striking example of parallel processing, towards which the electronically based computer is now striving, albeit with difficulty.

1.4 Characteristics of optical radiation

It could be argued that this section should come at the end rather than at the beginning of the book since by then the answer would be clear. Some general comments are apposite. Radiation is a form of energy that is transformable into other forms. We can have a light beam at one moment and a current of electrons later, by which time the original radiation has disappeared. The radiation energy is in motion, moving quite fast, and it carries with it information about the source that produced it and about what has happened to it in transit. It is thus a means of transferring information from one place to another. On being converted into another form—e.g. chemical energy as a result of interacting with the human retina, electrical energy in the form of photoelectric currents, magnetic energy in the form of a magnetic memory device—the result is the storage or display of the information carried by the light in a different form. The information-carrying aspect of light assumes enormous importance inasmuch as the number of waves passing a point in space per second is very large and the speed of the wave is also very high. These two features mean that the *amount of information* that can be transferred per unit time is also very large—far higher than that of all other methods which have evolved over the last century or so.

1.5   Waves and rays

The conflict in the seventeenth century on what light 'consisted of' was not surprising. Simple experiments gave strong evidence that light travelled in straight lines, as witnessed by the casting of shadows. There was little that could be described as 'wave-like'. On the other hand, the occurrence of interference effects ('fringes') was highly suggestive of a wave-like behaviour, as was the evidence that the edges of shadows—even if the light source were a minute point—were not in fact perfectly sharp. The light spread into the shadow in very much the way that water waves spread round, for example, a harbour wall. The pinhole camera illustrates this simple point. For a suitable size of pinhole, a reasonable picture (Fig. 1.1a) is obtained. Understandably, if the pinhole is made larger a fuzzier result (Fig. 1.1b) is obtained since light from the same point of the object can reach a finite area of the screen, so smudging the image. If the pinhole is made *smaller*, then the 'straight-line' notion indicates that the picture should be *sharper*, although less intense. In fact, the picture becomes fuzzier, as seen in Fig. 1.1(c), a consequence of the spreading, or diffraction, of the light passing through the pinhole.

The interference fringes such as appear in Fig. 4.6 show smooth undulations, consistent with the blending of wave-like disturbances. They may be photographed and show smooth, continuous gradations from light to dark. What happens, however, if we use beams of light of very *low* intensity? We obtain the results shown in Fig. 1.2. We do *not* get continuous gradations, but discrete spots. As the intensity increases, the spots become more numerous until they even-

Figure 1.1 Pinhole camera: (a) hole 0.10 mm diameter, (b) hole 1.0 mm diameter and (c) 0.06 mm diameter

tually merge so that we see continuous gradations. It would seem that although the light displays a wave-like character inasmuch as it produces interference effects, it can also behave in rather the way we consider that a particle behaves.

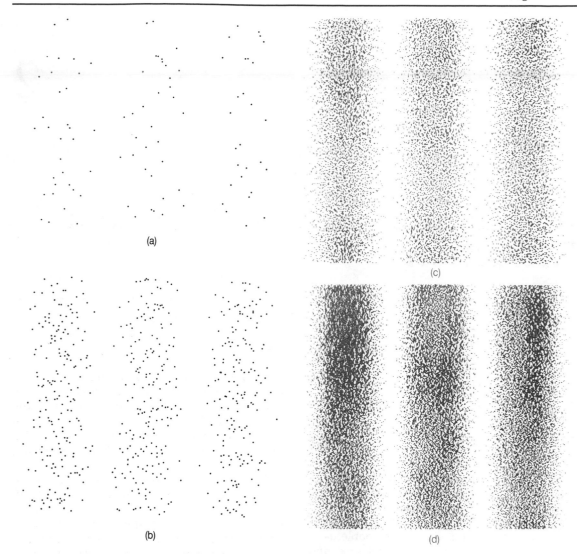

Figure 1.2 (a) to (d) Interference fringes for increasing values of intensity × time

1.6 Where to?

With the development of relativity and quantum mechanics we are able to account satisfactorily for all the experiments conducted so far on light (radiation) and it seems unlikely that present theories need a radical overhaul for the description of 'terrestrial' phenomena. As yet, however, we cannot know whether such theories are *truly* universal. Do they, for example, still apply to radiation in the neighbourhood of very dense masses? We are happy that a light beam grazing the limb of our Sun (density $\sim 1400 \, \mathrm{kg \, m^{-3}}$) will behave normally. We cannot be so sure that when grazing the limb of a pulsar (density $\sim 10^{18} \, \mathrm{kg \, m^{-3}}$) other, unexpected, effects might appear. Can we be sure that the not-inconsiderable distances that light travels in interstellar space do not themselves induce changes? Future cosmologists may be able to throw light on these matters.

Ray Optics (1): Reflection and Refraction

2.1 Introduction

In any situation in which light is to be examined or used it will be necessary to direct it, to focus it, to disperse it or to push it into an optical system or instrument. We do this with the help of of mirrors and lenses whose behaviour we can deduce from the fact that very narrow pencils of light (rays) travel in straight lines in uniform media and are reflected and refracted at surfaces according to simple laws. These enable us to establish the way in which mirrors and lenses, and systems of these components, act on light beams, thus enabling light to be used effectively. It could be said that the existence of subjects such as astronomy and biology rests squarely on our ability to handle beams of light in an effective way; the success or failure of a laser fusion system depends similarly on beam manipulation.

2.2 Principle of least time

If we examine the path of a light ray through an optical system we could ask: 'Why does it follow the path observed? Is there any simple, fundamental aspect that dictates that path? Why *that* path, rather than any other?'

Although rules for handling light rays evolved empirically over many centuries, it was Fermat (1601–1665) who established the basic principle, known as the 'principle of least time', that governed ray behaviour. He postulated that for all possible geometrical paths by which light might go between points A and B, the time taken for the actual path is a minimum. (He regarded this as an excellent example of the essential economy of nature! In fact, in some circumstances the time is a *maximum* rather than a minimum).

We can formulate Fermat's principle in a simple, elegant manner. The time taken for light to travel an infinitesimal distance ds in vacuum is $c\,ds$, where c is the speed of light in vacuum—a universal constant. In a material medium, the light velocity is c/n, where n is the refractive index of the medium. Thus the time for light to travel a distance ds in the medium is $c\,ds/n$ and the total time for a transit from A to B is

$$t_{AB} = \int_{A}^{B} \frac{c\,ds}{n} \qquad (2.1)$$

If AB is the actual light path, the time for infinitesimally differing neighbouring paths will be the same, to the first order, and we may express this as

$$\delta \int_A^B \frac{c\,ds}{n} = 0 \qquad (2.2)$$

where the δ, as in the calculus of variations, represents the change in the integral for an infinitesimal change in path. Equation (2.2) covers the case of both minima and maxima.

We draw two immediate conclusions from Fermat's principle, namely:

1. In a homogeneous medium (for which n is constant), rays are straight lines. The shortest distance between points A and B is the straight line AB.
2. Ray paths are reversible. If the integral for A to B satisfied eqn (2.2), then so does the integral from B to A.

The laws of reflection and refraction may be deduced from Fermat's principle (see Appendix 2.A).

2.3   Sign convention

We aim at establishing formulae that will enable us to determine the behaviour of the components (e.g. lenses and mirrors) in an optical system. In such a system, rays may be reflected or refracted at surfaces which may be plane, concave or convex. Unless we adopt a *sign convention*, we find that we need *different* formulae to cover different cases, a possible source of error and confusion. With a suitable sign convention, a single formula serves, for example, all cases of refraction at a surface.

The direction of light incident on a surface is taken as positive. Distances are measured *from* the surface (Fig. 2.1). For light travelling from left to right, a Cartesian system is used:

upwards and to the right are positive. Angles are positive if their tangents are positive and vice versa.

2.4   Reflection and refraction

Figure 2.2. shows what happens when a ray in a medium of index n_1 meets the boundary with a medium of index n_2. There are reflected and refracted rays, and the angles of incidence, reflection and refraction are as shown. The three rays and the normal to the surface at the point of incidence are coplanar. The relations between the angles are given by

$$n_2 \sin \theta_2 = n_1 \sin \theta_1 \qquad \text{(Snell's law)} \qquad (2.3)$$

and

$$\theta_1' = -\theta_1 \qquad (2.4)$$

The above relations apply both to plane *and* to *curved* surfaces.

From (2.4) we may write

$$n_1 \sin \theta_1' = (-n_1)\sin \theta_1 \qquad (2.5)$$

and comparing this with (2.3) suggests that reflection in medium n_1 at a surface is equivalent to refraction at a surface where the second medium has an index $(-n_1)$. Thus instead of deriving results separately for refraction and reflection, we need only deal with refraction and substitute $(-n_1)$ for n_2, resulting in equations describing reflection.

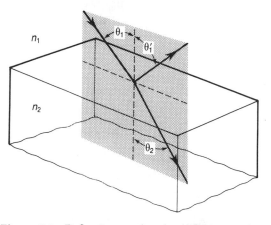

Figure 2.2 Reflection and refraction at a plane surface

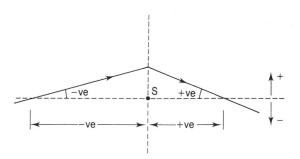

Figure 2.1 Sign convention

Thus $n_1/n_2(\equiv n_{12})$ is called the relative index of refraction between medium 1 and medium 2, n_1 and n_2 being the absolute indices. The refractive index of air at STP is about 1.0003. Thus the relative index from air to a medium differs by only 3 parts in 10^4 from the absolute index. Except for calculations of high accuracy, the difference between the relative and absolute index for the case of air and medium can be ignored.

2.5 Refraction by a prism

Figure 2.3 represents a ray passing through a prism of angle α between the two faces intersected by the ray, which lies in a plane perpendicular to the prism edge. The deviation δ of the ray is given by

$$\delta = \theta_1 - \theta_2 + \theta_4 - \theta_3 = (\theta_1 + \theta_4) - (\theta_2 + \theta_3) \quad (2.6)$$

As θ_1 is varied it is found that δ goes through a single minimum value (δ_m) which occurs when $\theta_1 = \theta_4$. (Reversibility of the ray direction ensures this. If a minimum occurred at $\theta_1 \neq \theta_4$, then one would also occur at θ_4, for a reversed ray.)

If $\theta_4 = \theta_1$, then $\delta_m = 2(\theta_1 - \theta_2)$ and $\alpha = 2\theta_2$, leading to $\alpha + \delta_m = 2\theta_1$. From Snell's law (eqn 2.3) we find, for a prism of index n in a medium of index n',

$$\frac{n}{n'} = \frac{\sin \frac{1}{2}(\alpha + \delta_m)}{\sin \frac{1}{2}\alpha} \quad (2.7)$$

Thus measurement of α and δ_m gives n/n'. For a measurement in air, the result is the refractive index of the prism material up to an accuracy of three decimal places.

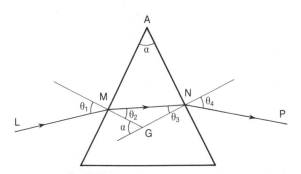

Figure 2.3 Refraction through a prism

2.6 Refraction by a thin prism

If α does not exceed a few degrees and θ_1 is small enough for the approximation $\sin \theta_1 \approx \theta_1$, to hold, then θ_2, θ_3 and θ_4 do not exceed a few degrees. For such angles, $n'\theta_1 = n\theta_2$ and $n\theta_3 = n'\theta_4$. Putting $n' = 1$ (air), eqn (1.6) gives

$$\delta = (n - 1)(\theta_2 + \theta_3) = (n - 1)\alpha \quad (2.8)$$

showing that the deviation depends only on n and α and is *independent of the angle of incidence*. The extent to which such a prism produces a deviation is called the power of the prism. A prism that produces a deflection of the beam of 1 cm at a distance of 1 m has a power of 1 dioptre (Fig. 2.4). One that produces a deflection of d cm at L m has a power of d/L dioptres.

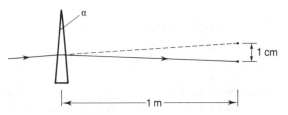

Figure 2.4 Refraction by a thin prism: power 1 dioptre

2.7 Dispersion

Newton showed that when a beam of white light is incident on a prism, the emergent light is dispersed into a spectrum. The deviation increases in the colour sequence red, orange, yellow, green, blue and violet (Fig. 2.5). Since colour is correlated with wavelength, and since the deviation δ is a function of refractive index, this shows that refractive index is a function of wavelength. The dispersion v of a material is defined as

$$v = \frac{n_Y}{n_F - n_C} \quad (2.9)$$

where n_C is the refractive index for light of wavelength 656 nm (red), n_F that for 486 nm (blue/green) and n_Y that for 587.6 nm (yellow). Values of v for a few materials are given in Table 2.1.

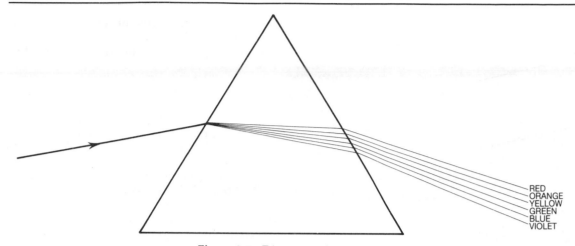

Figure 2.5 Dispersion by a prism

Table 2.1

Material	n_Y	v
Crown glass	1.5172	0.00573
Barium flint	1.5685	0.00797
Dense barium flint	1.6177	0.006923
Fused quartz	1.4585	0.00466

2.8 Total reflection

The refracting property of a medium is referred to as its *optical density*. The higher the refractive index, the higher the optical density. If light is incident on an interface from the denser of two media, then the angle of refraction is larger than the angle of incidence θ_1. As θ_1 is increased from zero, a point will be reached where $\theta_2 = \pi/2$: the refracted ray skids along the surface. The angle of incidence θ_c (critical angle) at which this occurs is given by

$$\theta_c = \text{arc } \sin\left(\frac{n_2}{n_1}\right) \qquad (2.10)$$

(from 2.3: $\theta_2 = \pi/2$).

For $\theta_1 > \theta_c$ there is no refracted ray and it is observed that incident light is totally reflected. We really *do* mean 'totally'. For a smooth interface between transparent media, no loss of intensity is observed even after more than 100 000 reflections, as a result of which light transmission through very long (tens of kilometres) optical fibres is possible (Chapter 14), thus providing us with an excellent means of long-distance communication.

Total reflection is employed in reflecting prisms for producing deviation of light beams—generally through 90° or 180° (Fig. 2.6). For a prism made by cutting the corner off a cube, as in Fig. 2.7, a beam incident on the diagonal face will, after total reflections, reemerge in a direction exactly parallel to the incident direction (and *not only* at normal incidence). A collection of such prisms was left on the Moon at the first Moon landing. We now send light pulses to the Moon, measure the time interval before they return, and so determine the Earth–Moon distance with very high accuracy.

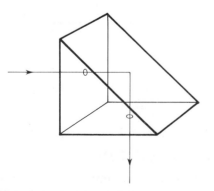

Figure 2.6 Total reflection at a prism: for this application the refractive index of the prism must exceed $(\sin 45°)^{-1} = \sqrt{2}$

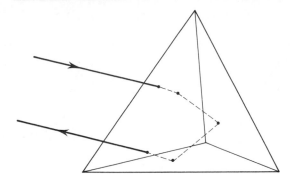

Figure 2.7   The corner-cube prism reflector

2.9   Fibre optics

If a light ray enters the end face of a glass cylinder or fibre, the refracted ray may strike the cylinder wall at an angle of incidence greater than critical, in which case total reflection will occur. The reflected ray will then again strike the wall at $\theta > \theta_c$ and total reflection will continue until the ray emerges from the opposite end of the cylinder. The reflection is total *only* if the surface is perfectly clean and smooth: dirty fingermarks play havoc. If, however, the fibre is coated with a layer of refractive index lower than that of the core, then the total reflection occurs inside the composite fibre (Fig. 2.8), so that dirt or disturbance at the outermost surface becomes irrelevant. The plain fibre is useless from a practical standpoint. The act of cladding transforms it into an immensely useful device, to be discussed in Chapter 14.

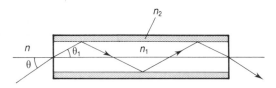

Figure 2.8   Propagation along a clad fibre

2.10   Numerical aperture of a coated fibre

If the index of a fibre core is n_1 and that of the cladding n_2, then for a ray that intersects the fibre axis ('meridional ray')

$$n \sin \theta = n_1 \sin \theta_1 \qquad (2.11)$$

where n is the index of the surrounding medium (Fig. 2.8). If the refracted ray strikes the n_1/n_2 interface at the critical angle, then $\theta_2 = \pi/2 - \theta_1$ and

$$n_1 \cos \theta_1 = n_2 \qquad (2.12)$$

Hence

$$n \sin \theta = (n_1^2 - n_1^2 \cos^2 \theta_1)^{1/2}$$
$$= (n_1^2 - n_2^2)^{1/2} \qquad (2.13)$$

where $n \sin \theta$ is termed the *numerical aperture* of the fibre (NA), by analogy with the theory of the microscope. The amount of light collected and transmitted by a fibre is proportional to $(NA)^2$, for practical values of NA.

The ray theory of fibres given above is a reasonable description for cases where the core diameter is greater than about ten times the light wavelength. For smaller diameters, the fibre needs to be treated as a waveguide (Chapter 14).

In cases where the ray description is valid, it would appear that, in the event of the fibre being bent, light could strike the core/cladding interface at less than the critical angle, in which case the light would leak out of the fibre. In practice, the curvature would need to be brutal for this to occur and the loss is negligible if the radius of curvature is more than ~ 10 times the fibre diameter. Since typical fibre diameters are of the order of tens of micrometres (or less), this forms no practical limitation.

2.11   Imaging through fibre bundles

If a bundle of fibres is made in such a way that their relative positions are accurately preserved throughout the length of the bundle, then when an image is projected on one end of the bundle, an identical image will occur at the opposite end (Fig. 2.9). Such a fibre bundle is termed 'coherent'—an unfortunate name, in view of the use of this word to characterise *light* properties (Chapter 5). Bundles in which relative fibre positions are not maintained are termed 'incoherent'; they may be used to transmit light, but not images.

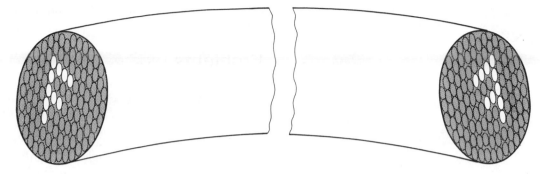

Figure 2.9   Imaging by a coherent fibre bundle

2.12   Intrascopes

There are situations in which it is necessary to observe what goes on in an inaccessible or hostile environment. One may well wish to observe the inside of a stomach (without making a large hole) or an active region of a nuclear reactor. Optical fibre bundles make this possible. With a combination of incoherent and coherent fibre bundles, the incoherent bundle serves to transmit light to the region of interest, and the coherent bundle relays images at the business end of the system to a viewing position where they can be examined. For medical use, in cystoscopes, gastroscopes, etc., one is limited to the maximum outside diameter of anything that is to be inserted into the interior of the human body. Hopkins has built systems capable of giving high-quality images using fibre bundles with outside diameters of ~2.5 mm.

Optical fibres play crucial role in optical communications. This is discussed in Chapter 14.

Appendix 2.A   Fermat's principle

2.A.1

For propagation in uniform media, rectilinear propagation and the laws of reflection and refraction are easily deduced from Fermat's principle. We illustrate the first and last of these, leaving the case of reflection as an exercise.

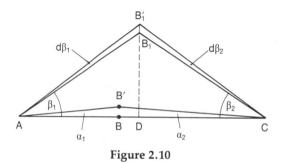

Figure 2.10

Figure 2.10 considers two possible paths, ABC and AB_1C for light paths from A to C. For the path AB_1C, where α_1, α_2 are sufficiently small that we can write $\cos\alpha_{1,2} = 1 - \alpha_{1,2}^2/2$, the difference in the paths AB_1C and ABC is approximately $(AB\,\alpha_1^2 + BC\alpha_2^2)$; to a *first* order of small quantities this is zero.

For the paths AB_1C and $AB_1'C$, the path difference is given by

$$(AD\ \sec\beta_1 + DC\ \sec\beta_2)$$
$$- [AD\ \sec(\beta_2 + d\beta_1) + DC\ \sec(\beta_2 + d\beta_2)]$$
$$= AD\ \sin\beta_1\ \sec^2\beta_1\ d\beta_1 + DC\ \sin\beta_2\ \sec^2\beta_2\ d\beta_2$$
$$(2.14)$$

Since β_1 and β_2 have the same sign then so do the two terms in (2.14). *Except* in the case $\beta_1 = \beta_2 = 0$, the path difference in (2.14) is to *first* order in $d\beta_1$, $d\beta_2$ and condition (2.2) of Section 2.2 is violated.

2.A.2

The case of refraction is illustrated in Fig. 2.11. We need to take account of the

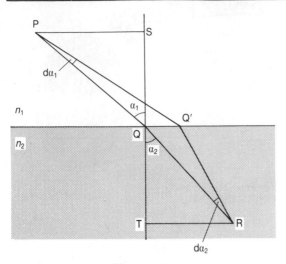

Figure 2.11

optical paths in the two media so the contributions to the integral of eqn (2.2) are

PQR $: n_1QS \sec \alpha_1 + n_2QT \sec \alpha_2$

PQ'R $: n_1QS \sec(\alpha_1 + d\alpha_1)$
$$+ n_2QT \sec(\alpha_2 + d\alpha_2)$$

To first order the difference is

$n_1QS \sin \alpha_1 \sec^2 \alpha_1 \, d\alpha_1$
$$+ n_2QT \sin \alpha_2 \sec^2 \alpha_2 \, d\alpha_2$$

Since

$QQ' = QS \sec^2 \alpha_1 \, d\alpha_1 = -QT \sec^2\alpha_2 \, d\alpha_2$

then the path difference is given by QQ' $(n_1 \sin \alpha_1 - n_2 \sin \alpha_2)$.
This vanishes provided

$$n_1 \sin \alpha_1 = n_2 \sin \alpha_2 \qquad (2.15)$$

that is provided Snell's law of refraction is satisfied.

Problems

2.1 Derive the law of refraction from Fermat's principle

2.2 Derive the law of reflection from Fermat's principle

2.3 With reference to Section 2.5, show that there is only one angle of minimum deviation.

2.4 Find the value of v for a glass whose refractive index is given by

$$n^2 - 1 = \frac{A_1\lambda^2}{\lambda^2 - \lambda_1^2} + \frac{A_2\lambda^2}{\lambda^2 - \lambda_2^2}$$

where λ is the wavelength, $A_1 = 0.5306$, $A_2 = 4.3356$, $\lambda_1 = 17500$ nm and $\lambda_2 = 1060$ nm.

2.5 The angle of a prism is nominally 60% and the index of refraction is about 1.5. Find an approximate value for the angle of minimum deviation.

2.6 Accurate measurement gives an angle α of $59°59'10''$ and the angle of minimum deviation is $39°29'20''$. Calculate limits for the refractive index. How many figures are significant? Is it necessary to include a correction for the refractive index of the air?

2.7 An optical fibre has a core refractive index of 1.525 and that for the cladding is 1.517. Calculate the numerical aperture of the fibre. What lens focal length would you use to focus a 0.4 mm diameter parallel beam of light into the fibre core?

3

Ray Optics (2): Paraxial Rays

3.1 Introduction: coaxial systems

One of the most important areas in which the behaviour of light rays needs to be considered is that involving curved (often spherical) refracting or reflecting surfaces. The simple biconvex lens, consisting of a piece of transparent material bounded by two spherical surfaces, has the property of forming *images*. If an illuminated transparency is placed squarely on the axis of such a lens, and sufficiently far away, then a plane can be located on the opposite side of the lens where the relative intensity distribution is the same as that leaving the transparency. This is referred to as the *image* of the object. It will appear identical to the object, although it may be larger or smaller. In this case the image will be seen on a screen placed in the appropriate plane. The light rays from each part of the object converge, after passing through the lens, at a single point on the screen. Sometimes an image can be seen by looking into a lens but it cannot be projected on to a screen—a so-called 'virtual' image (see Section 3.9).

The image produced by a simple lens such as that described would not be perfect: a closer correspondence with the object can be obtained by using several lenses in the form of a *coaxial* system. Thus a typical modern camera lens will have the form shown in Fig. 3.1. The centres of the lenses and the centres of curvatures of all the lens surfaces lie on a straight line—a *coaxial* system. The system of Fig. 3.2 would produce poor images and is of no interest: coaxial alignment of the lenses of a system is essential for good performance.

How do we arrive at the design of the lens of Fig. 3.1(a)? We need to be able to trace rays through the system, to determine how each of the components behaves. The *positions* of images formed by such systems can be found by considering rays that make angles with the axis for which the

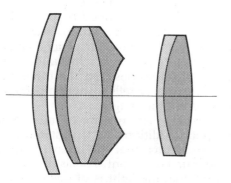

Figure 3.1 Section of modern camera lens

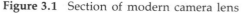

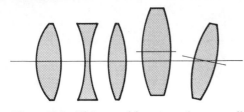

Figure 3.2 This would not work too well

approximations

$$\sin\theta = \theta, \qquad \tan\theta = \theta \quad \text{and} \quad \cos\theta = 1 \tag{3.1}$$

hold, where θ is the angle between the ray and the axis—the so-called *paraxial* rays. The treatment is termed 'Gaussian optics', after K. F. Gauss (1777–1855).

In this chapter we shall see how the behaviour of coaxial systems may be characterised and introduce the concepts and quantities needed in order to design and assess such systems. The mathematical treatment appropriate to paraxial rays entails the use of 2×2 matrices and is described in Section 3.23. Before embarking on this, however, we first need to gain insight into functions of the different surfaces and components of a system from a geometrical viewpoint.

3.2 Paraxial rays

Optical instruments generally use cones of rays which are much too wide for the paraxial approximation to be applied. They do this both to collect light efficiently and because, when diffraction is taken into account (Chapter 6), rays of wide angles are usually needed to give very sharp images. Nevertheless, it is desirable to begin by considering paraxial rays because the positions and sizes of images formed by paraxial rays are sufficiently good approximations to enable us to understand the general layout of most optical instruments including microscopes and telescopes. On the other hand, paraxial ray theory tells us nothing about the quality of optical images.

In this chapter we make one further assumption additional to that stated in eqn (3.1). We assume that the index (n) for any one medium is constant. This is equivalent to assuming that the light is of one colour, i.e. that it is confined to a narrow range of wave-

lengths so that effects due to dispersion are negligible. This is satisfactory in relation to laser light and to light from some other sources (e.g. sodium lamps) where the range of wavelengths, though large compared with that of laser light, is still small enough to make the variation of n very small. In Chapter 4 we shall consider what happens when neither of the above assumptions is valid.

3.3 Refraction at a single spherical surface

Consider the case of Fig. 3.3(a), in which an incident ray JK, in the direction of A_1, is refracted at a convex surface and meets the axis at A_2. By the sign convention (Section 2.3) the distances r, l_1, l_2, h and the angles ϕ, α_1, α_2 are all positive. (The latter are exaggerated in the diagram, for clarity.)

For small angles,

$$n_1\theta_1 = n_2\theta_2 \tag{3.2}$$

so that

$$n_1(\phi - \alpha_1) = n_2(\phi - \alpha_2) \tag{3.3}$$

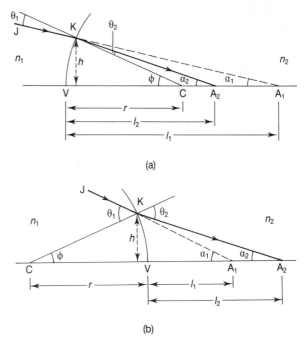

(a)

(b)

Figure 3.3 (a) Refraction at a convex spherical surface. (b) Refraction at a concave spherical surface

or

$$n_2\alpha_2 - n_1\alpha_1 = (n_2 - n_1)\phi \qquad (3.4a)$$

Since, within the paraxial approximation, $\alpha_1 = h/l_1$, $\alpha_2 = h/l_2$, $\phi = h/r$, then

$$\frac{n_2}{l_2} - \frac{n_1}{l_1} = \frac{n_2 - n_1}{r} \equiv P \qquad (3.5)$$

The quantity P is called the *power* of the surface: if r is in metres, the power is in *dioptres* (D). From (3.4a) and (3.5) we see that

$$n_2\alpha_2 - n_1\alpha_1 = Ph \qquad (3.4b)$$

Now examine Fig. 3.3(b) for a concave surface. If the sign convention were ignored, we should have $n_1(\phi + \alpha_1) = n_2(\phi + \alpha_2)$, but for this case ϕ is negative, while α_1, α_2 are positive. Thus eqn (3.3) applies in this case. Also, r is negative and l_1, l_2 positive, with the result that eqn (3.5) also applies. It is possible, when refraction occurs at a concave surface, that the refracted ray will not cross the axis. Provided the sign convention is used, eqns (3.3) to (3.5) still apply (Problem 3.1).

Note that eqn (3.5) does not contain h, so that all rays directed towards A_1 will arrive at A_2. A_1 and A_2 are termed *conjugate points* in respect of the surface VK.

3.3.1   Conjugate planes

When we are dealing with only one refracting surface any line through the centre of curvature may be regarded as an axis. If A_1 and A_2 in Fig. 3.4 are conjugate and B_1D_1 and B_2D_2 are small arcs of circles produced by rotating the line A_1CA_2 about the centre C, then all points on B_1D_1 have conjugates on B_2D_2. Within the paraxial approximation B_1D_1 and B_2D_2 may be replaced by small areas of planes tangent to the spheres at A_1

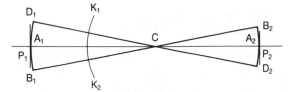

Figure 3.4 Although strictly a spherical object surface would appear to give a spherical image surface, for paraxial rays we have essentially plane-to-plane imaging

and A_2. These are called *conjugate planes* (P_1, P_2).

3.4   Magnification

The geometry of Fig. 3.3 enables the *positions* of conjugate points to be obtained. Light from any point on the plane P_1 (Fig. 3.4) will form an image on P_2. We need to know what size such an image will be. This emerges from Fig. 3.5 where, since $y_1/l_1 = \theta_1$ and $y_2/l_2 = \theta_2$, we have

$$\frac{n_1 y_1}{l_1} = \frac{n_2 y_2}{l_2} \qquad (3.6)$$

from which, since $\alpha_1 = h/l_1$ and $\alpha_2 = h/l_2$,

$$n_1 y_1 \alpha_1 = n_2 y_2 \alpha_2 \qquad (3.7a)$$

If a ray passes through many spherical refracting surfaces, (3.7a) applies to all the surfaces and intermediate images and

$$n_1 y_1 \alpha_1 = n_2 y_2 \alpha_2 = \cdots = n_f y_f \alpha_f \qquad (3.7b)$$

when n_f, y_f, α_f apply to the final image.

We define *transverse magnification* M_T as y_f/y_1 and *angular magnification* M_α as α_f/α_1. From (3.7b) we see that $M_T M_\alpha = n_1/n_f$. If the incident and final media have the same refractive index, then

$$M_T = \frac{1}{M_\alpha} \qquad (3.8)$$

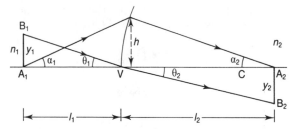

Figure 3.5 Imaging with a convex surface

3.5   Reflection at a spherical mirror

Figure 3.6 shows the reflection of a ray at a concave spherical mirror. The relationship between the object point A_1 and the image A_2 may be obtained by putting $n_2 = -n_1$, in accordance with the discussion of Section 2.4. Equation (3.2) shows that $\phi_2 = -\phi_1$:

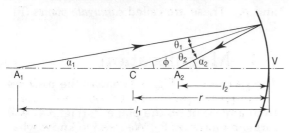

Figure 3.6 Imaging with a concave surface

Snell's law is transformed into the law of reflection. Equation (3.5) leads to

$$\frac{1}{l_2} + \frac{1}{l_1} = \frac{2}{r} = P \qquad (3.9)$$

This relation may be verified by putting $\phi_2 = -\phi_1$ and applying the procedure of Section 3.3 to Fig. 3.6.

If the object is at an infinite distance to the left, $1/l_1$ is zero and the image is at a point F distant $r/2 = 1/P$ from the mirror. This point is called the focus of the mirror and a plane through F (which is conjugate to a parallel plane at an infinite distance to the left) is called the focal plane. A set of rays parallel to FV may be regarded as coming from an infinitely distant point on the line FV and, after reflection, all these rays are directed towards F. A bundle of parallel rays making an angle α_1 with FV will, after reflection, be directed towards a point B in the focal plane (Fig. 3.7). The ray which passes through C is incident perpendicularly on the mirror and returns along its own path. Hence, we have

$$y = \alpha_1 f = \frac{\alpha_1}{P} \qquad (3.10)$$

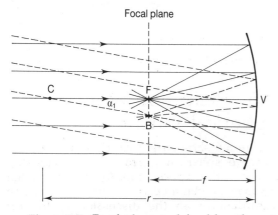

Figure 3.7 Focal plane and focal length

where $f = r/2$ is called the *focal length* of the mirror and depends *only* on the radius of curvature of the mirror. It is of interest to explore the variation of the position of the image as the object is brought in from a point at a great distance to the left of the mirror. As A_1 moves in, A_2 moves to the left from F and the object and image meet at C. If A_1 is brought still nearer to the mirror, A_2 moves to the left and when A_1 is at F, A_2 is at infinity. If A_1 continues to move towards the mirror A_2 reappears at an infinite distance to the right and moves in so that the object and image meet again at the mirror.

3.6 Reflection and refraction at a plane surface

For reflection at a plane surface ($r = \infty$), eqn (3.9) gives $l_2 = -l_1$. This is the well-known result that when reflection occurs at a plane mirror, the image is as far behind the mirror as the object is in front. (Note that this applies for rays at *all* angles between the ray and mirror normal, and not only those where the paraxial approximation applies.)

The effect of refraction at a plane surface manifests itself in the apparent depth of an underwater object viewed from above the water. In eqn (3.5) we put $n_1 = n$ (water) and $n_2 = 1$ (air), leading to

$$l_2 = \frac{l_1}{n} \qquad (3.11)$$

The *apparent depth* is less than the real depth by a factor $1/n$, or 0.75 for water, for which $n = 1.33$. For a block of glass with $n = 1.5$, the apparent depth is two-thirds of the actual depth.

3.7 Thin lenses in air

When a ray of light passes through a lens (Fig. 3.8), its distance from the axis changes by an amount that depends both on its inclination to the axis and on the thickness of the lens at the point of entry. For paraxial rays the inclination is small. We neglect this and also assume that the effect of finite lens thickness can be neglected. We thus need

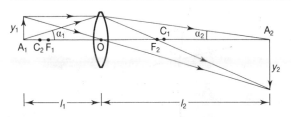

Figure 3.8 Imaging by a convex lens

to consider the effect of two spherical surfaces—air/glass and glass/air—separated by a negligibly small distance. By using eqn (3.5) twice, we can relate the final image distance l_2 to the object distance l_1. For the first surface, $n_1 = 1$, $n_2 = n$ (eqn 3.5) and the image forms at distance l', where

$$\frac{n}{l'} - \frac{1}{l_1} = \frac{n-1}{r_1}$$

For the second surface, $n_1 = n$, $n_2 = 1$ and the object distance is l'; therefore

$$\frac{1}{l_2} - \frac{n}{l'} = \frac{1-n}{r_2}$$

whence

$$\frac{1}{l_2} - \frac{1}{l_1} = (n-1)\left(\frac{1}{r_1} - \frac{1}{r_2}\right) \equiv P \qquad (3.12)$$

Note that, from (3.7a),

$$M_T = \frac{y_2}{y_1} = \frac{l_2}{l_1} = \frac{\alpha_1}{\alpha_2} = \frac{1}{M_\alpha} \qquad (3.13)$$

and that, for the case of Fig. 3.8, all of these quantities are *negative*.

3.8 Positive and negative lenses

A lens is called a *positive lens* or a *negative lens* according to the sign of its power. A beam of rays parallel to the axis coming from the left will meet at F_2 where $OF_2 = f$. The value of f is obtained by putting $1/l_1 = 0$ in eqn (3.12) and we have

$$\frac{1}{f} = (n-1)\left(\frac{1}{r_1} - \frac{1}{r_2}\right) = P \qquad (3.14)$$

where $f = 1/P$ is called the focal length of the lens. If $OF_1 = -f$, then rays from F_1 are parallel to the axis after they have passed through the lens.

3.9 Real and virtual images

When rays from an object actually meet in an image point (as in Fig. 3.8) we have a *real image* which may be seen on a screen placed in the image plane or recorded on a photographic plate. In the situation of Fig. 3.9, the image is *virtual*. An observer looking into the lens from the right of Fig. 3.9 would see an image I of the object O even though no rays actually pass through the image. (The lens of the eye will form a real image on the retina. Similarly, a camera, suitably focused on the image plane, will record an image of I.)

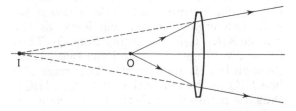

Figure 3.9 Formation of a virtual image when the object is closer than the focal point

3.10 General treatment of coaxial systems

With the help of Sections 3.3 to 3.7 the behaviour of any coaxial system can be analysed. Each surface or lens will produce an image (real or virtual) of an object. The positions and magnifications can be calculated and so the overall performance of the system determined. In practice we are generally interested in the overall behaviour of the system, rather than that at each interface. What do we need in order to specify the overall performance of an optical system?

Suppose K_1V_1 and K_fV_f (Fig. 3.10) are the initial and final surfaces of a coaxial system which may have any number of elements. We know that there will exist a point F_1 such that all rays through F_1 will emerge parallel to the axis. Similarly, rays parallel to the axis from the left will converge to (or diverge from) a point F_2. The planes through F_1 and F_2 perpendicular to the axis are the *first* and *second focal planes* respectively. (They may be at infinity: we consider this later.) The lines

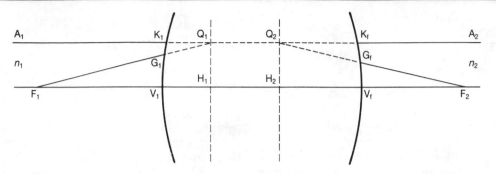

Figure 3.10   Principal planes and principal points

through F_1, G_1 and K_f, A_2 intersect in Q_1, so defining a plane perpendicular to the axis and intersecting it in H_1—and likewise for Q_2, H_2. Furthermore, a bundle of rays A_1K_i, F_1G_1 converging on Q_1 will leave as the bundle K_fA_2, G_2F_2, diverging from Q_2. Thus Q_2 is the image of Q_1 and the two planes through H_1, H_2 are *conjugate*. The image of a line H_1Q_1 is the line of the *same length* H_2Q_2; thus the planes through H_1, H_2 are planes of unit (± 1) magnification. They are termed *principal planes*.

For a thin lens, the idea of focal length is simple: it is the distance from the focal point to the lens. A collection of lenses forming a coaxial system will converge or diverge an incident parallel beam and will possess a focal length, but *where does one measure it from*? Suppose that we can see only an incident bundle of rays parallel to A_1K_1 and an emergent cone of rays through F_2. We would know that this form of focusing could be produced by placing a thin converging lens in the plane containing H_2 and Q_2, where Q_2 is

the intersection between the directions of the incident (A_1K_1) and emerging (G_2F_2) rays. This defines the axial point H_2, known as a *principal point*. The focal length of our equivalent converging lens would need to be H_2F_2. The other principal point H_1 can be similarly located by considering a bundle of rays parallel to A_2K_f, from the right, converging to F_1 on the left. Thus the principal points provide logical positions from which to measure the focal lengths of any coaxial system. In Fig. 3.10 the distance to F_2 is the *second focal length*. (Note the sign convention in Fig. 3.10 where f_1 is negative and f_2 positive.)

3.11   Cardinal points

F_1, F_2, H_1 and H_2 are all called *cardinal points*. When they are known the size and positions of the image of an object may be found by geometrical construction, as shown in Fig. 3.11 or by the formulae given in Section 3.7. In Fig. 3.11 a ray parallel to the axis and passing through B_1 meets the second prin-

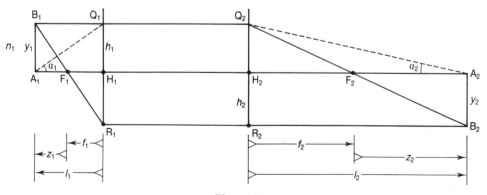

Figure 3.11

cipal plane in Q_2 and its final direction passes through F_2. Thus the direction Q_2F_2 is the final direction of one ray from B_1. Another ray from B_1 passes through F_1 and meets the *first* principal plane in R_1 and emerges parallel to the axis after passing through R_2 (where $H_1R_1 = H_2R_2$). The two rays meet in B_2 which is the image of B_1. Also A_2B_2 is the image of A_1B_1.

The cardinal points do not always occur in the order $F_1H_1H_2F_2$ shown in Fig. 3.11 but the above geometrical construction and the following algebraic formulae apply in all cases.

3.12 Magnification

From the similar triangles $H_2Q_2F_2$ and $R_2Q_2B_2$

$$\frac{f_2}{l_2} = \frac{y_1}{y_1 - y_2} \tag{3.15}$$

(remembering that, in the figure, y_2 is negative). Similarly,

$$\frac{f_1}{l_1} = \frac{-y_2}{y_1 - y_2} \tag{3.16}$$

so that, by addition,

$$\frac{f_1}{l_1} + \frac{f_2}{l_2} = 1 \tag{3.17}$$

and by division

$$\frac{f_1}{f_2} = -\frac{l_1 y_2}{l_2 y_1} \tag{3.18}$$

From the figure,

$$y_1 = l_1\alpha_1 = l_2\alpha_2 \tag{3.19}$$

Applying eqn (3.7a) to the initial and final directions of the ray $A_1Q_1Q_2A_2$, with eqns (3.18) and (3.19) we obtain

$$\frac{n_2}{f_2} = -\frac{n_1}{f_1} = P \tag{3.20}$$

As before, P is the power of the system. Also,

$$M_T = \frac{y_2}{y_1} = \frac{n_2 l_2}{n_1 l_1} \tag{3.21}$$

The similar triangles in the figure lead to Newton's equations:

$$z_1 z_2 = f_1 f_2 = \frac{-n_1 n_2}{P} \tag{3.22}$$

where z_1 and z_2 are the distances of the object and image planes from the corresponding focal points. Also,

$$M_T = \frac{n_1}{n_2} \frac{1}{M_\alpha} \tag{3.23}$$

In the following sections we apply these equations to special cases.

3.13 Newton's formula

If the initial and final media are the same, then $n_1 = n_2 = n$, and

$$f_2 = -f_1 = \frac{n}{P} \tag{3.24}$$

$$\frac{1}{l_1} - \frac{1}{l_2} = \frac{1}{f_1} = -\frac{P}{n} \tag{3.25}$$

Equation (3.8) applies to the magnification and Newton's equation reduces to

$$z_1 z_2 = f_1^2 = -\frac{n^2}{P^2} \tag{3.26}$$

The equations for a system of lenses in air are obtained by putting $n = 1$ in eqns (3.25) and (3.26).

3.14 Catadioptric systems

These systems include an odd number of mirrors so that a ray enters and emerges on the same side of the system. The appropriate equations are derived by putting $n_2 = -n_1 = -n$ in the equation of Section 3.11. We than have

$$f = f_2 = f_1 = -\frac{P}{n} \tag{3.27}$$

$$\frac{1}{l_1} + \frac{1}{l_2} = \frac{1}{f} \tag{3.28}$$

and

$$z_1 z_2 = f^2 = -\frac{n^2}{P^2} \tag{3.29}$$

These equations are valid when the object and image are in a medium of the same refractive index. If some lenses and a mirror are placed under water, to form a coaxial system with a vertical axis, the above equations apply when both the object and image are in water and when both are in air. They do not apply when one is in air and the other in water.

3.15   Calculation of the power of a system

Suppose that a ray parallel to the axis is incident upon a coaxial system and meets successive surfaces at heights h_1, h_2, ..., h_s. Let P_1, P_2, ..., P_s be the powers of the surfaces and P the power of the system as a whole. Then repeated use of eqn (3.4b) gives a series of equations:

$$n_2\alpha_2 - n_1\alpha_1 = P_1h_1$$
$$n_3\alpha_3 - n_2\alpha_2 = P_2h_2, \qquad \text{etc.}$$

and, by addition and putting $\alpha_1 = 0$ since the ray is initially parallel to the axis,

$$n_f\alpha_f = \sum_{r=1}^{s} P_r h_r \qquad (3.30)$$

where α_f is the angle between the emergent ray and the axis. However, in Fig. 3.11, the angle between the emergent ray and the axis corresponding to the ray B_1Q_1 is the angle $Q_2F_2H_2$ which equals h_1/f_2 and, by using eqn (3.20),

$$n_f\alpha_f = h_1 P \qquad (3.31)$$

so that

$$P = \frac{1}{h_1} \sum_{r=1}^{s} P_r h_r \qquad (3.32)$$

3.16   Two thin lenses in air

Suppose two thin lenses of powers P_1 and P_2 are separated by a distance a. A ray parallel to the axis meets the first lens at a height h_1 and is deviated through an angle P_1h_1. It meets the second lens at a height $h_2 = h_1(1 - aP_1)$ and eqn (3.31) gives

$$P = \frac{1}{h_1}[P_1h_1 + P_2h_1(1 - aP_1)]$$

so that

$$P = P_1 + P_2 - aP_1P_2 \qquad (3.33)$$

When the power is known, we need to find the positions of the foci (relative to some

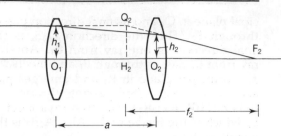

Figure 3.12   Two thin lenses in air

identifiable point such as the position of one of the lenses). The principal points can then be found because $H_1F_1 = f_1$ and $H_2F_2 = f_2$. In Fig. 3.12,

$$\frac{O_2F_2}{H_2F_2} = \frac{h_2}{h_1} = 1 - aP_1 \qquad (3.34)$$

so that

$$O_2F_2 = (1 - aP_1)\frac{1}{P} \qquad (3.35)$$

A similar procedure gives the position of F_1.

3.17   Thick lens

Suppose that a thick lens made of glass of index n has two surfaces of radii r_1 and r_2 and that its thickness is t. P_1 and P_2 are the powers of the two surfaces. Then the deviation of a ray, initially parallel to the axis at a height h_1, is P_1h_1/n. Following the procedure of Section 3.16 we then obtain

$$P = P_1 + P_2 - \frac{t}{n}P_1P_2 \qquad (3.36)$$

or

$$P = (n-1)\left(\frac{1}{r_1} - \frac{1}{r_2}\right) + \frac{n-1}{n}\frac{t}{r_1r_2} \qquad (3.37)$$

and eqns (3.35) and (3.37) are replaced by

$$O_2F_2 = \left(1 - \frac{t}{n}P_1\right)\frac{1}{P} \qquad (3.38)$$

Comparison of eqns (3.33) and (3.35) with eqns (3.36) and (3.38) shows that the thick lens behaves as though it had a *reduced thickness* t/n (cf. Section 3.6).

Equation (3.37) may be used to test the accuracy of the thin lens formula (eqn 3.14). Suppose a manufacturer is asked to make a lens whose power is 4 D and whose thickness is 5 mm with a glass of index 1.5. If a choice is made to put equal curvatures on the two faces ($r_2 = -r_1$), then use of eqn (3.14) leads to $r_1 = 0.25$ m $= 25$ cm. Equation (3.37) shows that the true power of the lens with this radius is $(4 - 0.027)$ D. The error is negligible in relation, for example, to a spectacle lens. On the other hand, consider a diamond 'loup' which is required to have a power of 40 D. Suppose the thickness is 8 mm and, as before, $r_1 = r_2$ and $n = 1.5$. Then eqn (3.14) leads to $r_1 = 0.025$ m $= 2.5$ cm but eqn (3.37) shows that if this radius is used the power actually obtained is $(40 - 4.3)$ D so that the error is over 10%.

3.18 Nodal points

These are two conjugate points N_1 and N_2 on the axis of a coaxial system which have the property that a ray incident on the system and directed towards N_1 will emerge parallel to its original direction and pass through N_2 giving unit angular magnification. In Fig. 3.13, the distances l_{N1} and l_{N2} from the corresponding principal points are equal. Equation (3.17) then gives

$$l_{N1} = l_{N2} = f_1 + f_2 \qquad (3.39)$$

(where for the case of Fig. 3.13, f_1 is negative), so that, for a system in air, $f_1 = -f_2$ and $l_{N1} = l_{N2} = 0$, i.e. the nodal points coincide with the corresponding principal points.

Figure 3.13 also shows that if P_1 and P_2 are the distances of any pair of conjugate planes from the corresponding nodal points, then

$$M_T = \frac{y_2}{y_1} = \frac{P_1}{P_2} \qquad (3.40)$$

Consider an object at a great distance from the system which subtends an angle $\omega = y_2/P_1$. The image is formed very near the

second focal plane whose distance from N_2 is $f_2 - (f_1 + f_2) = -f_1$ and

$$\frac{y_2}{f_1} = \frac{y_1}{P_1} = \omega$$

so that using eqn (3.27),

$$P = \frac{n\omega}{y_2} \qquad (3.41)$$

Even when the nodal points do not coincide with the principal points (i.e. for systems where the initial and final media are not the same), the nodal points can be found, when the principal and focal points are known, by using eqn (3.39). Conversely, if the nodal points and focal points are known the principal points can be found. Only four independent pieces of information are needed to characterise the system.

3.19 Experimental location of the cardinal points

It may occasionally be desirable to locate the cardinal points of a system relative to some well-defined point such as one of the outer surfaces. There is a variety of methods for doing this. One will now be described. Figure 3.14(a) shows an arrangement in which a fine pinhole P in a white screen is placed on the axial line to the left of the system and a mirror M nearly normal to the

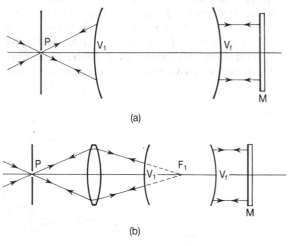

(a)

(b)

Figure 3.14 (a) Location of focal points. (b) How to do this when the focal point is inside the system

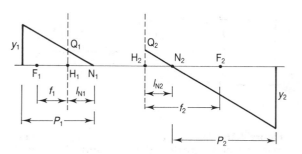

Figure 3.13 Nodal points

axis is on the right. The pinhole is strongly illuminated from the left. The distance between the pinhole and the system is altered until a sharp image of the pinhole is formed (near to it) on the screen. The pinhole is then at a focus of the system. The light emerging on the right falls on the mirror and is reflected almost along its own path. The distance of the focus from one surface of the system can be measured. The other focus can be located by turning the system round.

This method fails when a focus is inside the system but the modification shown in Fig. 3.14(b) can then be used. An auxiliary lens L_1 is used to direct the rays from P to a point within the system. When they are directed towards F_1, the parallel rays emerging from the system will return along their own paths to a point near P.

The nodal points may be found by mounting the system on a rotation table with a slide that permits any point on the axis to be placed over the centre of rotation. Suppose that a parallel beam is incident from the left and originally the system is oriented so that the axis coincides with the beam (Fig. 3.15). The light is focused at H, which then coincides with the second focus F_2 of the system. Now suppose that the system is rotated through an angle θ about a line normal to the plane of the paper and passing through C. Then the parallel beam is focused at a point H' in the focal plane. Also the ray A_1N_1 must emerge in a direction parallel to A_1N_1 and passing through N_2. It must also pass through H'. Then $HH' = z_N\theta$ where z_N is the distance of N_2 from C. The system is moved along the slide until a small rotation produces no movement of the image and this

happens when the centre of rotation coincides with N_2. The position of N_1 can be located in a similar way. For a system in air the principal points H_1 and H_2 coincide with the nodal points N_1 and N_2. When the initial and final media are not the same the distance N_2F_2 is f_1 and $N_1F_1 = f_2$ (Section 3.16) so that the distances $F_2H_2 = f_2$ and $F_1H_1 = f_1$. Thus H_1 and H_2 are located.

3.20   Telescopic systems

We now consider systems that have zero focusing power. A parallel beam entering the system emerges as a parallel beam, though usually the emergent beam differs from the entrant beam both in width and in direction (Fig. 3.16a). Such systems may be regarded as having focal points at infinitely great distances to the left and right of the system. They have no principal or nodal points and most of the discussion of Sections 3.10 to 3.13 is irrelevant to them except that the magnification relation $M_T M_\alpha = n_1/n_f$ is still valid. In Fig. 3.16(b) the lateral magnification is negative and its magnitude is less than 1. The angular magnification is also negative and its magnitude is greater than 1. In Fig. 3.17 the lateral magnification is positive and less than 1. The angular magnification (not shown) is also positive and greater than 1.

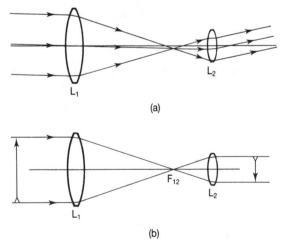

(a)

(b)

Figure 3.16 Telescopic system: (a) angular magnification greater than 1 and (b) lateral magnification less than 1

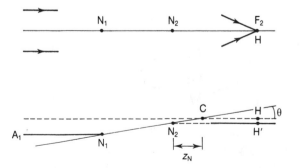

Figure 3.15 Location of nodal points

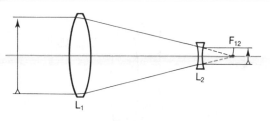

Figure 3.17

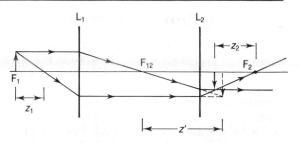

Figure 3.18   Longitudinal magnification

Equation (3.33) implies that if two lenses in air are separated by a distance equal to the algebraic sum of their focal lengths then the power of the combination is zero. Figures 3.16 and 3.17 show two systems of this type. For these systems, the lateral magnification is constant and equal to f_2/f_1 and the angular magnification f_1/f_2, in agreement with eqn (3.8). The combination of two positive lenses (Fig. 3.16) has a negative magnification. Images are always inverted and angles also change sign. This is acceptable in astronomical telescopes but not in terrestrial telescopes for which an erect image is required. This is given by the Galilean telescope shown in Fig. 3.17. However, binoculars usually employ lens systems that are equivalent to that shown in Fig. 3.16 and use reflecting prisms to give a correctly oriented image (Section 18.32).

3.21   Longitudinal magnification

Although they are generally used in connection with incident beams of parallel rays, telescopic systems do form images of objects at finite distances. In Fig. 3.18, suppose that an object is at a distance z_1, from the primary focus of L_1 and that this lens forms an image at a point distant z' from the point F_{12} which is the second focus of L_1 and the first focus of L_2. The second lens forms the final image at a distance z_2 from its second focus. Then, using Newton's formula twice,

$$z_1 z' = -f_1^2 \quad \text{and} \quad z_2 z' = -f_2^2$$

so that

$$\frac{z_2}{z_1}\frac{x_2}{x_1} = \frac{f_2^2}{f_1^2} = M_T^2 \tag{3.42}$$

This ratio is sometimes called the longitudinal magnification.

3.22   More on telescopic systems

Suppose that two dioptric systems are placed on the same axis and situated so that the second focus of one coincides with the first focus of the other. The distance from the second principal plane of one to the first principal plane of the other is then equal to the algebraic sum of f_1 and f_2, so that the combination is telescopic and the angular magnification is f_1/f_2. Conversely, all telescopic systems may be regarded as combinations of two dioptric systems. Originally telescopes were designed for astronomical use and later they were used for viewing distant terrestrial objects as in theodolites and telescopic gunsights. Recently telescopic systems have found a new application for expanding or compressing a nearly parallel beam emitted by a laser. Equation (3.8) implies that if the beam is compressed any small angular divergence is increased.

3.23   Matrix methods

In 1913, an astronomer, R. A. Sampson, showed that matrix algebra could be applied to ray optics but his work never became widely known and was nearly forgotten. In 1915, T. Smith independently discovered this application and it has now become recognised that matrix methods can be very useful for complex problems. In the following paragraphs we describe their application to Gaussian optics. Matrices will also appear in connection with polarised light (Sections 7.11 and 7.12).

3.24   System matrices

A ray may be specified by giving its angle (α) to the axis and the height y at which it cuts any chosen plane, which we may call a *reference plane* (Fig. 3.19). Note that, for the uninterrupted ray which does not meet a discontinuity, there is no change in α, so in Fig. 3.19 we have $\alpha_2 = \alpha_1$ and $\alpha_4 = \alpha_3$. For planes on either side of, and very close to, a refracting surface the value of α changes but that of y does not. Thus $y_3 = y_2$. By applying the two rules

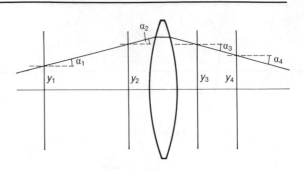

Figure 3.19 Relevant parameters for matrix formulation of ray-tracing

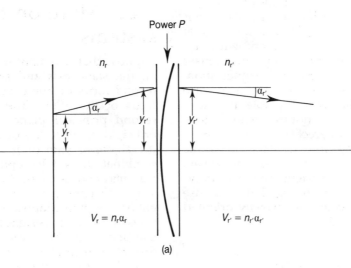

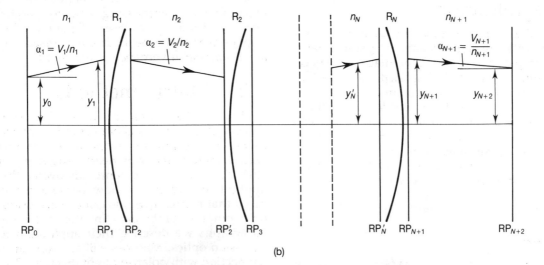

Figure 3.20   (a) Relevant planes for matrix formulation. (b) The refraction matrix

1. no change in α on transit between refracting surfaces,
2. no change in y on passing a refracting surface,

we can determine the position and direction of an emerging ray for any given incident ray. This is all that is needed to describe the behaviour of the system as a whole, in the paraxial approximation.

Suppose α_r is the angle between the axis and the ray at any surface r and z_{rs} is the distance between this surface and the next refracting surface. We can obtain expressions suitable for use in any system by defining $V_r = n_r\alpha_r$ and $T_{rs} = z_{rs}/n_r$ where V_r is the optical cosine and T_{rs} is the reduced distance. From Fig. 3.20(a) we see that

$$y_{r'} = y_r + T_{rs}V_r, \qquad V_{r'} = V_r \qquad (3.43)$$

$$V_s = Py_{r'} + V_{r'}, \qquad y_s = y_{r'} \qquad (3.44)$$

where P is the power of the refracting surface. We may write these equations in matrix form as

$$\begin{pmatrix} y_{r'} \\ V_{r'} \end{pmatrix} = \begin{pmatrix} 1 & T_{rs} \\ 0 & 1 \end{pmatrix}\begin{pmatrix} y_r \\ V_r \end{pmatrix} \quad \begin{array}{l}\text{(translation: no}\\\text{refraction)}\end{array} \quad (3.45a)$$

for the translation A to B and

$$\begin{pmatrix} y_s \\ V_s \end{pmatrix} = \begin{pmatrix} 1 & 0 \\ P & 1 \end{pmatrix}\begin{pmatrix} y_{r'} \\ V_{r'} \end{pmatrix} \quad \begin{array}{l}\text{(refraction: no}\\\text{translation)}\end{array} \quad (3.45b)$$

We may therefore write

$$\begin{pmatrix} y_s \\ V_s \end{pmatrix} = \begin{pmatrix} 1 & 0 \\ P & 1 \end{pmatrix}\begin{pmatrix} 1 & T_{rs} \\ 0 & 1 \end{pmatrix}\begin{pmatrix} y_r \\ V_r \end{pmatrix} \qquad (3.46)$$

giving the position and direction in terms of these quantities for the ray at A. We label the 2×2 translation matrix as $T(r, s)$ and the refraction matrix as $R(r, s)$. Note the order of the matrices.

3.25 Going through the system

Look at the system depicted in Fig. 3.20(b), made up of N refracting surfaces R_1 to R_N. A ray crosses a plane RP_1 at a reduced distance z_{12}/n_1 from the first surface at a height y_1 and with optical cosine V_1. The values of y and V at the plane RP_1 are given by

$$\begin{pmatrix} y_1 \\ V_1 \end{pmatrix} = T(0, 1)\begin{pmatrix} y_0 \\ V_0 \end{pmatrix} \qquad (3.47a)$$

and at RP_2,

$$\begin{pmatrix} y_2 \\ V_2 \end{pmatrix} = R(1, 2)\begin{pmatrix} y_1 \\ V_1 \end{pmatrix} = R(1, 2)T(1, 2)\begin{pmatrix} y_0 \\ V_0 \end{pmatrix} \qquad (3.47b)$$

For the ray crossing the final plane RP_{N+1}, we have

$$\begin{pmatrix} y_{N+2} \\ V_{N+2} \end{pmatrix} = T(N+1, N+2)R(N, N+1)T(N-1, N)$$

$$\times R(N-1, N)\cdots R(1, 2)T(1, 2)\begin{pmatrix} y_0 \\ V_0 \end{pmatrix} \qquad (3.48)$$

or

$$\begin{pmatrix} y_{N+2} \\ V_{N+2} \end{pmatrix}$$

$$= \left[\prod_{s=N}^{0} T(s+1, s+2)R(s, s+1)\right]T(1, 2)\begin{pmatrix} y_0 \\ V_0 \end{pmatrix} \qquad (3.49)$$

$$= \begin{pmatrix} M_{11} & M_{12} \\ M_{21} & M_{22} \end{pmatrix}\begin{pmatrix} y_0 \\ V_0 \end{pmatrix} \qquad (3.50)$$

The matrix $\begin{bmatrix} a & b \\ c & d \end{bmatrix}$

is unimodular if $ad - bc = 1$. The translation and refraction matrices are, on inspection, seen to be unimodular and the product of any number of unimodular matrices is itself unimodular. Thus $M_{11}M_{22} - M_{21}M_{12} = 1$, which serves as a check on the possible correctness of the final product matrix.

The elements of the final matrix are readily seen to be related to the useful physical quantities associated with a refracting system.

1. Imagine a whole system replaced by a single refractive surface of power P at the plane RP_1. The final value of V (V_{N+2}) will be changed from the initial value (V_0) by $M_{21}y_0$ (eqn 3.50); thus $P = M_{21}$.
2. M_{12} is found to contain $z_{N+2,N+1} \equiv z_f$, the distance of the final reference plane from the last surface of the system. If $M_{12} = 0$, the value of y_{N+2} ($= M_{11}y_0$) is *independent* of V_0. Thus all rays through the point y_0 in the plane RP_1 pass through y_{N+2}, regardless of direction—another way of saying that the point at y_{N+2} is the image of that at y_0. When $M_{12} = 0$, then y_{N+2}/y_0 is equal

to M_{11}: the ratio in question is the transverse magnification. Thus the position of the image is the value of z_f that makes M_{12} vanish and the value of M_{11} gives the transverse magnification.

3. If $M_{22} = 0$ then all rays from the point at y_0 will leave the system in the same direction, which makes an angle $M_{21}y_0$ with the axis. The plane RP_1 is then the first focal plane of the system. The M_{22} term contains z_{01}; putting $M_{22} = 0$ enables z_{01}, the distance of the first focal plane to the first refracting surface, to be found.

4. If $M_{21} = 0$, then the power of the system is zero: the system is telescopic. The angular magnification V_{N+1}/V_0 is given by M_{22}.

3.26 Analysis of the thin lens by matrix methods

Consider a thin lens in air and, as before, let r_1 and r_2 be the radii of the surfaces and l_1 and l_2 the object and image distances (Section 3.7). We have, using the reference planes shown in Fig. 3.21,

$$\begin{pmatrix} y_2 \\ V_2 \end{pmatrix} = \begin{pmatrix} 1 & l_2 \\ 0 & 1 \end{pmatrix}\begin{pmatrix} 1 & 0 \\ P_2 & 1 \end{pmatrix}\begin{pmatrix} 1 & 0 \\ 0 & 1 \end{pmatrix}\begin{pmatrix} 1 & 0 \\ P_1 & 1 \end{pmatrix}\begin{pmatrix} 1 & -l_1 \\ 0 & 1 \end{pmatrix}\begin{pmatrix} y_1 \\ V_1 \end{pmatrix} \quad (3.51)$$

The first step is to multiply the three central matrices to obtain a system matrix

$$S = \begin{pmatrix} 1 & 0 \\ P & 1 \end{pmatrix} \quad (3.52)$$

where $P = P_1 + P_2$ is the power of the lens as a whole. We might have written down (3.52) directly by collapsing the two inner reference planes of Fig. 3.21 and regarding the lens as a single refracting element (as in Section 3.8).

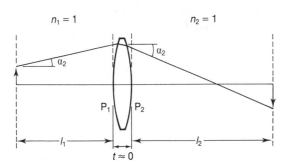

Figure 3.21 The thin lens by the matrix method

Completing the multiplication gives

$$\begin{pmatrix} y_2 \\ V_2 \end{pmatrix} = \begin{pmatrix} 1 + l_2 P & l_2 - l_1 - l_1 l_2 P \\ P & 1 - l_1 P \end{pmatrix}\begin{pmatrix} y_1 \\ V_1 \end{pmatrix} \quad (3.53)$$

Putting $M_{12} = 0$ gives the value of l_2 corresponding to a given value of l_1 in accord with eqn (3.12). Substituting the value of P from this equation into M_{11} gives the magnification equal to l_2/l_1, as in eqn (3.17).

3.27 Was it worth it?

The process of multiplying the matrices in Section 3.26 takes at least as long as working out the algebra in Section 3.7. In relation to the problem which we have used as an illustration the matrix method has no obvious advantage. This is because we have been using a sledgehammer to crack a nut. The advantages of the matrix method do not appear until it is applied to a system such as a compound lens with five components. There would then be ten refracting surfaces and eleven translations. The calculation of cardinal points for such a system by the methods of Sections 3.10 to 3.18 would be extremely cumbersome, as would also the calculation of the elements of a system matrix. If, however, the radii and separations were given numerically a computer would very quickly multiply the 2×2 matrices and find the elements of the system matrix. Then given the position and size of an object a small further computation would produce the position and size of the image. A large number of object points could then be treated in a time limited only by the rate of printout. The computer could also find the cardinal points of the systems.

3.28 Electron- and charged-particle optics

Although this book deals with the optics of electromagnetic waves, we may mention that analogues of the kind of optical systems discussed above exist for beams of charged particles, as manifested in the existence of the electron microscope. Analysis of the forces on an electron approaching a circular, current-carrying coil paraxially reveals that the coil has an imaging action similar in some respects to a converging lens. A feature special to the charged-particle case is, however, that the image is *rotated* with respect to the object about the system axis.

Further reading

W. Brouwer, *Matrix Methods in Optical Instrument Design*, W. A. Benjamin, New York, 1964.
A. Gerard and K. G. Burch, *Introduction to Matrix Methods in Optics*, John Wiley and Sons, London and New York, 1975.
H. Kogelnik, *Applied Optics*, **4**, 1562 (1965).
H. Kogelnik and T. Li, *Applied Optics*, **5**, 1550 (1966).
E. L. O'Neill, *Introduction to Statistical Optics*, Addison-Wesley, Reading, Mass. and London, 1963.

The second of the above references gives an introduction to matrix methods and is the most suitable for anyone who is not familiar with these methods. It also contains the application to lasers. For those who already have practical experience of matrix algebra, the second reference gives a compact treatment of Gaussian optics and an introduction to applications to aberrations.

Problems

3.1 Draw a scale diagram for a ray approaching in air a concave glass surface ($n = 1.5$) of radius of curvature 100 mm in a direction to strike the axis 400 mm beyond the surface. Locate the image point and show that eqn (3.5) correctly describes this case.

3.2 A glass sphere of radius 10 mm and refractive index 1.54 has a black inclusion at its centre. Locate the image of the speck when viewed from outside, (a) in air and (b) in water.

3.3 What are the transverse and angular magnifications of the images in Problem 3.2?

3.4 How far should your face be from a concave mirror of radius 400 mm if an enlarged virtual image is to be seen at a distance of 250 mm from your eye? What is the magnification produced?

3.5 What focal length lens would be needed to project a 35 mm slide on to a screen in a room 10 m long if the final image is to subtend an angle of $25°$ at an observer beside the projector? If the same projector were used in a room 5 m long, (a) what change in slide-lens distance would be needed and (b) what would be the angular subtense of the image seen by the observer next to the projector?

3.6 Before Valhalla was flooded, a camera was left to photograph an object at a distance of 10 m. Specify a lens to be added to the camera lens to enable a photograph to be taken after the deluge. (Refractive index of water = 1.33.)

3.7 A converging lens is placed between a source and a screen, 1 m apart. Sharp images are seen on the screen when the lens is in two positions, 0.2 m apart. Calculate the focal length of the lens. Generalise this to the case of a source–screen distance l and a lens shift d. Will the situation described above always occur?

3.8 A coaxial system consists of a convex lens ($f_1 = +300$ mm) 100 mm from a concave lens ($f_2 = -180$ mm) in air. Determine the cardinal points of the system.

3.9 A triplet lens consists of a converging lens ($f_1 = 100$ mm) followed, at 20 mm, by a diverging lens ($f_2 = -120$ mm). The third lens should be 25 mm beyond the diverging lens, but is missing. It is known that the second focal point should be 60 mm beyond the missing lens. Determine the focal length of the missing lens and also that of the combination.

3.10 Use the matrix method to determine l_2 in terms of l_1 for a lens of thickness t. Show that your result reduces to eqn (3.12) for a thin lens.

3.11 For what refractive index will a sphere in air have its focal points on the surface?

3.12 Locate the principal and focal points for a sphere of radius R and refractive index n, in air.

4

Wave Theory (1): Interference

4.1 Periodicity in time and space

Wave motion involves periodic variation in time, in space, or both. Figure 4.1 with OX as abscissa may be regarded as the instantaneous picture of a wave on a rope. It shows, for one particular time, the variation of displacement with position. The same curve with OT as abscissa shows the variation with time at one particular place. So long as we consider only one space dimension there is a symmetry between the two periodicities. We have a *temporal frequency* (ν_t) which is the number of waves per unit time and a *spatial frequency* (ν_s) which is the number of waves per unit distance. From these we define:

$$\varkappa = 2\pi\nu_s; \qquad \omega = 2\pi\nu_t$$

$$\lambda = \frac{1}{\nu_s} = \frac{{}^*2\pi}{\varkappa}; \qquad T = \frac{1}{\nu_t} = \frac{2\pi}{\omega} \qquad (4.1)$$

$\varkappa$ and ω are called *circular frequencies*; they are the numbers of waves in two units of space and time respectively. λ, called the *wavelength*, is the distance and T, called the *period*, is the distance/time between successive maxima. We shall frequently be dealing with waves of sinusoidal profile, although other profiles are possible. The important general feature of a wave is that if we make

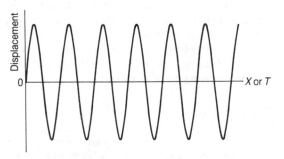

Figure 4.1 Periodic (sinusoidal) wave

a linear superposition of two or more waves, the resultant is itself a wave. The restriction to a consideration of sinusoidal waves is less stringent than may appear; a periodic function of any profile can (with only somewhat esoteric exceptions) be represented as a linear superposition of sinusoidal waves.

4.2 Progressive waves

A progressive wave may be represented by

$$E = E_0 \cos(\omega t - \varkappa x + \delta) = E_0 \cos\phi \quad (4.2)$$

where E is the quantity which fluctuates, E_0 is the *amplitude* and $\phi = \omega t - \varkappa x + \delta$ is the *phase*. δ is a constant. So long as we consider only one wave we may choose origins of x and t so that $\delta = 0$.

If we fix attention on one time, eqn (4.2)

28

reduces to

$$E = E_0 \cos(\varkappa x + \delta') \qquad (4.3a)$$

If we fix attention on one place, eqn (4.2) becomes

$$E = E_0 \cos(\omega t + \delta'') \qquad (4.3b)$$

Equations (4.3a) and (4.3b) both have the form of Fig. 4.1.

In the experiments described later in this book we shall see that the behaviour of light beams is precisely that which we expect for waves. If then we represent our light wave by eqn (4.2), to what does the E refer? The work of Maxwell showed that associated with the light wave there exist oscillating electric and magnetic fields. The two are linearly related so that we may characterise the light wave by either. We can usefully think of the E in eqn (4.2) as representing the electric field associated with the light wave. Since an electric field has a direction, as well as a magnitude, then we note that the E should be a vector and not a scalar quantity. For light propagation in materials whose properties are the same in all directions ('isotropic materials'), correct results are obtained by taking E to be the magnitude of the electric vector. For anisotropic materials the vector nature of E must be taken into account.

4.3 We cannot follow

In eqn (4.2) we have used ω and $\varkappa$ rather than ν_t and ν_s to obtain the simplest equation and also because ω and $\varkappa$ are generally used in books on wave mechanics and solid state physics. For light waves both ν_s and ν_t are numerically large compared with our everyday units of length (m) and time (s). For green light ν_s is about 2×10^6 waves m^{-1} and ν_t is about 6×10^{14} Hz (1 Hz = 1 wave s^{-1}). We cannot observe the fluctuating amplitude of light waves directly. Normally experiments involving light measure the energy received over items of at least a microsecond. This is so whether we use the eye, a photographic plate or a photocell as the detector, none of which can respond quickly enough to observe the individual waveform. Thus we usually measure the average value of the energy delivered over at least 100 million waves. For any type of wave, the energy carried by the wave is proportional to the square of the amplitude. By a suitable choice of units we may write the energy W as E^2.

If we wish to observe the wave properties of light we can make two beams of light interact to produce amplitude fluctuations in time or in space with frequencies that are small fractions of ν_t and ν_s (Section 4.5ff), and are hence directly observable.

4.4 Phase velocity of a progressive wave

If x changes by $1/\varkappa$ while t changes by $1/\omega$, the value of E, given by eqn (4.2) is unchanged. Thus the wave, as a whole, advances with a velocity v, where

$$v = \frac{\omega}{\varkappa} = \frac{\nu_t}{\nu_s} = \lambda \nu_t \qquad (4.4)$$

The *refractive index* n of a medium is defined as the ratio c/v, so that

$$v = \frac{c}{n} \qquad (4.5)$$

What happens when a wave crosses a boundary, e.g. from a vacuum to a medium? Does the wave at the boundary belong to the vacuum or to the medium? The answer is 'both' so the number of oscillations per second for the wave proceeding into the medium must be the same as that arriving from the vacuum side. Thus ν_t does not change when light passes from a vacuum into a material and it follows from (4.4) that

$$n\lambda_m = \lambda_v \qquad (4.6)$$

where λ_v and λ_m are the wavelengths in vacuum and in the medium respectively. In this book we use λ_{m1}, λ_{m2}, etc., where wavelengths in more than one medium are involved. Where only one medium is involved and no confusion can arise, λ is used. The value of c is approximately 3×10^8 m s^{-1} (Chapter 22).

4.5 Waves in three dimensions

In the three-dimensional propagation of waves there are surfaces, known as *wave surfaces*, over which the phase is the same at all points. The waves are called plane waves, spherical waves, cylindrical waves, etc., according to the shape of the wave surfaces. Any line normal to a wave surface is a wave normal. For plane waves:

$$E = E_0 \cos[\omega t - \varkappa(\alpha x + \beta y + \gamma z) + \delta] = E_0 \cos \phi \tag{4.7}$$

where α, β, γ are the direction cosines of the wave-normals. This may also be written

$$E = E_0 \cos(\omega t - \boldsymbol{\varkappa} \cdot \boldsymbol{r} + \delta) \tag{4.8}$$

where $\boldsymbol{\varkappa}$ is a vector with magnitude $\varkappa$ and components $\alpha\varkappa$, $\beta\varkappa$, $\gamma\varkappa$ and $\boldsymbol{r}$ is a radius vector from the origin. For the features of the most general wave, see Appendix 4.A. For a uniform spherical wave radiating from the origin,

$$E = \frac{E_0}{r} \cos(\omega t - \varkappa r + \delta) = \frac{E_0}{r} \cos \phi. \tag{4.9}$$

The mean value of the wave energy arriving at an area dS normal to the radius r is given by

$$\overline{E^2} = \frac{E_0^2}{r^2} \overline{\cos^2 \phi} = \frac{E_0^2}{2r^2}$$

The total energy flowing across the sphere of radius r is $4\pi r^2 \overline{E^2} = 2\pi E_0^2$, which is independent of r. Thus the total energy flow across any spherical surface surrounding a small isotropically emitting source of waves at the origin is the same, as required by the law of conservation of energy. Equation (4.9) implies that E increases indefinitely when r approaches zero, but sources of zero volume do not exist and, for the present, we may specify that eqn (4.9) is valid only when r exceeds λ.

4.6 Superposition of light waves

When two or more light waves pass simultaneously through the same volume, the resultant E_r of the amplitude at any point is the sum of E_1, E_2, E_3, etc. The value of E_r is thus the sum of the values of the individual amplitudes acting alone, so that

$$E_r = E_1 + E_2 + E_3 + \cdots = \sum_{m=1}^{N} E_m \tag{4.10a}$$

for N waves.

This assumption is called the *principle of superposition*. It is found to be valid for all experiments on light derived from the sun or from ordinary sources such as electric lamps, arc lights, etc.

For material media exposed to such light waves, the electric field of the combined light wave produces a polarisation which is linearly proportional to the electric field, so that we have, for linear media,

$$P = \chi E = \chi \sum_{m=1}^{N} E_m \tag{4.10b}$$

where χ is the susceptibility (see Chapter 15). This simple relation does not hold if the amplitudes E_m are very large—such as can be easily produced by pulsed lasers. In this situation we obtain a resultant polarisation which we need to represent by

$$P = \chi \left(\sum_{m=1} E_m + \sum_m \sum_n \alpha_{mn} E_m E_n \right.$$
$$\left. + \sum_m \sum_n \sum_p \alpha_{mnp} E_m E_n E_p + \cdots \right) \tag{4.10c}$$

i.e. terms of the second and higher orders have to be taken into account. The ratio of a second-order term to a first-order term is typically $\alpha_{11} E_1^2 / E_1 = \alpha_{11} E_1$. Now the coefficients α_{11}, etc., are all of such a magnitude that it is not until E_1 is extremely large compared with that of ordinary light beams that the ratio $\alpha_{11} E_1$ becomes comparable with unity. However, a laser may produce a pulse of one terawatt ($= 10^{12}$ W) which can be concentrated into an area of 10^{-11} m^2 giving, for a short time, a flux density of 10^{18} times the flux density ($1\,\mathrm{kW\,m^{-2}}$) of sunlight at the Earth's surface. At power densities of this order, the above high-order effects can become very significant.

The optics for which eqn (4.10a) is valid is called *linear optics*. The very interesting range of phenomena opened up by lasers, for which eqn (4.10c) is needed, is known as *non-linear optics*. This is considered in Chapter 15. It is still true that most optical theory and most instruments are concerned with linear optics and in other chapters of this book we assume that eqn (4.10a) is valid unless otherwise stated.

4.7 Temporal beats

Suppose two plane waves of equal amplitude E, but with different phases, interact. Then

$$E_r = E_0(\cos \phi_1 + \cos \phi_2)$$
$$= 2E_0 \cos \tfrac{1}{2}(\phi_1 - \phi_2)\cos \tfrac{1}{2}(\phi_1 + \phi_2) \quad (4.11)$$

If the directions are the same and $\delta_1 = \delta_2 = 0$, eqn (4.2) gives

$$E_r = 2E_0 \cos \tfrac{1}{2}[(\omega_1 - \omega_2)t - (\varkappa_1 - \varkappa_2)x]$$
$$\times \cos \tfrac{1}{2}[(\omega_1 + \omega_2)t - (\varkappa_1 + \varkappa_2)x] \quad (4.12)$$

This resultant is shown in Fig. 4.2. There is a beat wave (shown by the dotted lines) with a temporal frequency $\tfrac{1}{2}(\omega_1 - \omega_2)$. The light energy fluctuates with a frequency $(\omega_1 - \omega_2)$.

This may be shown experimentally by dividing a beam of light into two, reflecting one beam from a moving mirror and then combining the two beams on a detector. If the direct beam at the detector has a frequency ω_1, the beam received from the moving mirror will have a frequency higher or lower than ω_1 depending on whether the mirror has a component of velocity towards or away from the detector—the well-known Doppler effect. The two beams are then recombined and generate an electrical signal which follows the energy fluctuations. The speed of the moving reflector may be adjusted to give a beat frequency of about 1 MHz, which can be measured by standard radio-frequency methods. If the speed v of the moving mirror is known, $\omega_1/\omega_2 = (c \pm v)/c$, and since $\omega_1 - \omega_2$ is also known ω_1 and ν_t can be calculated. It is found that, for red light, $\nu_t = 4.5 \times 10^{14}$ Hz.

4.8 Spatial beats: interference fringes

Now consider the interaction of two waves of the same amplitude and temporal frequency but with slightly different directions. If the planes of constant phase are normal to the xy plane and the wave-normals make small angles $\pm \theta$ with the x axis (Fig. 4.3) then, in (4.8), $\alpha = \cos \theta$, $\beta = \pm \sin \theta$ and $\gamma = 0$, so that we have

$$E_1 + E_2 = E_0\{\cos(\omega t - \varkappa x \cos \theta - \varkappa y \sin \theta)$$
$$+ \cos(\omega t - \varkappa x \cos \theta + \varkappa y \sin \theta)\} \quad (4.13)$$

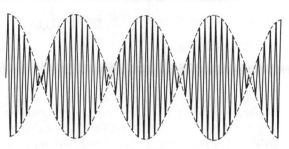

Figure 4.2 Beat wave

If we confine our attention to the plane for which $x = 0$ and put $\sin \theta = \theta$ because θ is small, eqn (4.13) reduces to

$$E_1 = E_0 \cos(\omega t - \varkappa y\theta) \quad (4.14a)$$

and

$$E_2 = E_0 \cos(\omega t + \varkappa y\theta) \quad (4.14b)$$

so that

$$E_r = 2E_0 \cos \omega t \cos \varkappa y\theta \quad (4.15)$$

and

$$W = E_r^2 = 2E_0^2(1 + \cos 2\varkappa y\theta)\cos^2 \omega t \quad (4.16)$$

Thus the intensity on the plane $x = 0$ is a series of fringes (perpendicular to the plane of the paper in Fig. 4.3) with a spatial frequency $\nu_f = 2\theta\varkappa/2\pi = 2\theta\nu_s$, which is equal to the angle between the wave-normals multiplied by ν_s.

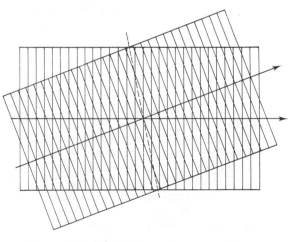

Figure 4.3 Intersecting plane wave-fronts

4.9 Interference in thin films

Suppose that a parallel beam of monochromatic light is incident normally upon one of a pair of glass plates inclined at a small angle θ (Fig. 4.4). Then the angle between the wave reflected from the upper surface AB of the air film and that reflected from the lower surface AC is 2θ. An observer viewing the plates from above sees fringes of spatial frequency $\nu_f = 2\theta\nu_s$. If $2\theta = 0.001$ radians, ν_f is found to be about 2000 fringes m^{-1} with green light so that ν_s is 2×10^6 waves m^{-1} and λ is 5×10^{-7} m. The fringes can be seen with the naked eye and are conveniently measured with a low-power microscope.

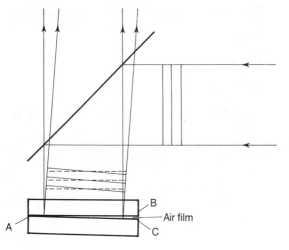

Figure 4.4 Interference from a wedge

4.10 Newton's rings

Now suppose that a parallel beam of light is incident on the air film between the curved surface of a weak planoconvex lens and a plane (Fig. 4.5). If e_P is the thickness of the film at the point P, the path difference for the two waves reflected from either side of the film is $2e_P$. This produces a phase difference $2\varkappa e_P$. All points on a circle through P will have the same path difference. Let d_P be the diameter when $2\varkappa e_P = 2p\pi$; then since $(d_P/2)^2 = e_P(2R - e_P) \doteq e_P R$, then

$$d_P = 2(p\lambda R)^{1/2} \qquad (4.17)$$

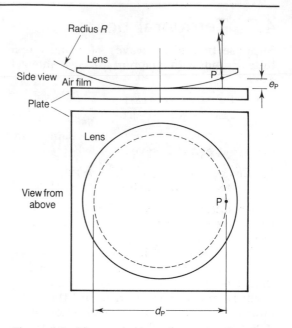

Figure 4.5 Newton's rings: lens on a flat plate

If we consider only the path difference we should expect to see bright fringes when p is an integer and the centre (for which $p = 0$) should be bright. In fact the centre is dark (Fig. 4.6) and the bright fringes are seen when $2e_P$ is an odd number of half-wavelengths. This is because there is a phase change of π on reflection at the lower surface (where the reflected ray is in the less dense medium) and none on reflection at the upper surface (where the reflected ray is in the more dense medium). This change of phase is also found in the Lloyd's mirror experiment (Section 4.14). (The need for such phase changes is seen if the gap between two surfaces is reduced to zero, which results in solid glass—no interface and hence no reflection.).

4.11 Optical path

With $R = 4$ m and $\lambda = 5 \times 10^{-7}$ m eqn (4.17) gives 1 mm for the diameter of the first bright ring (seen when $p = \frac{1}{2}$). With the naked eye a few fringes can be seen and many more with a low-power microscope. Newton, who first studied these fringes, sometimes used surfaces of very small curvature ($R = 16$ m) and made surprisingly accurate measurements of the

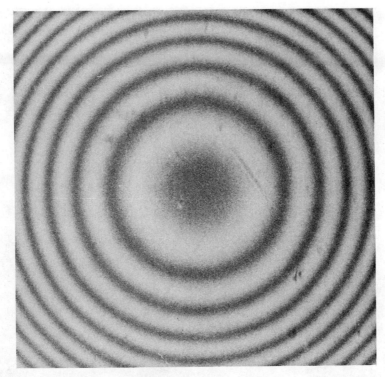

Figure 4.6 Newton's rings

diameter of several rings with the naked eye. He showed that the values of e_P for a succession of bright rings were in arithmetic progression. He also found that if a film of liquid was placed between the two glass surfaces the diameters of the corresponding rings were reduced in the ratio $1 : \sqrt{n}$. This is because the wavelength in the medium has been reduced in the ratio $1 : n$ (eqn 4.6) and substitution in eqn (4.7) gives a lower value of d_P. Passage through a distance e in a medium of index n produces the same effect on the phase as a distance ne in a medium of index 1; the distance ne is called the *optical path*. If we use the vacuum wavelength λ_v and, for a ray that passes through different media, add up the optical paths we obtain the phase difference by multiplying the total optical path by $\varkappa_v$ (the value of $\varkappa$ for the wave in vacuum).

4.12   Contour fringes

We may note that both with the air wedge and with Newton's rings the fringes give a contour map of the upper surface if the lower surface is plane. In the case of the wedge the upper surface is a hill ascending at constant rate and the fringes are straight lines. In the second case the contours are circles. The contours are at height intervals of $\lambda/2$. If a piece of imperfectly plane glass is placed on an optical flat (whose surface is plane to within $\approx 0.03\lambda$), fringes that are contours of the 'hills' and 'valleys' on the imperfect glass are obtained. The fringes originally used by Fizeau (1819–1896) are called *contour fringes* or *fringes of constant thickness* (Fig. 4.7).

4.13   Young's experiment

In both of the above experiments the two beams of light are produced by division of the amplitude of an incident monochromatic wave through partial reflection at the upper surface. We now consider an experiment (Fig. 4.8a) in which two sources are obtained by selecting different parts of a wave-front by two apertures in a diaphragm. In Fig. 4.8(a), S_0, S_1 and S_2 are slits perpendicular to the

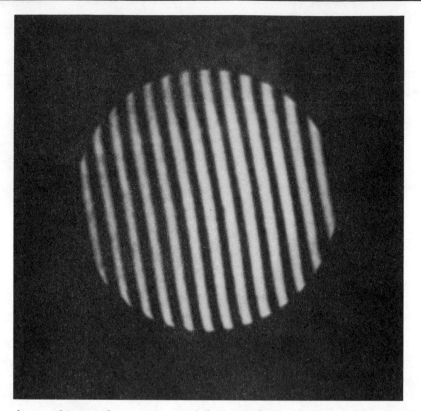

Figure 4.7 Interference fringes of constant optical thickness formed between an optical flat and a glass surface: if the fringes were *perfectly* straight, parallel and equidistant, the surface would also be flat. It is very slightly curved at the edges

plane of the paper; x_1 and x_2 (each about 1 m) are large compared with $2h$ (about 1 mm). For the present assume that the slits are very narrow (width a small fraction of 1 mm), and we explore the variation of intensity with y over a range small compared with x. Consider first the case where S_1 and S_2 are small circular apertures. The trace of the spherical waves W_1, W_2 in the plane of the diagram of Fig. 4.8 will, for distances close to the axis, be almost straight lines, crossing at an angle $\theta = 2h/x_2$. Thus fringes of spatial frequency $\nu_f = \theta \nu_s$ will be obtained along the line OQ, and

$$\nu_f = \frac{2\pi}{x_2} \nu_s = \frac{2h}{x_2 \lambda} \qquad (4.18)$$

If the circular apertures are extended above and below the plane of the diagram, to produce long slits, the above argument will apply to each element of the slit, with the

result that parallel straight-line fringes will be seen on a screen placed at Q.

For an experiment using a mercury lamp and a green filter as source, the distance between successive fringes $(1/\nu_f)$ is found to be ~ 0.5 mm when $x_2/2h = 10^3$ so that (for green light) we again obtain $\nu_s = 2 \times 10^6$ waves m^{-1} or $\lambda = 5 \times 10^{-4}$ mm.

4.14 Fresnel's biprism and Lloyd's mirror

Two other arrangements which effectively give slit sources side by side are depicted in Fig. 4.9 which show (a) Fresnel's biprism and (b) Lloyd's mirror. With Lloyd's mirror the lowest fringe is dark owing to the change of phase on reflection in the less dense medium (Section 4.10).

The fringes obtained with thin films were

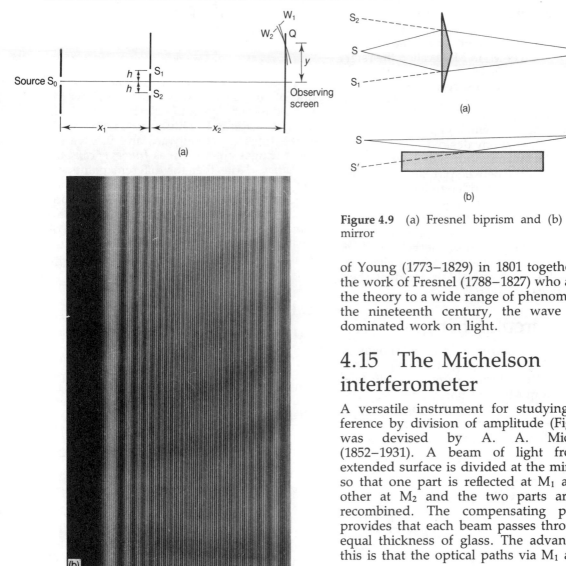

(a)

(b)

Figure 4.8 (a) Young's experiment and (b) Young's fringes

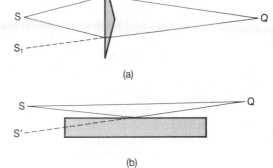

(a)

(b)

Figure 4.9 (a) Fresnel biprism and (b) Lloyd's mirror

of Young (1773–1829) in 1801 together with the work of Fresnel (1788–1827) who applied the theory to a wide range of phenomena. In the nineteenth century, the wave theory dominated work on light.

4.15 The Michelson interferometer

A versatile instrument for studying interference by division of amplitude (Fig. 4.10) was devised by A. A. Michelson (1852–1931). A beam of light from an extended surface is divided at the mirror M_0 so that one part is reflected at M_1 and the other at M_2 and the two parts are then recombined. The compensating plate C provides that each beam passes through an equal thickness of glass. The advantage of this is that the optical paths via M_1 and M_2

observed by Huygens (1629–1695) and Hooke (1635–1703) in the middle of the seventeenth century and Newton studied the rings in 1666. Unfortunately these phenomena did not lead Newton to the wave theory though he recognised the periodicity in space and in time and even considered the possibility of vibrations. In the eighteenth century the wave theory was neglected. It did not gain acceptance until the experiment

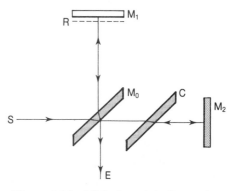

Figure 4.10 Michelson interferometer

can be made the same for all wavelengths. For this position only, fringes can be obtained with white light. The plane R which is the reflection of M_2 in M_0 is called the reference plane. Interference between the two beams seen by the observer at E occurs as though one beam had been reflected from M_1 and the other from R. The mirrors are provided with screws which enable R to be set parallel to M_1 or at any desired small angle to it. M_1 can be moved, normal to itself, slowly. Thus the distance between M_1 and R and the angle can be altered independently.

If M_1 is set at a small angle to R and a parallel beam is incident, straight-line fringes as described in Section 4.9 are produced, and these fringes move across the field if M_1 is traversed. If M_1 is set accurately parallel to R and a solid cone of light is incident then circular rings are seen

4.16 Circular fringes

The formation of these rings is shown in Fig. 4.11. A beam arriving at P behaves as if it were partly reflected from R (really it is reflected from M_2) and partly from M_1. The path difference between the two parts is $AB + BC - AE = 2e/\cos\theta - 2e\tan\theta'\sin\theta = 2e(1 - \sin^2\theta)/\cos\theta = 2e\cos\theta$, where e is the

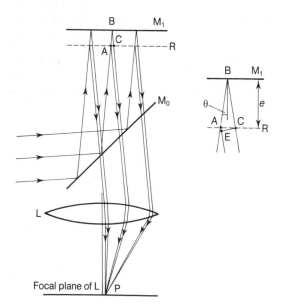

Figure 4.11 Ray paths in Michelson interferometer

distance from M_1 to R. If e is constant, bright rings which subtend angles θ_1, θ_2, ..., θ_P are seen where

$$2e\cos\theta_P = p\lambda \qquad (4.19)$$

Since both beams are reflected at the same kind of surface we do not have to allow for any phase change on reflection. If M_1 is brought nearer to R (reducing e), θ_P increases so the rings move outward and vice versa. These fringes are known as *fringes of constant inclination*. Note that for different points on one fringe e and θ are both constant. With fringes of constant inclination e is constant for the whole field and each fringe corresponds to a different value of θ; for fringes of constant thickness (contour fringes) θ is constant over the whole field and each fringe corresponds to a different value of e.

4.17 Interference with a number of waves

The calculation of a resultant by adding cosine terms, used in Sections 4.11 to 4.16, is easy when only two beams are involved. The following procedure is more convenient when there are many waves. We represent one wave by the expression:

$$E = E_0 \exp i(\omega t - \varkappa x + \delta) = E_c\, e^{i\phi_0} \qquad (4.20)$$

where $\phi_0 = \omega t - \varkappa x$ and $E_c = E_0\, e^{i\delta}$ is called the *complex amplitude* (see Appendix 4.B). Then

$$W = EE^* = E_c E_c^* \qquad (4.21)$$

so that a real number is available for comparison with measurements of W.

The resultant E_r for a number of waves of the same frequency is

$$E_r = (E_{c1} + E_{c2} + \cdots)e^{i\phi_0}$$

and

$$E_{rc} = \sum_r E_{cr} \qquad (4.22)$$

i.e. *the complex amplitude of the resultant is the sum of the complex amplitudes of the component waves.*

4.18 Grating spectra

Let us now consider the superposition of waves from a number N of slits of equal spacing d, instead of the two slits used in Young's experiment. This arrangement of apertures is usually called a diffraction grating but in the present context it is an arrangement for multiple-beam interference. Each slit is assumed to be narrow, the width being much smaller than the distance between the slits. Such gratings were at one time made by ruling with a fine point on glass that had been silvered. We shall describe modern gratings later (Section 6.13).

When parallel light falls normally on the grating (Fig. 4.12), each aperture scatters light into the region beyond the grating. Light emerging in a particular direction arrives at the focal plane of the lens L at Q. The phase difference between the waves that arrive at O from successive slits is constant. If the amplitudes at Q are all the same and equal to E and the phases are $0, \delta, 2\delta, \ldots, (N-1)\delta$, then

$$\delta = \varkappa d \sin \theta \qquad (4.23)$$

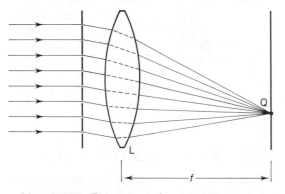

Figure 4.12   Formation of grating spectrum

The complex amplitude of the resultant is

$$E_{rc} = E_0(1 + e^{-i\delta} + e^{-2i\delta} + \cdots + e^{-(N-1)i\delta}) \quad (4.24)$$

Although the phases are in *arithmetic* progression the terms of eqn (4.24) form a *geometric* progression whose sum is

$$\frac{E_{rc}}{E_0} = \frac{1 - e^{-N\delta}}{1 - e^{-i\delta}} \qquad (4.25a)$$

and

$$|E_{rc}|^2 = E_0^2 \left(\frac{1 - \cos N\delta}{1 - \cos \delta} \right) = E_0^2 \frac{\sin^2(N\delta/2)}{\sin^2(\delta/2)} \qquad (4.25b)$$

When $\delta = 2m\pi$ (where m is an integer, $-2, -1, 0, 1, 2$, etc.),

$$\sin \theta = \frac{\delta}{\varkappa d} = \frac{m\lambda}{d} \qquad (4.26)$$

When δ is small we may substitute the angles for the sines and as δ approaches zero $|E_{rc}|^2$ approaches $N^2 E_0^2$. The same value is obtained whenever δ approaches an integral multiple of 2π. The variation of $|E_{rc}|^2$ with δ is shown in Fig. 4.13.

Considering the case when $m = 1$, we see that white light is spread out into a spectrum. Previously we saw that refraction through a prism gave a spectrum and this established a relation between perceived colour and refractive index. The grating spectrum establishes a relation between perceived colour and wavelength in the most direct way, though this relation was implied by the fact that Newton's rings for red light have larger diameters than the rings for blue light. We find that $\lambda = 620-700$ nm is seen as red light, $500-560$ nm as green and $350-450$ nm as blue to violet.

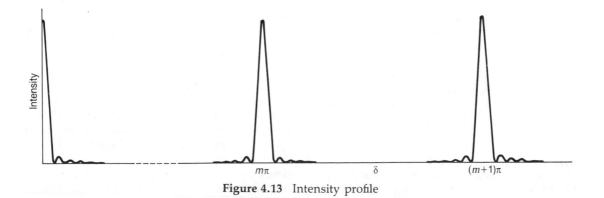

Figure 4.13   Intensity profile

4.19 Spectrum lines

With white light from an electric filament lamp the spectrum is found to be continuous. Light from electric arc and spark sources gives spectra containing a large number of narrow lines (Fig. 4.14). Simpler spectra are obtained with a sodium lamp which gives only two strong lines (at 589.0 and 589.6 nm) or with a low-pressure mercury lamp (strong lines at 597, 577, 546, 436 and 405 nm). In the range from the ultraviolet to the infrared part of the spectrum, hydrogen emits several series of lines, of which four are shown in Fig. 4.14(b). In contrast to emission sources, the white-light spectrum of sunlight is crossed by *dark* lines named after J. Fraunhofer (1787–1826) who also made some of the first diffraction gratings.

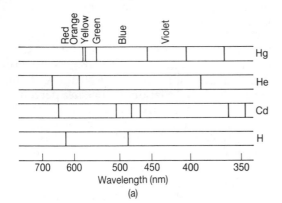

(a)

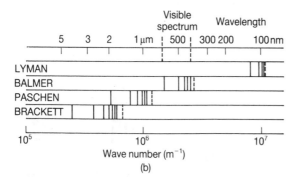

(b)

Figure 4.14 (a) Position of lines in spectra of some common gases and vapours. (b) Four of the series of hydrogen spectra

4.20 Why spectrum lines?

In modern physical theory an atom or molecule has a series of stationary states of energies ξ_1, ξ_2, etc. When a large collection of atoms changes from a state of high energy ξ_n to one of lower energy ξ_m a train of light waves is emitted. The frequency ν is given by

$$h\nu = \xi_n - \xi_m \qquad (4.27)$$

The energy difference corresponding to the transition $n \rightarrow m$ is called a *quantum*. h is a universal constant equal to 6.62×10^{-34} J s. The intensity of the emitted light is observed to fall exponentially with time with $I(t) = I_0 e^{-t/\tau}$. The time constant τ is usually less than 10^{-8} s so that there are more than a million waves in the train and the change of amplitude over a thousand waves is quite small. Thus in many experiments one of these wave trains behaves like an indefinitely long wave of constant amplitude, i.e. like a truly monochromatic wave. They give sharp lines in the spectrum.

When light falls upon a gaseous atom quanta of energy may be absorbed, raising the energy from ξ_m to ξ_n and, since (4.26) is still valid, dark lines are produced in a spectrum of white light. The Fraunhofer lines in the Sun's spectrum are caused by absorption of light from the inner region of the Sun (*photosphere*) by a cooler outer layer called the *chromosphere*.

4.21 Free spectral range

So far we have discussed the spectrum for $m = 1$. Other spectra corresponding to $m = 2$, 3, 4, etc., exist and the mth-order line of λ_1 will coincide with the $(m + p)$th-order line for λ_2 if

$$m\lambda_1 = (m + p)\lambda_2 \qquad (4.28)$$

and, if $p = 1$,

$$\lambda_1 - \lambda_2 = \frac{\lambda_1}{m + 1} \qquad (4.29)$$

The difference of wavelength between two lines whose successive orders coincide is called the *free spectral range*. The eye has very little sensitivity (see Fig. 16.1) outside the limits 400–700 nm and if the ultraviolet is excluded (by a filter or by the absorption of glass lenses) no overlapping occurs in the first order. When $m = 1$ the free spectral range covers the whole visible spectrum. We shall find a very different situation at high orders of interference (when m may be 10^5 or 10^6).

4.22 Interference of multiple beams by repeated reflections

In Sections 4.9 to 4.16, we considered the effects of interference between *two* beams of light of equal amplitude, the result of which is the formation of fringes with an intensity distribution given by eqn (4.16). Since $(1 + \cos 2\varkappa\theta y) = 2 \cos^2(\varkappa\theta y)$, such fringes are termed 'cos^2' fringes. The widths at half-maximum of the light and dark regions are equal. In Sections 4.18 to 4.21, we saw that the combination of large numbers of waves, from the slits in a grating, could give rise to fringes in which the bright fringes are very narrow compared with the dark.

An alternative way of producing large numbers of beams is to arrange for interference to occur between beams arising between surfaces that are silvered and so have a large reflection coefficient (but with *some* transmission).

Let us contrast the behaviour of a pair of unsilvered surfaces with that of high-reflecting surfaces. For unsilvered surfaces the reflected light will be dominated by two beams of nearly equal amplitude. For glass of refractive index 1.5, 4% of the incident light energy is reflected; Fig. 4.15(a) shows the amplitudes of the first few beams reflected and transmitted in this case. For the reflected light, the only two beams of consequence are almost equal in amplitude and the argument of Section 4.8 applies. Note that in the transmitted light, one beam dominates. Although fringes may be seen in transmission (e.g. when slides are projected) the contrast is very low.

Now look at the case of highly reflecting plates, where we assume that each surface has an energy reflectance ρ and transmittance σ. In practice, in contrast to the few per cent reflectance of uncoated glass, values of ρ approaching unity may be obtained.

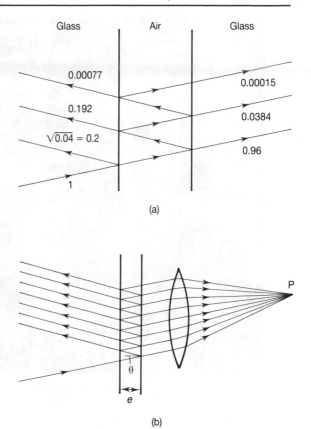

Figure 4.15 (a) Two-beam interference and (b) Multiple-beam interference

We are generally interested in the transmitted light in such a system. For Fig. 4.15(b) the phase difference between successive emerging beams is given by

$$\delta = 2\varkappa e \cos \theta + \varepsilon \qquad (4.30)$$

Although for a dielectric surface ε will be zero or π, for a metal film the value will be somewhere between these two limits.

The resultant transmitted amplitude is given by

$$E_r = \sigma E_0(1 + \rho e^{-i\delta} + \rho^2 e^{-2i\delta} + \cdots)$$

$$= \frac{\sigma E_0}{1 - \rho e^{-i\delta}} \qquad (4.31)$$

Thus the energy flux at the point P in the focal plane of the lens in Fig. 4.15(b) is proportional to

$$E_r E_r^* = \frac{\sigma^2 E_0^2}{1 + \rho^2 - 2\rho \cos \delta}$$

Figure 4.16   Multiple-beam fringes in transmission

which may be written

$$E_r E_r^* = \frac{\sigma^2 E_0^2}{(1-\rho)^2} \frac{1}{1 + L \sin^2(\delta/2)} \quad (4.32)$$

where

$$L \equiv \frac{4\rho}{(1-\rho)^2}$$

for $\rho = 0.8$, $L = 80$. If $\rho = 0.95$—easily obtained with dielectric multilayers (Section 4.30)—L is over 1500.

From eqn (4.32) we see that maxima in the interference pattern occur for $\delta = 2\pi \times$ integer. The half-maximum value occurs when $L \sin^2(\delta/2) = 1$. Writing $\delta_{1/2} = 2m\pi \pm \delta$ and $\sin^2(\delta_{1/2}/2) \doteq (\delta_{1/2}/2)^2$, we see that

$$\delta_{1/2} = \frac{2}{L^{1/2}} \quad (4.33)$$

Successive fringes occur at intervals if δ equal to 2π. Thus the ratio of fringe separation to fringe width, a convenient measure of fringe sharpness, is given by

$$F \equiv \frac{2\pi}{2\,\delta_{1/2}} = \frac{\pi L^{1/2}}{2} \quad (4.34)$$

F is called the *finesse*. For $\rho = 0.8$, the finesse is 14; for $\rho = 0.95$, $F = 60$. The multiple-beam fringes are seen as sharp, bright lines against a dark background Fig. 4.16).

4.23   Fabry–Perot etalon

In Fig. 4.17 two highly reflecting plates, both accurately plane, are supported by a separator which is made so that they are accurately parallel and have a fixed separation. This arrangement is called a *Fabry–Perot etalon* after C. Fabry (1867–1945) and A. Perot (1863–1925). If a weakly convergent beam of light is incident upon the etalon and the transmitted light is focused by the lens L_2 then bright narrow circular rings are seen in the focal plane. The order of interference at the centre may be written as $(m + f)$, where m is an integer and f is a fraction. Usually m is a large number. It is about 10^5 when the separation (e) of the plates is 2.5 cm. From measurements of e with a travelling microscope, m can generally be determined to

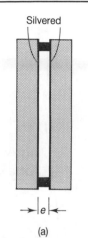

(a)

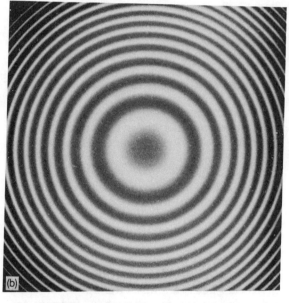

(b)

Figure 4.17   Fabry–Perot etalon

within ± 3. Rings occur at $\theta_1, \theta_2, \ldots, \theta_n$, where

$$\frac{2e}{\lambda} = m + f = \frac{m}{\cos \theta_1} = \frac{m - n + 1}{\cos \theta_n} \quad (4.35)$$

and if θ_n is small enough for us to put $\cos \theta_n = 1 - \frac{1}{2}\theta_n^2$ we find

$$\theta_n^2 = \frac{(f + n - 1)\lambda}{e} \quad (4.36)$$

If θ_n is plotted against $n - 1$, the intercept on the axis gives f; alternatively the method of least squares may be used to give a more accurate value. The measurements of

$\theta_1, \ldots, \theta_n$ are usually good enough to determine f with an accuracy of ± 0.01–0.02.

4.24 Method of exact fractions

If the fractions f_1, f_2, f_3, ... are determined for three wavelengths and the integers m_1, m_2, m_3 are known to within ± 3, then taking the seven possible values of m gives seven possible values of e from which possible values of $m_2 + f_2$ and $m_3 + f_3$ can be computed. It is usually found that only one value of e gives the measured fractions for all three wavelengths, and this is accepted as the correct value. In the rare event that more than one value of e, within the possible range, gives three correct fractions a fourth wavelength must be used. Thus the ratios of the wavelengths and the thickness of the etalon are known to within a few parts in 10^7, using an etalon of separation 2.5 cm. Longer etalons may be used if higher accuracy is required. The limit is set by the width of the rings due partly to imperfection in planeness and parallelism of the plates and partly to the fact that no real source can produce a *perfectly* monochromatic line.

4.25 Analysis of hyperfine structure

It is often found that a line which, in a grating spectrograph, appears to be single and fairly sharp shows several rings (of one order) when tested by an etalon. This is due to two causes. Firstly, many elements have significant fractions of more than one isotope and the energy levels depend to a small extent on the atomic weight. Secondly, the orientation of nuclear spin with respect to the resultant magnetic moment of the electrons has also a small effect on energy levels. The etalon gives very accurate values of the wavelength differences and hence (through eqn 4.27) of the energy differences. The isotope separations can be distinguished by using material in which one or more isotopes have been enriched to a known extent. The other separations then give important information on nuclear spin.

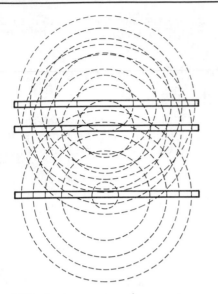

Figure 4.18 Appearance of spectra when Fabry–Parot etalon is inserted in spectograph

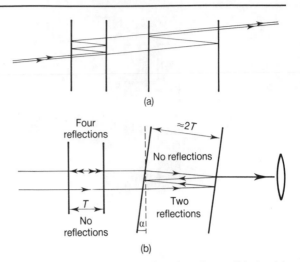

Figure 4.19 (a) Two inclined etalons. (b) As (a) with axis of first etalon parallel to incident beam

The possibility of analysing complicated lines depends on the power of the etalon to separate close lines. This depends on the sharpness of the rings and this increases with e because $\delta_{1/2}$ is independent of e (eqn 4.33). Also, eqn (4.30) shows that the change in $\cos \theta$ required for a given value of δ is inversely proportional to e. However, the order of interference also increases with e so that the free spectral range decreases (eqn (4.29), and if e is too great the lines of one order become confused with those of another. There will thus be an optimum value of e. In order to measure many lines in one experiment, the light from an etalon may be focused on the slit of a spectrograph. The rings then appear as horizontal divisions on the lines of a spectrum (Fig. 4.18) and their diameters may be measured.

4.26 Comparison of two etalons

If light passes successively through two etalons, it is possible for interference to occur between light that has been reflected $2n$ times in one etalon and $2m$ times in the other, as shown in Fig. 4.19(a) which shows the simplest case where $m = 1$ and $n = 2$. If the thicknesses of the etalons are me_0 and ne_0

respectively then the total path difference is $mne_0 - nme_0 = 0$ and a fringe of zero order should be seen when the etalons are exactly parallel to one another. If the ratio is not quite m/n but $(m + \Delta)/n$ (where Δ is very small compared with m) then zero-order interference will be obtained when

$$(m + \Delta)n = mn \cos \alpha \qquad (4.37)$$

if the light is incident normally on the first etalon and at an angle α on the second.

With white light the fringe of zero order is easily distinguished from the other fringes because it is the only one for which the maxima for all wavelengths coincide. It thus appears as a clear white fringe (or black if there has been a phase difference of π by reflection) whereas all other fringes are coloured. By making one etalon normal to the incident light (and to the telescope; see Fig. 4.19b) and tilting the other till the fringe of zero order is in the centre of the field the angle α can be measured. If two etalons have been made with the intention that one shall have, for example, exactly twice the separation of the other, α will be a very small angle and its measurement will give the true ratio to better than 1 part in 10^7. This method is good when m and n are less than about 10. For much higher values of m and n the zero-order fringe is obscured by stray light from rays that have suffered other numbers of reflections. These fringes were first observed by Brewster (1781–1868) in 1815. Despite its age, this method remains a simple, powerful, accurate means of comparing etalon lengths.

4.27   Channelled spectrum

If a parallel beam of white light passes normally through an etalon and is then concentrated on the slit of a spectrograph sharp bright fringes appear at wavelengths for which $2e = n_1\lambda_1 = (n_1 + 1)\lambda_2$, etc. The corresponding frequencies (ν_{s1}, ν_{s2}, etc.) are

$$\frac{n_1}{2e}, \qquad \frac{n_1 + 1}{2e}, \qquad \text{etc.} \qquad (4.38)$$

i.e. the fringes are at constant frequency intervals. These fringes may be used to calibrate the spectrograph if e is known.

4.28   Multiple-beam contour fringes

Tolansky (1907–1973) saw that the contour fringes (Section 4.12) could be made very much sharper by coating both the optical flat and the surface under test with highly reflecting metal films. Fringes produced in this way are shown in Fig. 4.20. A very small deviation from straightness or a break in a fringe can be seen and this enables steps on a surface of less than 0.5 nm to be detected. The topography of surfaces could thus be explored with an accuracy that revealed features of molecular dimensions. Tolansky used both fringes of transmission (as in the Fabry–Perot interferometer) and fringes of reflection (as in the experiments of Newton and Fizeau). The multiple-beam fringes of reflection are complementary to the fringes of

Figure 4.20 Multiple-beam contour fringes seen in transmission

Figure 4.21 Multiple-beam contour fringes seen in reflection

transmission. They show sharp black lines on a light background (Fig. 4.21).

4.29 Where *are* all these fringes?

We obtain a clue to this question by considering how we can observe the different kinds of fringe that we have been discussing. Take the case of Young's fringes (Section 4.13). We find that we can observe fringes on the screen of Fig. 4.8(a) whatever the value of x_2 (although they might be a bit hard to see if we made x_2 too big). We describe such fringes as 'non-localised'. In contrast we find in the case of Newton's rings that we need to focus our eyes (or a magnifying glass) on the region between the lens and plate of Fig. 4.5. These fringes are 'localised' in the neighbourhood of the air film. In the third type of fringe we consider the Fabry–Perot etalon. Although we can see fringes when we look through an etalon we cannot obtain them on a screen in the manner of Fig. 4.8(a) unless we interpose a converging lens. When we do this the Fabry–Perot rings appear on a screen placed in the focal plane of the lens. The fringes are formed by the parallel bundles of rays leaving the etalon which appear to come from infinity. Thus these fringes are said to be 'localised at infinity'.

So far we seem simply to be looking at the results of experiments which illustrate that light waves can interfere with one another. Is this of any particular use? Can we make interference work for us?

4.30 Multilayer dielectric films

By evaporation in a good vacuum on to clean surfaces it is possible to coat glass with one or more thin layers of materials of different refractive indices. These layers adhere to the substrate so well that the final surface is at least as hard as glass. Interference of beams reflected at the different interfaces can be made to produce:

1. Surfaces of very low reflection
2. Surfaces of very high reflection and low absorption

3. Surfaces that divide a beam of light into two parts (one reflected and the other transmitted) with very little loss (called *beam-splitters*)
4. interference filters that select bands of the spectrum down to a few nanometres in width

4.31 Antireflection coatings

It is shown in Section 8.17 that, for light incident normally on a surface between two dielectrics of refractive indices n_1 and n_2, the amplitude reflection coefficient is

$$\rho^{1/2} = \frac{n_1 - n_2}{n_1 + n_2} \qquad (4.39)$$

If a layer of optical thickness $\lambda/4$ and index n_c is deposited on the surface of glass of index n_g (Fig. 4.22) then the amplitudes of the waves reflected at the two interfaces are of opposite sign (if $n_c < n_g$) and are equal in magnitude if

$$\frac{n_g - n_c}{n_g + n_c} = \frac{n_c - 1}{n_c + 1} \qquad (4.40)$$

i.e. if $n_c^2 = n_g$. Thus, by destructive interference of the two waves, the reflection is zero. Strictly speaking, the effect of multiple reflections should be take into account. In fact, the result $n_c^2 = n_g$ still holds when an exact treatment is used.

For most glasses, it is usually not possible to find a suitable material whose index com-

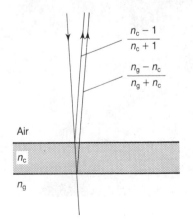

Figure 4.22   Single-film antireflection coating

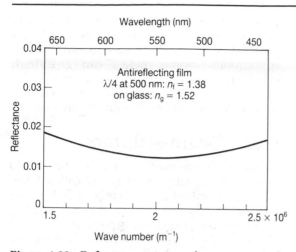

Figure 4.23 Reflectance curve for a practical single antireflecting film of MgF_2 on glass

plies with eqn (4.40) exactly. Also, even if the material had the correct index at every wavelength the thickness would be correct for only one wavelength. In practice the reflection coefficient follows a curve similar to that shown in Fig. 4.23. It is usual to choose the thickness to obtain the minimum reflection in the green region of the spectrum where the eye is most sensitive. The residual reflection is mainly red and blue which together give a purple which resembles the bloom on a ripe plum and so the process is often called 'blooming'. The overall effect is to reduce the energy reflection coefficient from about 0.04 (for a glass of index 1.5) to 0.015. This is a considerable advantage in an instrument such as binoculars where light will have suffered at least eight reflections. For uncoated surfaces, the overall transmission would be 75%; this is raised to over 90% by coating. Moreover, the unwanted reflections serve to degrade the final image. Coating with a single layer is not good enough for instruments with many surfaces and for them it is worth while incurring the additional expense involved in depositing two or more layers which reduce the energy reflection coefficient to about 0.0025.

4.32 High-efficiency reflection films

Mirrors were originally made by polishing metals or depositing thin layers of the metal on glass. Silver and aluminium both gave very high reflection coefficients (up to $\rho = 0.98$). Aluminum-coated mirrors, being more durable, are used for telescopes, although recoating once a year is needed. Rhodium-coated mirrors last much longer but are more difficult to make. Thin metallic films are sometimes used to divide a beam into two equal parts, but 35% transmission, 35% reflection and 30% absorption is a typical performance and when, as in the Michelson interferometer, each beam is transmitted *and* reflected the loss is significant.

Mirrors of very high reflection and low loss may be made by depositing on glass alternate layers of high and low index materials (Fig. 4.24). The films are of equal optical thickness ($\frac{1}{4}\lambda$) so that transmission twice

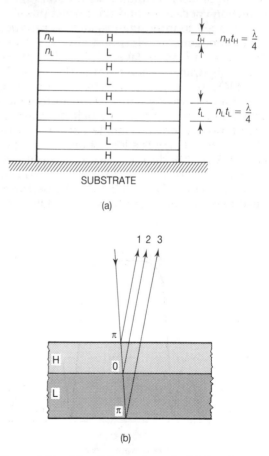

Figure 4.24 (a) Schematic of high-reflecting quarter-wave stack. (b) Waves from single element of quarter-wave stack

through any one produces a phase difference of π. There is also a phase difference of π whenever reflection occurs in the less dense medium (Section 8.17). Suppose that the phase of the incident wave at the surface is zero; then beam (1) (Fig. 4.24(b)) has a phase of π (due to reflection), beam (2) has no phase difference on reflection but has a phase difference to π due to transmission twice and beam (3) has a phase difference of 2π due to transmission and π due to reflection, making 3π, so that it is in phase with beams (1) and (2). If ρ is 0.04 at each reflection, the amplitude reflection coefficient ($\rho^{1/2}$) is 0.2, so that with about ten films, surfaces with $\rho = 0.995$ are produced. The transmittance is around 0.003, the residual loss being due not to absorption but to small inhomogeneities in the films which produce scattering. The demand of laser technologists for still higher reflection coefficients and the lowest possible loss has led to intensive research to produce very homogeneous films and also to the use of stacks with more layers. Since the phase change by transmission is proportional to $(2e \cos \theta)/\lambda$, the highest reflection is obtained only for one wavelength and for normal incidence but there is only a small decrease in reflectance over an angle of $10°$ (cos $10° = 0.985$). There is a loss of only a few per cent if λ is 15% different from the optimum so that a mirror designed for 500 nm will be fairly good from 575 to 425 nm. The performance falls off very rapidly when λ differs by

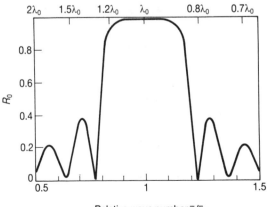

Figure 4.25 Reflectance versus wave number for high-reflecting quarter-wave stack of eleven layers

20% from the design value (Fig. 4.25). The high-efficiency films, often essential for lasers, do not entirely replace metals in other applications. Some metals can give high reflection from 100 nm in the far ultraviolet through the visible spectrum, infrared and microwave regions of the spectrum.

4.33 Beam-splitters

When stacks are made to divide a beam into two nearly equal parts they are called *beam-splitters*. A typical performance is 50% transmission with 49.5% reflection. Usually beam-splitters are designed for incidence at $45°$ so that the reflected beam is deflected through a right angle. However, they can be made for any ratio of splitting and for any angle of incidence.

4.34 Interference filters

Glass or gelatine filters transmit regions of the spectrum about 50 nm wide but many experiments require bands of width 5 nm or less. We have seen in Section 4.28 that a Fabry–Perot etalon gives a channelled spectrum with narrow lines whose separation increases if the etalon thickness is reduced. An etalon of suitable thickness will give lines at intervals of a little more than 50 nm (so that only one line in the visible spectrum passes an appropriate glass filter), and if the reflection coefficient is very high the lines will be very sharp. Figure 4.26(a) shows an arrangement in which two high-efficiency reflecting stacks surround a separator film S. If the spacer is of thickness $6\lambda_0$, the order of the pass-band at λ_0 is 12, since $2e = 12\lambda_0 = n_0\lambda$. There will be pass-bands corresponding to $n = 11$ and $n = 13$. If $\lambda_0 = 500$ nm, these will occur at 545 and 461 nm. The transmittance between successive orders will be small. A typical filter of this type has a transmittance curve such as that of Fig. 4.26(b). Narrower bands can be produced at some loss of maximum transmission.

Interference filters can also be made to transmit all wavelengths below a certain wavelength λ_0 and to reflect longer wavelengths. If λ_0 is about 650 nm the visible spectrum is transmitted and the infrared which contains most of the heat is removed

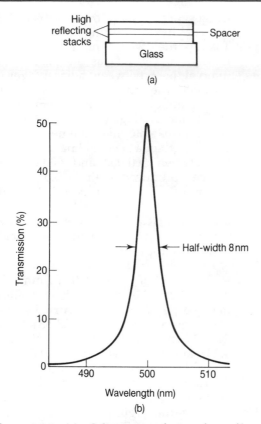

(a)

(b)

Figure 4.26 (a) Schematic of interface filter. (b) Transmittance of the Fabry–Perot type interference filter

from the beam. This is very useful in certain situations where overheating might destroy certain components.

Appendix 4.A   The wave equation

Equation (4.7) represents a simple form of progressive wave. A more general waveform is represented by

$$E = E_0 f[\omega t - \varkappa(\alpha x + \beta y + \gamma z)] \qquad (4.41)$$

where f is any smooth function. Equation (4.41) has the periodicity with time and with space which is characteristic of a wave. The phase is constant over the surface $\alpha x + \beta y + \gamma z = $ constant, i.e. it represents a *plane* wave. Differentiating twice with respect to time and twice with respect to space and

remembering that $\alpha^2 + \beta^2 + \gamma^2 = 1$ we have

$$\frac{\partial^2 E}{\partial x^2} + \frac{\partial^2 E}{\partial y^2} + \frac{\partial^2 E}{\partial z^2} = \frac{\varkappa^2}{\omega^2} \frac{\partial^2 E}{\partial t^2} \qquad (4.42)$$

and, using eqn (4.4),

$$\nabla^2 E = \frac{1}{v^2} \frac{\partial^2 E}{\partial t^2} \qquad (4.43)$$

Similarly, eqn (4.9) may be twice differentiated with respect to t and to r and yields

$$\frac{\partial^2 E}{\partial r^2} + \frac{2}{r} \frac{\partial E}{\partial r} = \frac{1}{v^2} \frac{\partial^2 E}{\partial t^2} \qquad (4.44)$$

which is eqn (4.42) in polar coordinates.

Equation (4.42) is known as the *wave equation* since its various solutions represent waves.

Appendix 4.B Representation of waves by complex quantities

We need two quantities—the amplitude and phase—to represent a wave. The same goes for a complex number, so representation of a wave by a complex number is certainly possible. It turns out not only to be possible but highly advantageous, since the manipulation of exponentials is often very much simpler than that of sines and cosines. The clue to the connection is that the expression for a wave,

$$E = E_0 \cos(\omega t - \varkappa x + \delta)$$

can be written as the real part of

$$E_0 \exp i(\omega t - \varkappa x + \delta).$$

Can we simply use a complex number for a wave with the understanding that the necessarily real quantities associated with a wave are given in the real part of the complex number? For many (but not all) purposes the answer is 'yes'.

The expression $E = E_0 \cos(\omega t - \varkappa x + \delta) \equiv E_0 \cos(\phi_0 + \delta)$ tells us that if we measure E when $\phi_0 + \delta = 0$ we shall get the result E_0 and that for $\phi_0 = 0$ we get $E_0 \cos \delta$. If we write

$$\begin{aligned} E &= E_0 \exp i(\omega t - \varkappa x + \delta) \\ &= E_0 \exp i(\phi_0 + \delta) \qquad (4.45) \\ &= (E_0\, e^{i\delta})e^{i\phi_0} \end{aligned}$$

then for $\phi_0 + \delta = 0$ we obtain $E = E_0$ and for $\phi_0 = 0$, $E = E_0\, e^{i\delta}$, of which the real part is $E_0 \cos \delta$. Thus the prescription appears to work.

The expression $E_0\, e^{i\delta}$ is referred to as the complex amplitude. It may be represented on an Argand diagram. The time variation of a wave at a point x is given by examining the value of E with increasing ϕ_0. Since $E = (E_0\, e^{i\delta})e^{i\phi_0}$, this represents a vector $(E_0\, e^{i\delta})$ in the Argand diagram, corresponding to $\phi_0 = 0$, which rotates in an anticlockwise direction with angular frequency ω.

The complex expressions for waves may be combined (added or subtracted); the real part of the resultant complex number will give the correct result for the combination of real expressions. The multiple-beam problem, discussed in Section 4.22), would be practically impossible without the use of complex representation.

Problems

4.1   A mercury lamp emits light of wavelengths 404.6 and 546.1 nm, measured in vacuum. Calculate the values of ν and ω for these lines.

4.2   Two flat plates form a wedge, in contact at one end and 0.014 nm apart at the other. When illuminated normally with light of wavelength, λ, fringes 0.975 mm are seen. Determine λ.

4.3   In a Newton's rings experiment, with a spherical surface of radius 1 m, the diameter of the tenth dark ring is found to be 4.85 ± 0.01 mm. Calculate the wavelength of the light used.

4.4   In the arrangement of Problem 4.3, the space between the plane and spherical surfaces is filled with water ($n = 1.33$). What will be the radius of the tenth dark ring?

4.5   In demonstrating Young's experiment with a helium—neon laser ($\lambda = 633$ nm) the slits are 0.8 mm apart and the

fringes are formed on a screen 5 m away. What is the fringe separation?

4.6   Taking the width of the principal maxima of Fig. 4.11 as one half of the interval in δ between adjacent zeros, calculate the ratio of the line width to separation between orders for a 40 000-line grating.

4.7   An approximate measurement of the wavelength of a spectral line shows it to be between 500.000 and 500.005 nm. Fringes produced with a Fabry—Perot etalon of plate separation 5.0000 mm have a fraction f (Section 4.23) of 0.7 ± 0.03. Calculate the unknown wavelength to as many significant figures as appropriate.

4.8   A Fabry—Perot etalon is known to have a separation of 10.000 ± 0.001 mm, the best that could be done with a travelling microscope. For the mercury lines of wavelength 404.656, 435.834 and 546.073 nm, the fractions f (Section 4.23) are respectively 0.40, 0.85 and 0.19. The values of f are known to an accuracy of ± 0.05. Find the thickness of the etalon.

4.9   Criticise the argument leading to the result $n_c^2 = n_g$ in Section 4.31.

4.10 Show that if r_1 and r_2 are respectively the amplitude reflection coefficients at successive surfaces of a parallel-sided film, the amplitude R of the reflected beam is given by

$$R = \frac{r_1 + r_2\, e^{-2i\delta}}{1 + r_1 r_2\, e^{-2i\delta}}$$

where $\delta = 2\pi n d/\lambda$; d is the film thickness, n the refractive index and normal incidence is assumed.

4.11 An interference filter for the mercury green ($\lambda = 546$ nm) line has the smallest possible spacer thickness. At what wavelengths above and below 540 nm will adjacent transmission peaks occur? Where would the peaks occur if the spacer thickness were doubled? Are your answers exact?

Wave Theory (2): Wave Groups

5.1 Finite wave trains

The experiments described in Chapter 4 give convincing support to a wave theory of light but leave certain important questions unanswered. Why do we not observe interference fringes whenever a room is illuminated by two independent sources of light? Why, in Young's experiment, was it necessary to use a slit S_0 yet in the observation of Newton's rings an extended source of light is satisfactory? With the Michelson interferometer (Fig. 4.10) it is found that if the distance between M_1 and R is increased to several centimetres, the fringes gradually become less clear and, so far, there is no obvious reason why this should be so. In this chapter we shall see that these queries arise because, in writing eqn (4.2) without qualification we assumed that it was valid (and that E_0 was constant) for all values of x and t. Thus we represented light by infinite unbroken wave trains of constant amplitude, lasting for ever and stretching from $-\infty$ to $+\infty$. We need to know *how* E_0 varies with time and space for a real source. We cannot measure the electric field of a light wave directly—the oscillations are too rapid for any detector to record. We can, however, examine plausible models of how E_0 can be expected to behave and use them to predict the results of experimental measurements.

We need to assume that E_0 depends on x and t, viz.

$$E_0 \equiv E_0(x, t) \qquad (5.1)$$

If we think of a source as being switched on at time t_0 and off at t_1, then E_0 at the source is zero for $t < t_0$ and $t < t_1$; the duration of the wave train is $t_1 - t_0$. If the source radiates at a fixed frequency then the time variation of E_0 at a given value of x will be sinusoidal during the emitting period (left-hand diagram of Fig. 5.1a). Another possibility is that E_0 will rise to a maximum value and that the oscillations will decay exponentially with time (Fig. 5.1b). How will we know whether one or other of these models represents a real light wave?

We shall see below that any wave train of finite duration may be represented as a superposition of the infinite wave trains which we used with such abandon in Chapter 4. Each of the superposed waves will have its own distinct frequency, so that the energy in the light beam is spread over a range of wavelengths—a *spectrum*. This is exactly the kind of measurement we can make with a spectrometer. Can we therefore work backwards from our measurements on the spectra of light sources and deduce how the electric field is behaving? We can, with the help of a theorem first propounded by J. B. Fourier (1768–1830).

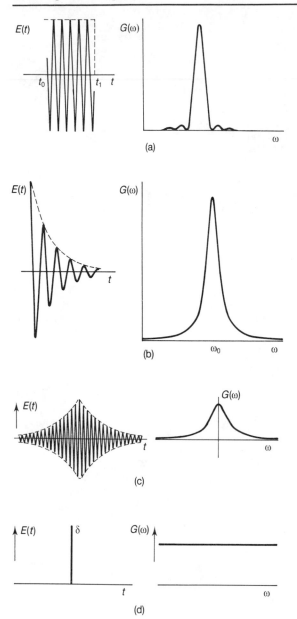

Figure 5.1 Wave trains, temporal distribution and spectral distribution., (a) Sharply limited wave train (b) damped harmonic waves, (c) Gaussian wave group, (d) delta function

5.2 Fourier transforms

If a time-varying light field $E(t)$ can be represented as a superposition of infinite wave trains with frequencies ν and angular frequencies $\omega \equiv 2\pi\nu$, then we can define a

function $g(\omega)$ by

$$g(\omega) = \int_{-\infty}^{\infty} E(t)e^{-i\omega t} \, dt \qquad (5.2)$$

where $g(\omega)$ is called the *Fourier transform* of $E(t)$. In general $g(\omega)$ is complex and the real quantity $g(\omega)g^*(\omega) \equiv G(\omega)$ is called the *power spectrum* of $E(t)$. If the light beam represented by $E(t)$ were collected on a detector which responded to a band of frequencies of width $d\omega$ at frequency ω, then the detector reading would be $G(\omega) \, d\omega$. Alternatively, if the light were sent into a spectrometer and a detector which was equally sensitive to all frequencies were moved along the spectrum, a curve of $G(\omega)$ against ω could be obtained.

With $g(\omega)$ defined by eqn (5.2), an inverse relation exists, namely

$$E(t) = \frac{1}{2\pi} \int_{-\infty}^{\infty} g(\omega)e^{i\omega t} \, d\omega \qquad (5.3)$$

Equations (5.2) and (5.3) define a Fourier transform pair.† (*Note* the opposite signs in the exponents.) If $E(t)$ is to represent the light field, then it is a real quantity. In this case, eqn (5.3) implies that

$$g(\omega) = g^*(-\omega)$$

so that in (5.3) the integral from $-\infty$ to 0 is equal to the integral from 0 to $+\infty$. Hence we may write

$$E(t) = \frac{1}{\pi} \int_{0}^{\infty} g(\omega)e^{i\omega t} \, d\omega \qquad (5.4)$$

and we need consider only positive frequencies.

In many cases of interest in optics $E(t)$ has the form

$$E(t) = E_0(t) \cos \omega_0 t = \tfrac{1}{2}E_0(t)(e^{i\omega_0 t} + e^{-i\omega_0 t}) \qquad (5.5)$$

where $E_0(t)$ varies only a very little in the period of one wave. In most optical cases, it varies little over the time of some millions of

---

† There is no general agreement on whether the integral defining the Fourier transform should have a positive or negative exponent. Equation (5.2) *could* be written with a positive sign in which case (5.3)—the inverse transform—would have a negative exponent. It is unimportant which is adopted but the signs in the pair *must* be opposite.

Table 5.1 Fourier transforms

$E(t)$	$g(\omega)$	$G(\omega)$
$0\ (-\infty < t < 0)$ $Ae^{-\gamma t/2}\cos\omega_0 t$ $(0 \leqslant t < \infty)$	$\dfrac{A}{2[i(\omega-\omega_0)+\gamma]}$	$\dfrac{A^2}{4[\omega-\omega_0)^2+\gamma^2]}$
$A\exp\left(-\dfrac{\sigma^2 t^2}{4}\right)\cos\omega_0 t$	$\dfrac{A\pi^{1/2}}{\sigma}\exp[-\sigma^2(\omega-\omega_0)^2]$	$\dfrac{A^2\pi}{\sigma^2}\exp[-2\sigma^2(\omega-\omega_0)^2]$
$0\ (-\infty < t < -t_1)$ $A\cos\omega_0 t\,(-t_1 < t < t_1)$ $0\ (t_1 < t < \infty)$	$\dfrac{A\sin(\omega-\omega_0)t_1}{\omega-\omega_0}$	$\dfrac{A^2\sin^2(\omega-\omega_0)t_1}{(\omega-\omega_0)^2}$
Delta function (pulse at $t=0$)	Constant	Constant
$\sum\limits_{n=-N}^{N}\delta(t-nt_1)$	$\dfrac{\sin(N+\frac{1}{2})\omega t_1}{\sin(\frac{1}{2}\omega t_1)}$	$\dfrac{\sin^2(N+\frac{1}{2})\omega t_1}{\sin^2(\frac{1}{2}\omega t_1)}$

waves. The optical frequency then acts as a carrier frequency which is modulated by $E_0(t)$. In this case,

$$g(\omega)=\tfrac{1}{2}\int_{-\infty}^{\infty}E_0(t)\{\exp[-i(\omega-\omega_0)t]$$
$$+\exp[-i(\omega+\omega_0)t]\}\ dt \quad (5.6)$$

In all the cases in which we are interested the integral of the second term in the bracket is negligible compared with the integral of the first term. In such cases we can therefore find the transform of $E(t)$ by substituting $(\omega-\omega_0)$ for ω in the transform of $E_0(t)$. Figure 5.1 shows some typical modulated waves. The left-hand figures represent $E_0(t)$. The power spectrum is shown to the right. Table 5.1 gives some useful Fourier transforms. All the transforms between t and ω have corresponding transforms between x and $\varkappa$.

5.3 Damped harmonic waves

The waves emitted by free atoms can be represented as damped waves starting at $t=0$, so that $E_0(t)$ is zero from $-\infty$ to 0 and

$$E_0(t)=A\exp(-\gamma t) \quad (5.7)$$

for $0 < t < \infty$.

The energy (proportional to $[E_0(t)]^2$) falls to half its maximum value when $\gamma t_{1/2}=\log_e 2$, i.e. when $t_{1/2}=0.7/\gamma$. The Fourier transform (Table 5.1) is

$$g(\omega)=\frac{A}{2[i(\omega-\omega_0)+\gamma]} \quad (5.8)$$

and

$$G(\omega)=\frac{A^2}{4[(\omega-\omega_0)^2+\gamma^2]} \quad (5.9)$$

This has its maximum when $\omega=\omega_0$ and falls to half-maximum when $\omega-\omega_0=\pm\gamma$. The full width between half-value points is 2γ. For strong lines in the spectrum of a gas-discharge lamp, γ has a value of 10^7 to $10^8\ \text{s}^{-1}$, when the pressure of the source is low so that collisions are infrequent. Since ω_0 for visible light is about $4\times10^{15}\ \text{s}^{-1}$ the line width γ is a very small fraction of ω_0. Collisions between atoms effectively increase γ since the emitting atom effectively starts a new (unrelated) wave at each collision during its normal emitting period. In a high-pressure mercury arc the lines are very wide and at higher pressure still (in a xenon arc), (5.7) is no longer valid and a continuous spectrum (with $G(\omega)$ nearly constant) is produced. At moderate pressures in a discharge tube source γ may be 10^{10} or $10^{11}\ \text{s}^{-1}$ so that the duration of a wave train is about 10^{-11} to

10^{-10} s, corresponding to a length of 3–30 mm.

5.4   Gaussian wave groups

Suppose

$$E_0(t) = A \, \exp\left[-\left(\frac{\sigma t}{2}\right)^2\right] \cos \omega_0 t \quad (5.10)$$

Then

$$G(\omega) = \frac{A^2 \pi}{\sigma^2} \exp\left[-\frac{(\omega - \omega_0)^2}{\sigma^2}\right] \quad (5.11)$$

This is called a Gaussian wave group.

The energy is half its maximum value when $(\sigma t_{1/2}/2)^2 = \log_e 2$, i.e. when $t_{1/2} = \pm 1.66/\sigma$ so that the length of a wave train is $3.3/\sigma$ in time and $3.3c/\sigma$ in distance. Similarly $G(\omega)$ has half its maximum value when $(\omega - \omega_0)^2 = \sigma^2 (\log_e 2)/2$, i.e. when $\omega - \omega_0 = \pm 0.59\sigma$, so that the width between half-value points is 1.2σ.

Atoms in a discharge tube have different components of velocity in the direction of the observer and, owing to the Doppler effect (Section 4.7), the light received from different atoms will have a spread of different frequencies. This adds a new source of line broadening to that considered in Section 5.3. When the pressure is low the Doppler effect predominates. It leads to a Gaussian distribution with

$$\sigma^2 = \frac{2\omega_0^2 kT}{Mc^2} \quad (5.12)$$

where M is the mass of the emitting atom, T is the absolute temperature and k is Boltzmann's constant. For cadmium ($M = 112 \times 1.66 \times 10^{-27}$ kg) with $T = 300$ K, σ is about 2×10^9 s^{-1}. The width between half-value points for a line of wavelength 644 nm is $\omega - \omega_0 \approx 2.4 \times 10^9$ s^{-1} or $\lambda - \lambda_0 \approx 5 \times 10^{-4}$ nm. The length of a wave train is about 1.4 ns in time or 0.4 m in space.

In the outer parts of a line $G(\omega)$ from (5.11) falls off more rapidly than $G(\omega)$ from (5.9). When $\omega - \omega_0$ is 4 times the half-value the distribution from (5.9) has fallen to 0.06 times the maximum and that from (5.11) to less than 10^{-7} of the maximum value. This makes it possible to deduce the effects of damping ('natural broadening') and atomic motion ('Doppler broadening') in a single line. Doppler effects determine the shape of the line centre and damping effects that of the wings of the line.

5.5   Limited wave trains

From Table 5.1 and Fig. 5.1(c) we have that when $E_0(t) = A$ from $-t_1$ to $+t_1$ and zero at other times, giving a wave train of constant amplitude, then

$$G(\omega) = A^2 \left[\frac{\sin(\omega - \omega_0)t_1}{\omega - \omega_0}\right]^2 \quad (5.13)$$

The half-maximum value of $G(\omega)$ is obtained when $\sin(\omega - \omega_0)t_1 = 0.7(\omega - \omega_0)t_1$. Solving this equation numerically by computer gives $2.8/t_1$ for the full width $2(\omega - \omega_0)$ of the line.

A highly stabilised, continuous-wave laser gives a wave train of constant amplitude, of the order of tens of metres in length, so that t, in (5.13) is around 10^{-7} s. The associated line width should be $\sim 3 \times 10^7$ s^{-1}. In practice small vibrations, etc., broaden the line so that, although very narrow, it is not 'transform limited', i.e. it is not as narrow as (5.13) would predict. When pulsed lasers are used to give wave trains only 10^{-12} s long, the Fourier transform indicates a width of about 3×10^{12} Hz or $10^{-3}\omega_0$. In this case a transform-limited width may be achieved.

5.6   Very short pulses

Dirac defined a function that can describe a very short pulse occurring at times t_1. He considered the limit of a sequence of functions whose value at some time t_1 increased indefinitely while its width decreased indefinitely, but in such a way that the area under the curve (i.e. the integral) retained a finite value A. This is known as a delta function. The Fourier transform of a delta function is a constant (Fig. 5.1d) so an extremely short pulse gives essentially an equal energy spectrum. Obviously we cannot produce an indefinitely short pulse of light but lasers can give extremely short pulses. The shortest pulses that have been produced are only 3×10^{-14} s long, corresponding to about one period of the light wave.

The symmetry of eqns (5.2) and (5.3) is such that we must expect that when $g(\omega)$ is a delta function, i.e. when we have only one frequency present, the light will be a monochromatic wave of constant amplitude and we may regard this as a definition of a monochromatic wave.

5.7 Quasi-monochromaticity

In each of the three examples discussed in Sections 5.3 to 5.5, there is one parameter that defines the length of the wave train and another that defines the width of the frequency distribution. These parameters are inversely proportional to one another. It is always found that as the length of the wave train increases the width of the frequency distribution decreases. This would be expected since the longer the wave train the more closely it approaches the indefinitely long monochromatic wave train for which there is a precisely defined frequency. It is usual to speak of light that includes a range of frequencies which is very small compared with ω_0 as *quasi-monochromatic*.

5.8 Propagation of a wave group in a dispersive medium

What happens to a wave group traversing a medium whose refractive index depends on wavelength? From Section 4.2 we have, for the phase of a wave,

$$\phi = \omega t - \varkappa x + \delta$$

If we wish to follow the progress of one wave we note that a point of constant phase ϕ_1 moves a distance dx in time dt, such that

$$d\phi_1 = \omega\, dt - \varkappa\, dx = 0, \qquad \text{so}\ \frac{dx}{dt} = v = \frac{\omega}{\varkappa} \quad (5.14)$$

where v is the phase velocity.

If we consider a group of waves with a narrow frequency range and want to follow the point of maximum agreement of phase ϕ_2 we have $d\phi_2 = t\, d\omega - x\, d\varkappa = 0$. The group travels a distance x in time t and hence has

a velocity

$$U = \frac{x}{t} = \frac{d\omega}{d\varkappa} \quad (5.15)$$

In terms of v and λ this may be written as $U = -\lambda^2\, dv/d\lambda$. Thus when the medium is dispersive the velocity U of the group as a whole differs from the phase velocity v. By dropping a stone into a pond a group of waves that advances as a whole can be seen. Careful observation, however, shows that individual waves come up at the back of the group, advance through the group (because in this case $v > U$) and die out as they reach the front.

What happens to a Gaussian wave group in a dispersive medium? We write, following eqn (5.4):

$$E(x, t) = \frac{1}{\pi} \int_0^\infty g(\varkappa) \exp\left[i(\varkappa x - \omega t)\right]\, d\varkappa \quad (5.16)$$

(using the Fourier transform between x and $\varkappa$), where

$$g(\varkappa) = \frac{A\pi^{1/2}}{\rho} \exp\left[-\frac{(\varkappa - \varkappa_0)^2}{\rho^2}\right] \quad (5.17)$$

and $\rho = c\sigma$ because the length (in space) of the wave train is c times its duration (in time) (see Section 5.4). Using Taylor's theorem we may write

$$\omega = \omega_0 + U\, \Delta + \tfrac{1}{2} W\, \Delta^2 \quad (5.18)$$

where $\Delta = \varkappa - \varkappa_0$, $U = d\omega/d\varkappa$ and $W = d^2\omega/d\varkappa^2$. It is assumed that $\omega - \omega_0$ is small compared with ω_0.

It is shown in Appendix 5.A that, on substituting from (5.17) and (5.18) into (5.16) and integrating, we have:

1. When $W = 0$,

$$|E(x, t)|^2 = \frac{A^2}{4} \exp\left[-\frac{\rho^2(x - Ut)^2}{2}\right] \quad (5.19)$$

Thus, in this case the envelope of the wave group advances unchanged with velocity U.

2. When $W \neq 0$,

$$|E(x, t)|^2$$
$$= \frac{A^2}{4(1 + W^2 t^2 \rho^2)^{1/2}} \exp\left[-\frac{\rho^2(x - Ut)^2}{2(1 + W^2 t^2 \rho^4)}\right] \quad (5.20)$$

In this case the wave group still advances with velocity U but expands as it advances.

This expansion depends on the term $4W^2t^2$ and since W and t are generally small enough to make $W^2t^2\rho^2 \ll 1$, the expansion is slow. As the wave advances the maximum of the envelope must decrease because the total energy of the wave train is constant (Fig. 5.2). The spreading is small if ρ is small (so that the range of frequencies is small) or if W is small. This broadening of an advancing pulse is important in relation to optical-fibre communications. The rate of transmission of information is proportional to the number of pulses per second so it is desirable to have short pulses with short intervals between them. This means that only a small broadening over a distance of 100 km is acceptable. Also the range of frequencies is inversely proportional to the length of the pulse so that very short pulses imply that ρ cannot be extremely small. Thus it becomes very important to reduce W as far as possible.

The ideal fibre material would have both zero absorption and $W = 0$ at the appropriate wavelength. Glasses have been developed that approach the above ideal specification. They enable optical pulses to be transmitted over distances of the order of 100 km with acceptably low loss and with very small spreading. Pulse spreading may in fact be practically eliminated by the subtle use of non-linear properties of the transmitting medium (see Section 5.12).

The expression for U (eqn 5.15) can be put in various forms which are suitable for different applications. Since $\omega = v\varkappa$ we have

$$U = \frac{d\omega}{d\varkappa} = \frac{d(v\varkappa)}{d\varkappa} = v + \varkappa\frac{dv}{d\varkappa} \qquad (5.21)$$

Also, putting $n = c/v$:

$$U = \frac{c}{n}\left\{1 + \frac{\lambda}{n}\frac{dn}{d\lambda}\right\} \qquad (5.22)$$

and, since $c/U = d(n\omega/d\omega)$,

$$\frac{1}{U} = \frac{1}{v} + \frac{\omega}{c}\frac{dn}{d\omega} \qquad (5.23)$$

In these equations λ and $\varkappa$ represent the values of the wavelength and wave number *in the medium*.

5.9 Coherence

We cannot measure either the amplitude or phase of visible light waves directly because the frequency is too high for our detectors to respond. We can measure the phase *difference* between two beams provided this difference remains constant for a time which is longer than the response time of our instruments. We can also measure the energy (W) averaged over a certain area which depends on the properties of our instruments. With thermal sources of light we are able to observe interference provided we take appropriate precautions. Thus in Young's experiment we have to use the slit so as to restrict the size of the primary source. With laser light, on the other hand, it is very easy to observe interference and sometimes difficult to avoid it. We now wish to understand how this difference is related to the restrictions on what we can observe and to the basic differences between laser light and thermal light.

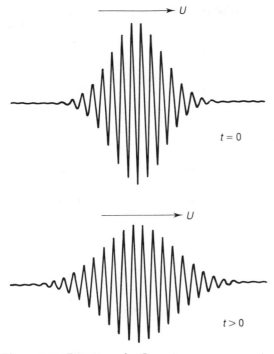

Figure 5.2 Progress of a Gaussian wave group in a dispersive medium.

5.10 Photometric summation

In Section 4.17, we showed that if $E = E_0 \, e^{i\delta}$ is the complex amplitude of a wave then the energy associated with a single wave is $W = EE^*$. For the resultant of a number (p) of waves (using eqn 4.22),

$$W = E_r E_r^* =$$
$$(E_1 + E_2 + \cdots + E_p)(E_1^* + E_2^* + \cdots + E_p^*) \quad (5.24)$$

W_r is the resultant energy.† Hence

$$W_r = \sum_{n=1}^{p} E_n E_n^* + \sum_{\substack{n=1 \\ n \neq m}}^{p} \sum_{m=1}^{p} E_m E_n^* \quad (5.25)$$

$$= \sum_{n=1}^{p} W_n + \sum_{\substack{n=1 \\ n \neq m}}^{p} \sum_{m=1}^{p} E_{0m} E_{0n}^* \exp i(\delta_m - \delta_n)$$
$$\quad (5.26)$$

where $E_n = E_{0n} \, e^{i\delta_n}$.

If a room is illuminated with light from a filament lamp or a fluorescent tube, then the phases of the light waves from different small areas of the filament or tube are varying rapidly and at random. The positive contributions to the second term in eqn (5.26) are on average equal to the negative ones, and averaged over a very short time the sum is zero. Equation (5.26) reduces to

$$W_r = \sum_{n=1}^{p} W_n \quad (5.27)$$

which is the ordinary *law of photometric summation*, i.e. the total light energy arriving at the point of observation is equal to the sum of the energies from all the small areas regarded as independent sources. This is true at any point in the room and at any time (provided the time average is taken). We observe interference only if we arrange that the phase differences between light from different sources remain constant. We can do this if we derive two sources (as in Sections 4.8 to 4.17) or more than two (as in Sections 4.18 and 4.22) from the same source. We

cannot obtain interference by the addition of light from independent thermal sources. Also the phase fluctuations in light from different parts of an extended thermal source are independent.

5.11 Visibility of interference fringes

In the preceding section we discussed a situation where no interference fringes are produced by the interaction of light from two (or more) sources (Fig. 5.3a). In Section 4.8 we derived equations for the energy distribution in fringes in which the intensity fluctuates between a maximum value and zero (eqn 4.16 and Fig. 5.3b). We now consider an intermediate case in which fringes are formed whose minima are not zero (Fig. 5.3c). Michelson introduced a term 'visibility' (V) to describe these fringes where

$$V = \frac{W_{max} - W_{min}}{W_{max} + W_{min}} \quad (5.28)$$

W_{max} and W_{min} are the maximum and minimum intensities respectively.

When $W_{min} = 0$, $V = 1$, there are fringes of the highest possible contrast. Also when $V = 0$, $W_{max} = W_{min}$ and there are no fringes at all. So the two extreme cases are included but we can also deal with intermediate cases. Note that V is a quantity that can be measured by passing a detector (covered with a narrow slit) along the fringes. From Fig. 5.3(b), fringes of visibility V may be regarded as the superposition of a fraction V of light that is giving fringes of visibility 1 upon a fraction $(1 - V)$ that is not contributing to the fringes but is being added by photometric summation.

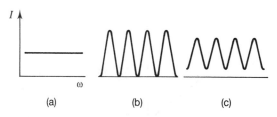

(a)   (b)   (c)

Figure 5.3 (a) No fringes ($V = 0$), (b) fringes for $V = 1$, (c) fringes for $V = 0.63$

---

† The energy associated with an electromagnetic wave is *proportional* to E^2 (Chapter 8). For the present discussion, the constant of proportionality is of no consequence and we put it equal to unity.

5.12 Mutual coherence function

Modern development of the theory of interference fringes started with Van Cittert (1934) and Zernike (1938). The theory was restated and extended by Wolf [1]. Hopkins [2] gave a particularly simple and powerful way of calculating V in special cases. What follows is due to these four authors.

Suppose that a thermal source S illuminates two pinholes P_1 and P_2 in an opaque screen (Fig. 5.4). We wish to calculate the visibility of fringes in the region of Q which is distant s_1 from P_1 and s_2 from P_2. The energy W_1 measured at P_1 is the mean value (measured over a time determined by the detector) of $E_1 E_1^*$ and similarly for W_2 at P_2. The complex amplitude at Q at a time t is given by

$$E_Q(t) = K_1' E_1\left(t - \frac{s_1}{c}\right) + K_2' E_2\left(t - \frac{s_2}{c}\right) \quad (5.29)$$

where K_1', K_2' are geometrical constants. From the discussion of Section 6.15, we note that these are pure imaginary quantities and put $K_{1,2}' = iK_{1,2}$ where $K_{1,2}$ are real. For a source of constant intensity the average values of E_1 and E_2 do not change with time and we may shift the origin to give

$$E_Q(t) = K_1' E_1(t + \tau) + K_2' E_2(t) \quad (5.30)$$

where

$$\tau = \frac{s_2 - s_1}{c} \quad (5.31)$$

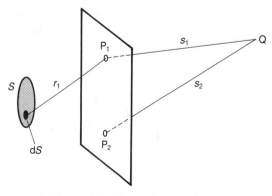

Figure 5.4 Young's experiment

Then

$$W_Q = K_1^2 W_1 + K_2^2 W_2 + 2K_1 K_2 \Gamma_{12}\left(\frac{s_2 - s_1}{c}\right) \quad (5.32)$$

where

$$\Gamma_{12}(\tau) = \tfrac{1}{2}[E_1(t + \tau)E_2^*(t) + E_1^*(t + \tau)E_2(t)] \quad (5.33)$$

Γ_{12} is called the *mutual coherence function*. When P_1 and P_2 coincide at P, $\tau = 0$ and

$$\Gamma_{11}(0) = |E_1(t)|^2 = W_1$$

is called the self-coherence. Similarly $\Gamma_{22}(0) = W_2$.

If it is arranged that parts of the *same* wavefront are presented at P_1 and P_2, e.g. by illuminating with a small source a long way from the pinholes, then as the point of observation is moved, τ will vary. The value of Γ_{11} will then be given by

$$\Gamma_{11}(\tau) = \tfrac{1}{2}[E_1(t + \tau)E_1^*(t) + E_1^*(t + \tau)E_1(t)]$$
$$= \mathrm{Re}\,[E_1(t + \tau)E_1^*(t)] \quad (5.34)$$

known as the autocorrelation function of $E_1(t)$.

5.13 Complex degree of coherence

The energy measured at Q when P_2 is covered is $K_1^2 W_1 = W_{1Q}$; when P_1 is covered, it is $K_2^2 W_2 = W_{2Q}$. We can define a normalized version of Γ_{12} by

$$\gamma_{12}(\tau) = \frac{\Gamma_{12}(\tau)}{\sqrt{W_1}\sqrt{W_2}} \quad (5.35)$$

which has the advantage that it does not depend on the absolute value of the energy from S. Then the energy at Q is, using (5.32),

$$W_Q = W_{1Q} + W_{2Q} + 2(W_{1Q}W_{2Q})^{1/2}\mathrm{Re}\,[\gamma_{12}(\tau)] \quad (5.36)$$

The complex quantity $\gamma_{12}(\tau)$ may be written $|\gamma_{12}|\exp\,i(\omega\tau + \delta_{12})$ and is known as the *complex degree of coherence*.

The real part is

$$\mathrm{Re}\,[\gamma_{12}(\tau)] = |\gamma_{12}|\cos\,(\omega_0\tau + \delta_{12}) \quad (5.37)$$

where δ_{12} is a constant. Inserting this value into (5.36) we have, for the case when

$W_{1Q} = W_{2Q},$

$$W_Q = 2W_{1Q}[1 + |\gamma_{12}| \cos(\omega_0\tau + \delta_{12})] \quad (5.38)$$

The maximum and minimum values of W_Q are $2W_{1Q}(1 + |\gamma_{12}|)$ and $2W_{1Q}(1 - |\gamma_{12}|)$. From (5.28) we see that, in this case, $|\gamma_{12}|$ is equal to the visibility of the fringes. Thus the light behaves as though a fraction $|\gamma_{12}|$ is coherent with a phase difference $\omega\tau + \delta_{12}$ and a fraction $(1 - |\gamma_{12}|)$ is non-coherent. For the fringes at the centre ($\tau = 0$) the phase difference of the sources P_1 and P_2 is δ_{12}. Zernike pointed out that the *degree of coherence* is equal to the visibility obtained under the most favourable conditions (i.e. in this case sources of equal intensity); $|\gamma_{12}|$ meets this criterion.

5.14   Coherence of a single beam

In the above discussion we have treated coherence as a mutual property of two sources such as P_1 and P_2 and of the light that comes from them. Nevertheless, it is not unusual to speak of single beams of light as being coherent, non-coherent or partially coherent. These terms may be applied in the following way. Suppose we have a nearly parallel beam of light (Fig. 5.5) and that P_1 and P_2 are two points on a surface of constant phase. Then if the beam has been derived from a thermal source (by putting a small-area source at the focus of the lens) it will be found that the degree of coherence is large when P_1 and P_2 are close together and decreases as the distance between them increases. If the degree of coherence between points on the edges of the beam is so high as to give good interference fringes then the beam may be said to be 'highly coherent' spatially. If good fringes can be obtained

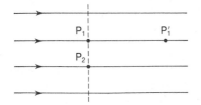

Figure 5.5   Parallel beam, partial coherence

between light from points such as P_1 and P_1' (e.g. by combining beams scattered from these regions) then the beam may be said to be 'temporally coherent', with a 'coherence length' P_1P_1' or a coherence time P_1P_1'/c.

The above definitions are not very precise because we have not said exactly what constitutes 'good' interference fringes and it is open to the reader to choose any value of V between, say, 0.25 and 0.75 as 'good' in the above sense. Fringes of visibility 0.6 look very clear and this is a reasonable value.

A beam of light from a gas laser operating under suitable conditions will have virtually complete spatial coherence across the beam and a coherence length that is more than 100 m, even when no special precautions are taken. Thus this kind of laser light is spoken of as 'completely' coherent. For light from thermal sources the 'coherence length' or coherence time depends on the length of the wave train. A single spectrum line selected from light emitted by a discharge at low pressure may have a coherence length of 10 cm or occasionally more. As the pressure is increased the coherence length falls so that in a high-pressure mercury arc it may be only a few wavelengths and, in a xenon arc, effectively zero.

5.15   Spatial coherence

The spatial coherence of light from a thermal source depends on the geometry of the arrangement that has produced the beam—especially on the angular size of the source. There is a theorem, known as the Van Cittert–Zernike theorem, which shows that the degree of spatial coherence ($|\gamma_{12}|$ when $\tau = 0$) is given by a formula that is the same as a well-known formula in diffraction theory, though the quantities that appear in the formula are different. This enables us to take over the results of calculations which were done a century ago in relation to diffraction. This formula is discussed in Appendix 5.B. Here we may assume the result for Young's experiment. For the case of parallel slits the degree of coherence of the light from slits S_1 and S_2 in Fig. 4.8 is

$$|\gamma_{12}| = \frac{\sin\theta}{\theta} \quad (5.39)$$

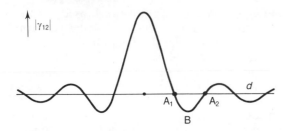

Figure 5.6 Visibility as a function of width of source S in Fig. 5.4

where $\theta = 2\pi hd/x_1$ and d is the width of the source S_0. The variation of $|\gamma_{12}|$ with d is shown in Fig. 5.6. $|\gamma_{12}|$ goes negative for some values of d. This is equivalent to a phase change of π and the fringes are displaced (by half a fringe width) from the positions of fringes formed by light from the centre of S_0.

5.16   Coherence—slit-width dependence

In a qualitative way we may explain the variation of coherence (and hence of the fringes formed at Q) as follows. When the slit S_0 is very narrow the fringes formed by different parts of it nearly coincide so that fringe pattern as a whole is very clear (Fig. 5.7). As the slit is widened the maxima from different

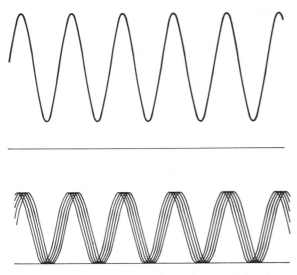

Figure 5.7   Visibility of interference fringes

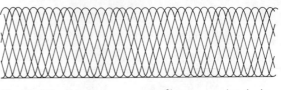

Figure 5.8   Fringes corresponding to a point A_1 in Fig. 5.6

parts of the source no longer coincide and when d corresponds to the point A_1 in Fig. 5.6 the different fringes are as shown in Fig. 5.8 and the overall visibility is zero. Now suppose the slit is opened a little further, exposing two small areas on either side of the region that gives no fringes. These areas will produce fringes that coincide but which are displaced by half a fringe width from the fringes seen when S_0 is very narrow. As these areas are increased the fringes get stronger at first but they are never very strong because they are superposed on the non-coherent background. After a width corresponding to point B in Fig. 5.6 they lose visibility by non-coincidence just as the first set did until at A_2 they have vanished completely. Further widening produces a third set of still weaker fringes.

5.17   Coherence and the Michelson interferometer

In the Michelson interferometer, and in most fringe patterns formed by division of amplitude, the wave-fronts are superposed so that P_1 interferes with an image of itself and P_2 with an image of itself (Fig. 4.10). The two beams that interfere are thus coherent at every point even though P_1 and P_2 are not coherent with each other. It is thus not necessary to use a small source. If the path difference becomes equal to the coherence length the visibility of the fringes falls because the wave trains no longer arrive at the detector at precisely the same time. As the path difference is increased the overlap decreases (Fig. 5.9) until fringes are no longer seen.

If in a Michelson interferometer the mirror M is replaced by an Amici prism (Fig. 5.10) one wave-front is reversed. Light from A interferes with light from B and fringes are

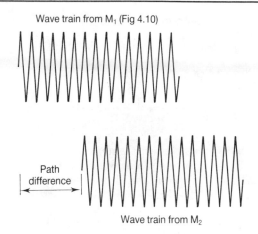

Wave train from M₁ (Fig 4.10)

Path difference

Wave train from M₂

Figure 5.9 Overlap of wave trains

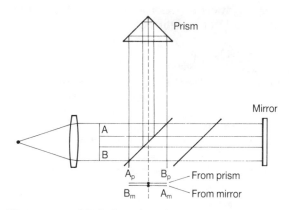

Prism

Mirror

A
B

A_p B_p — From prism
B_m A_m — From mirror

Figure 5.10 Michelson interferometer with Amici prism

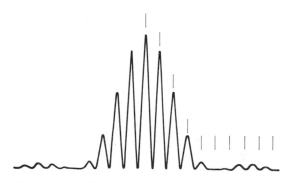

Figure 5.11 Fringes shown in arrangement of Fig. 5.10. Note that the minima in the side-lobes correspond to the maxima in the central region

seen only if a small source is used. The visibility of fringes in the centre of the field is then high because light waves from neighbouring points are interfering. The visibility gradually falls to zero following Fig. 5.6 until the fringes disappear. They reappear (with a displacement) as shown in Fig. 5.11.

5.18 Measurement of degree of coherence

The coherence of beams of light is of importance firstly because, unlike phase and amplitude, the degree of coherence *can* be measured. For example, the degree of coherence $|\gamma_{12}|$ for P_1 and P_2 in Fig. 5.4 can be obtained by covering P_1 and P_2 alternately to obtain W_1 and W_2 and then measuring the visibility of the fringes. To obtain the mutual coherence function we need to measure also the energy at P_1 and P_2 (eqn 5.35).

The coherence is also important because it is propagated with the beam of light which it characterises. The coherence for points in the exit pupil of an optical system is the same as the coherence for corresponding points in the entrance pupil. If coherent light is scattered, e.g. by ground glass, the degree of coherence is reduced. There is, however, no way of *increasing* the coherence by passing a beam through a system of lens or mirrors.

Appendix 5.A Gaussian wave group in a dispersive medium

Note that

$$\int_0^\infty \exp(ay - by^2)\,dy$$

$$= \exp\left(\frac{a^2}{4b}\right) \int_0^\infty \exp\left[-b\left(y - \frac{a}{2b}\right)^2\right]dy$$

$$= \exp\left(\frac{a^2}{4b}\right) \int_0^\infty \exp(-bz^2)\,dz$$

$$= \left(\frac{\pi}{4b}\right)^{1/2} \exp\left(\frac{a^2}{4b}\right) \tag{5.40}$$

Substituting from (5.17) and (5.18) into

(5.16),

$$E(x,t) = \frac{A}{\pi^{1/2}\rho} \int_0^\infty \exp\left\{-\frac{\Delta^2}{\rho^2} + i[(\varkappa_0 + \Delta)x\right.$$
$$\left. - (\omega_0 + U\Delta + W\Delta^2)t]\right\} d\Delta \quad (5.41)$$

where $\Delta = \varkappa_0 - \varkappa$, i.e.

$$E(x,t) = \frac{A}{\pi^{1/2}\rho} \exp i(\varkappa_0 x - \omega_0 t) \int_0^\infty$$

$$\exp\left[i(x - Ut)\Delta - \left(\frac{1}{\rho^2} + Wt\right)\Delta^2\right] d\Delta \quad (5.42)$$

1. For the case $W = 0$, $a = i(x - Ut)$, $b = 1/\rho^2$ and

$$E(x,t)$$
$$= \frac{A}{2} \exp\left[-\frac{\rho^2(x - Ut)^2}{4}\right] \exp i(\varkappa_0 x - \omega_0 t)$$
$$(5.43)$$

2. For $W \neq 0$, $a = i(x - Ut)$, $b = (1/\rho^2 + iWt)$ and

$$E(x,t) =$$

$$\frac{A \exp i(\varkappa_0 x - \omega_0 t)}{2(1 + iWt\rho^2)} \exp \frac{-\rho^2(x - Ut)^2}{4(1 + iWt\rho^2)}$$
$$(5.44)$$

$$|E(x,t)|^2 =$$

$$\frac{A^2}{4(1 + W^2 t^2 \rho^4)} \exp \frac{-\rho^2(x - Ut)^2}{2(1 + W^2 t^2 \rho^4)}$$
$$(5.45)$$

Appendix 5.B Calculation of $\Gamma_{12}(0)$ and $\gamma_{12}(0)$

The complex amplitude at P_1 associated with an element dS' of the source is

$$dE_1 = \frac{(W_s)^{1/2}}{r_1} dS \exp(-i\varkappa r_1) \quad (5.46)$$

where r_1 is the distance from P_1 to the element (Fig. 5.4). The energy at P_1 is given by

$$W_1 = \int \frac{W_s}{r_1^2} dS \quad (5.47)$$

Similar formulae apply for P_2. Then

$$dE_1\, dE_2^* = \frac{W_s}{r_1 r_2} \exp[-i\varkappa(r_1 - r_2)]\,dS \quad (5.48)$$

and the average value for $E_1 E_2^*$ is

$$\Gamma_{12}(0) = \int \frac{W_s}{r_1 r_2} \exp[-i\varkappa(r_1 - r_2)]\,dS \quad (5.49)$$

and, from eqn (5.35),

$$\gamma_{12}(0) = \frac{1}{(W_1 W_2)^{1/2}} \int \frac{W_s}{r_1 r_2} \exp[-i\varkappa(r_1 - r_2)]\,dS$$
$$(5.50)$$

Now consider an element dS' at a distance r_1 from a point source, which radiates a wave whose complex amplitude at dS' will be $A\,e^{-i\varkappa r_1}/r_1$ (Fig. 5.12). Regard dS' as a second point source of amplitude $A\,e^{-i\varkappa_1 r_1}/r_1$ radiating to a point Q distant r_2. The amplitude at Q will be

$$\frac{A\,e^{-i\varkappa r_1}}{r_1} \frac{e^{-i\varkappa r_2}}{r_2} dS'$$

For an aperture surrounding dS', the total amplitude at Q will be

$$a(Q) = \int_{\text{aperture}} \frac{A\,e^{-i\varkappa(r_1 - r_2)}}{r_1 r_2} dS' \quad (5.51)$$

which is identical in form with (5.50). Equation (5.51) represents the effect of the diffraction of a wave by an aperture. Thus the form of the propagation equation for $\gamma_{12}(0)$ is the same as that for diffraction of a wave, and the extensive results of diffraction theory can be taken over to deal with problems of coherence.

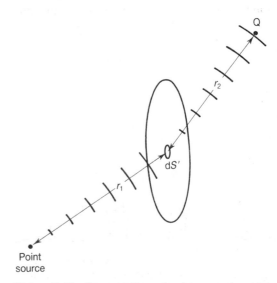

Point source

Figure 5.12 Propagation of coherence function

References

1. M. Born and E. Wolf, *Principles of Optics*, 3rd ed., Ch. 10, Pergamon Press, 1965.
2. H. H. Hopkins, *Proc. Roy. Soc. Lond.* A, **208**, 263 (1951).

Problems

5.1 Verify the expressions for $g(\omega)$ given in Table 5.1.

5.2 Draw the curves for a line with intensity distributions given by eqn (5.9) with $\gamma = 2 \times 10^7 \, s^{-1}$ for a range $\pm 10^8 \, s^{-1}$ about the centre of the mercury 546 nm line, normalised to unity at the line centre. On the same diagram, plot the normalised Doppler line (eqn 5.11) for mercury (atomic mass 200) at a temperature of 500 K.

5.3 With various tricks, light pulses of duration 6×10^{-15} s (6 femtoseconds) can be produced. Assuming the pulse to be rectangular, sketch the spectrum of such a pulse. Mark the range of the visible spectrum on your diagram. The wavelength of the light forming the pulse is 633 nm. Comment.

5.4 Over a limited wavelength range, the refractive index of certain glasses follows the relationship

$$n = A + \frac{B}{\lambda^2}$$

Derive an expression for the group velocity of light in this material.

5.5 The refractive index of quartz for $\lambda = 257$ nm is 1.50379; for $\lambda = 589.3$ nm it is 1.45848. Assuming the relation between n and λ given in Problem 5.4, determine the phase and group velocities for a wavelength of 404.6 nm.

5.6 Find the resultant amplitude, phase and energy of five waves of equal amplitude E_0 and with phases 0, $\pi/4$, $\pi/2$, $3\pi/4$ and π. Verify the result with a vector diagram.

5.7 Find the resultant energy of a very large ($\sim$ infinite) number of waves of amplitude E, $\frac{1}{2}E$, $\frac{1}{4}E$, $\frac{1}{8}E$, ... with phase angles 0, $\pi/2$, π, $3\pi/2$, ...

5.8 Young's fringes are formed in a system with slits of equal width. An absorber of transmittance 30% is placed over one slit. Calculate the visibility of the fringes formed.

5.9 Verify the relation stated in eqn (5.34).

6

Diffraction

6.1 Introduction

When a beam of light is partially obstructed by an obstacle (Fig. 6.1a), some of the light is diverted sideways and usually light and dark fringes are seen. The spreading of light in the neighbourhood of discontinuities is known as diffraction. It also occurs when the obstacle absorbs only part of the light (Fig. 6.1b) and when it changes the phase over part of the wavefront, as would be obtained by the insertion of a plate of glass of uneven thickness. Thus a small bubble of air in a lens or a local variation of refractive index causes diffraction effects.

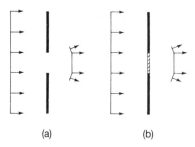
(a)   (b)

Figure 6.1 Diffraction of plane wave by (a) a clear aperture and (b) and absorbing aperture. The arrows indicate the amplitudes of the waves in the directions indicated. The alternation of intensity manifests itself as fringes

6.2 Calculating diffraction patterns

Can we calculate the intensity distribution in a diffraction pattern? The correct procedure for calculating diffraction effects is to solve the wave equation (eqn 4.42) subject to the boundary conditions imposed by the obstacle. This can be done only in the very simplest of cases. Thus Sommerfield [1,2] has obtained a solution for the simple case of diffraction at an infinite straight edge, but in general exact solutions cannot be obtained and approximations have to be made. In most cases the approximate methods used lead to results that agree closely with experimental measurements. What kind of approximations do we need to make in order to solve real problems? We assume first that an opaque plane obstacle completely absorbs that part of the wave-front which falls on it and leaves completely unaltered the part of the wave-front which misses the obstacle (St Venant's hypothesis). Once past the obstacle the unobstructed part of the wave-front is free to spread laterally—and does so!

The St Venant hypothesis implies the existence of abrupt discontinuities in the wave-front at the boundary, although studies of water waves and microwaves show that this does not occur. The amplitudes of waves close to boundaries *are* affected by the presence of the obstacles. We assume that for optical waves such effects are negligible. Since the analysis with this assumption leads to results that agree with the results of experiments, we conclude that this assumption is justified.

6.3 Huygens' principle

Huygens suggested that, since any point of a wave-front is a region of disturbance, it acts as a source of secondary waves and that the position of the wave-front at a later time is the envelope of the secondary waves. This

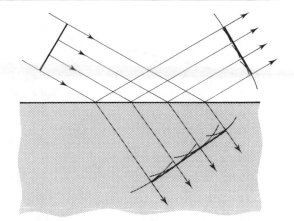

Figure 6.3 Reflection and refraction of waves in terms of Huygens' description

gives a qualitative explanation of rectilinear propagation (Fig. 6.2a) and of circular wave-fronts (Fig. 6.2b) which converge or diverge and also explains the sine law of refraction if we assume that the velocity in a material medium is c/n (Fig. 6.3). Huygens assumed that the secondary waves operated only where they touched their envelope in the forward direction and he ignored the backward wave, although there is no logical reason for this assumption. Despite these faults, Huygens correctly represented one significant fact. A beam of light carries information about the source and about anything that has modified it (e.g. by reflection or refraction). This information crosses any surface that intersects the beam. Kirchhoff showed that if we know both the disturbance and its rate of change on such a surface we can construct the wave surface at all later times.

6.4 Fresnel's equation

If each point on a wave-front behaves as a point source, we first need to specify how the amplitude of a wave from a point source will vary with distance. We know (1) that the energy at distance r from any point source (which radiates uniformly in all directions) varies as $1/r^2$ and also (2) that the energy flow in a wave varies as the square of the amplitude. If the wavelength is λ, then the phase will vary as $\exp(-2\pi i r/\lambda)$. If the point source on a wave-front is due to a small area

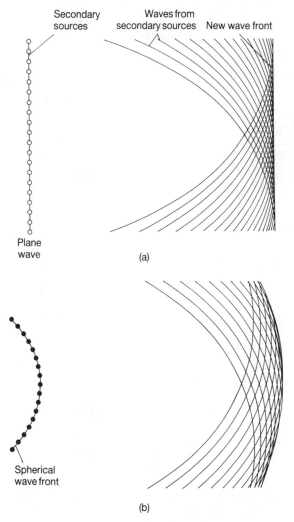

Figure 6.2 Huygens' reconstruction of (a) a plane wave-front and (b) a spherical wave-front

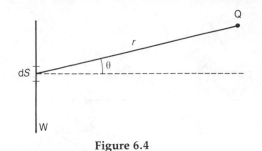

Figure 6.4

dS at which the amplitude is E_0, then the contribution to the amplitude of the wave at Q (Fig. 6.4) from dS is expected to have the form

$$dE_Q = \text{constant} \times \frac{E_0 \ dS \ \exp(-2\pi i r/\lambda)}{r} \quad (6.1a)$$

The constant may include a term that depends on θ, so we can put

$$dE_Q = \text{constant} \times F(\theta) \frac{E_0 \ dS \ \exp(-2\pi i r/\lambda)}{r}$$
$$(6.1b)$$

In the neighbourhood of $\theta = 0$, $F(\theta) = 1$.

In order that the results of using the above assumptions agreed with experimental observations, Fresnel found it necessary to put the constant in (6.1b) equal to i/λ. The factor 'i' indicates that the phase of the oscillation at Q differs by $\pi/2$ from that at dS [$\exp(i\pi/2) = i$].

It was subsequently established by Gouy that the phase of a spherical wave that passes through a focus changes by π. We can think of the diverging spherical wave from dS as having come from a converging wave focused on dS from the left; a $\pi/2$ phase change occurs when the approaching wave collapses into dS and a further $\pi/2$ when the secondary wavelet leaves dS. Thus for the element dS of Fig. 6.4, the field at Q is

$$dE_q = \frac{i F(\theta) E_0 \ dS}{\lambda r} \exp\left(\frac{-2\pi i r}{\lambda}\right)$$

$$= \frac{i F(\theta) E_0 \ dS}{\lambda r} \exp(-i\varkappa r) \quad \text{where } \varkappa = \frac{2\pi}{\lambda}$$
$$(6.1c)$$

In order to obtain the total field at Q, we should need to integrate (6.1c) over the whole area of the wave-front. If we were able

to do this we should find that the result satisfied the wave equation. In any real situation the integration is extremely difficult. We can, however, distinguish two extreme cases for which calculations are possible.

6.5 Near-field and far-field diffraction

The two extremes are shown in Fig. 6.5 In Fig. 6.5(a), $S_d S_0$ is small compared with the size of the aperture and a fringed image of the *aperture* is seen. In Fig. 6.5(b), $S_d S_0$ is large compared with the aperture size and a fringed image of the *source* is seen. The near-field case (a) is referred to as *Fresnel diffraction*; the far-field case (b) is *Fraunhofer diffraction*. Different mathematical approaches are needed for the two cases.

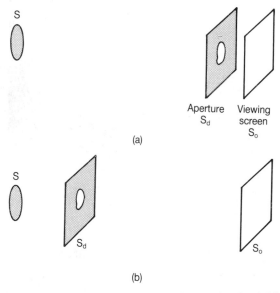

Figure 6.5 Near-field (Fresnel) and far-field (Fraunhofer) situations

6.6 Fraunhofer diffraction

The extreme case of Fraunhofer diffraction is where both source S and viewing screen S_0 are at infinite distances from S_d, a situation easily contrived by a pair of convex lenses, as seen in Fig. 6.6. The screen S_d receives plane waves from the lens and thinks that the source is at infinity. A point on S_0 records the plane wave component emerging from S_d

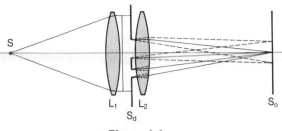

Figure 6.6

and believes the aperture to be an infinite distance away. In practice we consider first a point source S at the focus of L_1, so that a plane wave falls normally on S_d and we calculate the field distribution with S_0 at a *finite* distance from S_d. We can then readily determine the distribution over S_0 as its distance from the aperture tends to infinity.

We shall use eq (6.1c) for this case, illustrated in Fig. 6.7. We note that (6.1c) contains r both in the denominator and in the exponential term. For the far-field case, r will be enormously large compared with the wavelength and we may put $r = D$ in the denominator of (6.1c) with negligible error. The exponential term cannot be treated in such a cavalier fashion. A change in r of $\lambda/2$ can change the value $\exp(-2\pi i r/\lambda)$ from $+1$ to -1 so that an approximation closer than $r \approx D$ is essential.

In Fig. 6.7, the coordinates of P are $(x, y, 0)$ and those of Q are (x_0, y_0, z_0). Thus

$$r^2 = (x_0 - x)^2 + (y_0 - y)^2 + z_0^2 \qquad (6.2)$$

and

$$D^2 = x_0^2 + y_0^2 + z_0^2 \qquad (6.3)$$

so that

$$r = (D^2 - 2xx_0 - 2yy_0 + x^2 + y^2)^{1/2} \qquad (6.4)$$

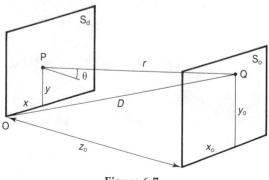

Figure 6.7

If the lateral distances x, y, x_0, y_0 are small compared with D, we may use the binomial approximation for r in (6.4) to obtain

$$r = D - \alpha x - \beta y + \frac{x^2 + y^2}{2D} - \frac{(\alpha x + \beta y)^2}{2D} \qquad (6.5)$$

where α, β are the direction cosines of the line OQ. When x and y are small compared with D, the last two terms of (6.4) may be neglected: indeed this may be regarded as the criterion for Fraunhofer diffraction to occur. If the distance from S_d to S_0 is more than ~ 100 times the size of the aperture in S_d, then the distribution over S_0 is virtually the same as that at infinity, with the arrangement of Fig. 6.6. Moreover, θ is confined to values sufficiently small that $F(\theta) = F(0) = 1$.

From (6.5) for the Fraunhofer case we have $r = D - \alpha x - \beta y$. From (6.1c), putting $2\pi/\lambda \equiv \varkappa$, we obtain

$$E_Q = \frac{iE_0}{\lambda D} e^{-i\varkappa D} \int\int \exp[i(\alpha \varkappa x + \beta \varkappa y)] \; dx \; dy \qquad (6.6)$$

where the integral is taken over the area of the aperture in S_d. We have thus obtained the complex amplitude E_Q of the wave diffracted in the direction (α, β) for a plane wave incident on S_d. It will be convenient to express the direction (α, β) in terms of $\varkappa_x = \alpha \varkappa$ and $\varkappa_y = \beta \varkappa$. Furthermore, if the amplitude of the wave emerging from the aperture is not uniform over S_d, but has a distribution given by $f(x, y)$, then (6.6) is readily generalised to

$$E(\varkappa_x, \varkappa_y)$$

$$= \frac{iE_0 e^{-i\varkappa D}}{\lambda D} \int\int f(x, y) \exp[i(\varkappa_x x + \varkappa_y y)] \; dx \; dy$$

$$= A \int\int f(x, y) \exp[i(\varkappa_x x + \varkappa_y y)] \; dx \; dy \qquad (6.7)$$

The power of the wave in the $(\varkappa_x, \varkappa_y)$ direction is given by $|E(\varkappa_x, \varkappa_y)|^2$. Hence the normalised angular power distribution is given by

$$P(\varkappa_x, \varkappa_y)$$

$$= \left| \frac{E(\varkappa_x, \varkappa_y)}{E(0, 0)} \right|^2$$

$$= \frac{|\int\int f(x, y) \exp[i(\varkappa_x x + \varkappa_y y)] \; dx \; dy|^2}{|\int\int f(x, y)|^2 \; dx \; dy} \qquad (6.8)$$

6 Insight into Optics

The function $f(x, y)$ enables any type of aperture to be dealt with. In regions where the aperture is completely transparent, $f(x, y) = 1$; where opaque, $f(x, y) = 0$. If the aperture contains a thin, transparent film, which changes the phase of the beam by $\phi(x, y)$ without changing the amplitude, then $f(x, y) = e^{i\phi(x, y)}$.

6.7 Fourier transforms

Equation (6.8) implies that a spatially non-uniform distribution of wave amplitude in the incident wave leads to a non-uniform *angular* distribution of power leaving the aperture. Moreover, the *larger* the area of the aperture, the *smaller* is the extent of the angular spreading. In Chapter 4 we considered the relation between the length or duration of a finite wave train and the associated distribution of wave number or frequency. The integral in eq (6.7) looks suspiciously like a Fourier transform (of eqn 5.2). In fact, $E(\varkappa_x, \varkappa_y)$ and $f(x, y)$ are connected by Fourier transform relations exactly as are $E(t)$ and $g(\omega)$ in Section 5.2.

6.8 Far-field (Fraunhofer) diffraction by a rectangular aperture

For a rectangle $2a \times 2e$, $f(x, y) = 1$ when x is between $-a$ and $+a$ and y is between $-e$ and $+e$. We obtain $P(\varkappa_x, \varkappa_y)$ by integrating eqn (6.7) (cf. Section 5.5):

$$P(\varkappa_x, \varkappa_y) = \left(\frac{\sin \varkappa_x a}{\varkappa_x a} \frac{\sin \varkappa_y e}{\varkappa_y e}\right)^2 \quad (6.9)$$

When a is large and e is small so that the aperture becomes a long narrow slit, there is essentially no variation in the x direction. The distribution in the y direction (Fig. 6.8a) is given by

$$W(\varkappa_y) = \text{sinc}^2(\varkappa_y e) \quad (6.10)$$

where *sinc* is a function here defined by sinc $\phi = \sin \phi/\phi$.[†]

When observing diffraction by a slit aperture it is usual to use a slit source. Each point

[†] Not universally: sometimes sinc x is defined as $\sin(\pi x)/(\pi x)$.

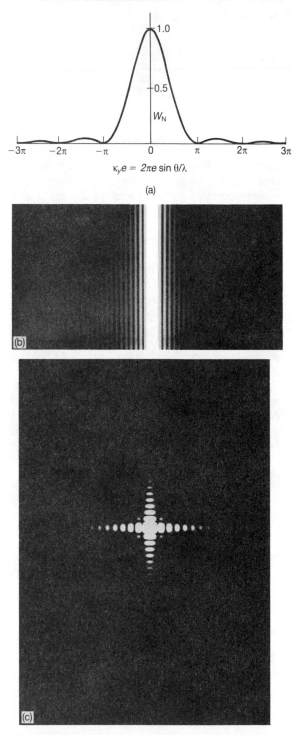

$\varkappa_y e = 2\pi e \sin \theta/\lambda$

(a)

Figure 6.8 Intensity distribution for infinitely long slit of width e: (a) calculated, (b) experimental, (c) diffraction pattern for rectangular aperture

of the source then produces a distribution similar to that given by eqn (6.10) in the plane perpendicular to the slit that passes through its image point on S_0. The light from different points on the source is non-coherent and therefore there is no interference on S_0 between light from different points of the source. The result is a set of fringes shown in Fig 6.8(b). The pattern for a rectangular slit is seen in Fig 6.8(c).

6.9 Far-field diffraction by a circular aperture

This case is important because in microscopes, telescopes, etc., the edges of lenses and mirrors commonly constitute circular apertures that limit the field. When aberrations have been made very small, diffraction at these boundaries sets the limit to the quality of images obtained.

Consider a circular aperture of radius R with a point source on the axis and lenses as in Fig. 6.6. The aperture in S_d is now a circle centred on the axis. In view of the symmetry we need to calculate the variation of W only along one radius, which we may choose as the OX direction. Then eqn (6.7) becomes

$$E_Q = A \int_{-R}^{R} dx \int_{-(R^2-x^2)^{1/2}}^{(R^2-x^2)^{1/2}} \exp(i\varkappa_x x)\, dy \tag{6.11}$$

$$= 2A \int_{-R}^{R} (R^2 - x^2)^{1/2} \exp(i\varkappa_x x)\, dx \tag{6.12}$$

Now change to variables ρ and χ, where $\chi = R\cos\chi$ and $\rho = \varkappa_x R$. Write $\exp(i\rho\cos\chi)$ as $\cos(\rho\cos\chi) + i\sin(\rho\cos\varkappa)$ and note that the second integral vanishes. We then have

$$E_Q' = 2AR^2 \int_0^\pi \sin^2\chi \cos(\rho\cos\chi)\, d\chi \tag{6.13}$$

The integral reminds us that Bessel's function of the first kind of order unity is defined by

$$J_1(\rho) = \frac{\rho}{\pi} \int_0^\pi \sin^2\chi \cos(\rho\cos\chi)\, d\chi \tag{6.14}$$

and since $J_1(\rho)/\rho$ tends to the value $\frac{1}{2}$ when ρ approaches zero then the angular power spectrum, normalised to the value at $\varkappa_x = 0$,

is

$$P(r') = 4\left[\frac{J_1(\rho)}{\rho}\right]^2 = 4\left[\frac{J_1(2\pi Rr'/\lambda f)}{(2\pi Rr'/\lambda f)}\right]^2 \tag{6.15}$$

where r' is the distance from the centre of the pattern on S_0 and f is the focal length of the lens.

Tables of Bessel functions are available [3]. From these the graph shown in Fig. 6.9(a)

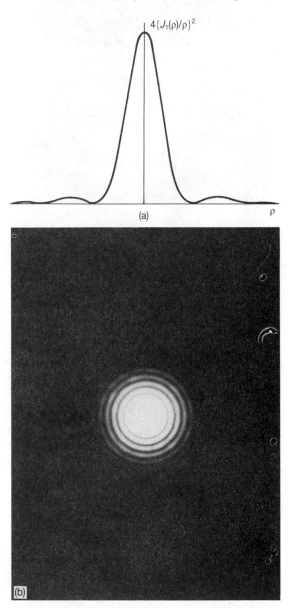

Figure 6.9 Intensity distribution for circular aperture of radius R, $\rho = (2\pi \cos\theta)/\lambda$: (a) calculated, (b) experimental

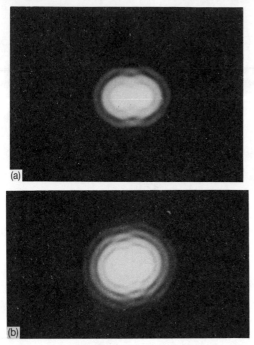

(a)

(b)

Figure 6.10

has been calculated. A photograph of the pattern is shown in Fig 6.9(b). Fig 6.9(a) looks similar to Fig. 6.8(a) but the relative sizes and positions of the subsidiary maxima are different. The first minimum in Fig. 6.9 occurs when $\rho = 3.83$, corresponding to a radius $r' = 1.22 f\lambda/2R$.

As two circular apertures are moved closer together, their diffraction patterns overlap (Fig. 6.10a) until the pattern can scarcely be distinguished (Fig. 6.10b) from that of a single aperture. This sets a limit to the ability to separate points of detail (the resolving power), to be further discussed in Chapter 19.

6.10 Apodised circular aperture

Hitherto we have assumed that the wave amplitude $f(x, y)$ is the same at all points of the aperture and we see that the diffraction pattern exhibits rings round the central spot. We can obtain a pattern *without* rings if we arrange that

$$f(x, y) = f(R) = \exp\left[-\pi\left(\frac{R}{\sigma}\right)^2\right] \quad (6.16)$$

where σ is a constant. This is a Gaussian function whose Fourier transform from $-\infty$ to ∞) is given in Table 5.1. From it we obtain

$$P_N(r') = B \exp[-\pi\sigma(r')^2] \quad (6.17)$$

where B is a constant and r' is the coordinate in the Fourier transform plane, i.e. in the plane of observation.

The irradiance falls smoothly from the centre outwards (Fig. 6.11). This calculation is not strictly accurate because we have assumed that eqn (6.16) is true for all values of R whereas $f(x, y)$ is zero outside the circular aperture. However, if the radius of the aperture is large compared with σ, $f(R)$ will have fallen to a very small value at the edges of the aperture and the error is small. Thus if a circular aperture is covered with a filter whose radial transmission varies as eqn (6.16), the far-field diffraction pattern contains no annular rings. This process is known as apodisation.

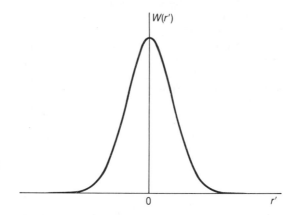

Figure 6.11 Intensity distribution for apodized circular aperture

6.11 Diffraction by a number of similar apertures

Suppose the diffraction screen S_d contains N apertures that are similar and similarly orientated (Fig. 6.12). In each aperture we choose a point (O_1, O_2, etc.) as a local origin. All the local origins are similarly placed in the apertures. Let the coordinates of a point P in the pth aperture be (x, y) relative to the local

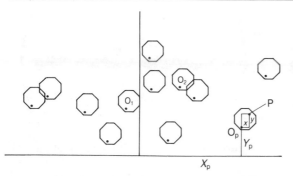

Figure 6.12 Random distribution of identical apertures

origin and let X_P, Y_P be the coordinates of O_P relative to a main origin at O. Then the coordinates of P relative to O are $(X_P + x)$ and $(Y_P + y)$ and eqn (6.7) gives

$$E_0(x, y) = A \sum_{p=1}^{N} \exp[i(\varkappa_x X_P + \varkappa_y Y_P)]$$

$$\int_S \exp[i(\varkappa_x x + \varkappa_y y)] \, dx \, dy \quad (6.18)$$

the integral being taken over the area of one aperture. The normalised angular power spectrum is

$$W = (GG^*)(gg^*) \quad (6.19)$$

where

$$G = \sum_{p=1}^{N} \exp[i(\varkappa_x X_P + \varkappa_y Y_P)] \quad (6.20a)$$

and

$$g = \frac{\pi}{2} \int \exp[i(\varkappa_x x + \varkappa_y y)] \, dx \, dy \quad (6.20b)$$

Thus the diffraction pattern depends on the product of two factors G and g: g depends on the size and shape of one aperture and is called the *form factor* and G depends on the way in which the elements are arranged and is called the *structure factor*.

6.12 Random and regular arrays

Two important cases arise: (1) a random arrangement and (2) a completely ordered arrangement on some kind of symmetrical grid. When the arrangement is random the relative phases of light beams going in a given direction are also random. We have photometric summation (as in Section 5.10) and $GG^* = N$. The distribution of light for all apertures is the same as that for a single aperture but is N times stronger. When the diffracting elements are arranged in a regular way there are some directions in which the phases differ by even multiples of π. Then GG^* becomes N^2 for these directions and very bright spots appear. The chief application of the study of regular arrangements is in X-ray crystallography where the problem becomes more complicated since there is usually a *three-dimensional* array of diffracting elements.

The random arrangement was applied in Young's Eriometer, used to determine the diameters of particles of approximately the same size. A parallel beam of quasi-monochromatic light is incident on a microscope slide covered with the particles, typically blood corpuscles. Since the refractive index of the cell material differs from that of the surrounding fluid, they behave in a similar way to apertures. If the corpuscles are all roughly circular a fairly sharp ring pattern would be obtained, and the angular diameter of the rings may be measured. In fact blood corpuscles are not circular and are oriented at random so the rings are broadened. It is nevertheless possible to measure the average size of the corpuscles with sufficient accuracy for medical diagnosis.

6.13 Diffraction gratings

The optical diffraction grating is the simplest possible arrangement of similar diffracting elements. It consists of a set of parallel apertures or lines on a plane surface. Originally gratings were made by ruling lines in a metal film (or a smoke deposit) on glass. If measurable angles of diffraction are to be achieved, the separation between the lines must be of the order of the wavelength of the light to be diffracted. This indicates that hundreds of lines per millimetre are needed. The art of ruling fine parallel lines developed fairly rapidly to a point where 200 lines mm^{-1} could be ruled. The structure factor for such a grating is obtained by substituting from eqn (4.23) into eqn (4.25b), and the form factor from eqn (6.10). Combining these equations gives

$$P(x_y) = \frac{\sin^2(N\delta/2)}{(N \sin \delta/2)^2} \, \text{sinc}^2(x_y e) \quad (6.21)$$

for the normalised power spectrum. The factor $1/N^2$ is included because when $\delta \to 0$, $\sin^2(N\delta/2) \to N^2\sin^2\delta/2$.

Substituting from eqn (4.23) for δ and putting $x_y = x \sin \theta = 2\pi \sin \theta/\lambda$ we have

$$P(x_y)$$
$$= \left[\frac{\sin(2\pi Nd \sin \theta/\lambda)}{N \sin(2\pi d \sin \theta/\lambda)} \, \text{sinc}(2\pi e \sin \theta/\lambda) \right]^2$$
$$(6.22)$$

The first term in the bracket is the structure factor and the second the form factor. The θ-dependence of the individual factors is shown in Fig. 6.13(a) and that of $P(x_y)$ in Fig. 6.13(b).

It is also possible to make *reflecting* gratings in which the lines reflect light and the spaces

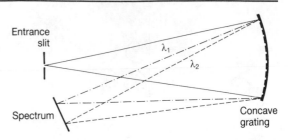

Figure 6.14 Concave diffraction grating

are dark. It is usual to make these in the form of spherical mirrors which refocus the light to give an image near to the plane of the slit source (Fig. 6.14). When the light is incident on a grating at an angle θ_1 and viewed in a direction θ_2 the value of δ is obtained by substituting $(\sin \theta_1 - \sin \theta_2)$ for $\sin \theta_1$ in eqn (4.23). A similar alteration is needed in respect of the form factor so that $(\sin \theta_1 - \sin \theta_2)$ must be substituted for $\sin \theta$ in both factors of eqn (6.22). For a reflecting grating, $(\sin \theta_1 + \sin \theta_2)$ must be substituted if we use the sign convention stated in Section 2.3.

6.14   Blazed gratings

Fig 6.13 shows the form factor, the structure factor and $W(x_y)$ for the kind of grating discussed above. The strongest maximum is the one for $m = 0$ (in eqn 4.26). This is in the same position for all wavelengths and so does not give a spectrum. The number of side spectra (corresponding to $m = \pm 1$, ± 2, etc.) that have any appreciable intensity depends on the form factor and hence on the width of the lines. If the lines are made narrow the fraction of the energy in the side spectra increases but the total energy diffracted into all lines decreases. Rayleigh suggested (in 1888) that reflecting gratings with the form shown in Fig. 6.15 should be made.

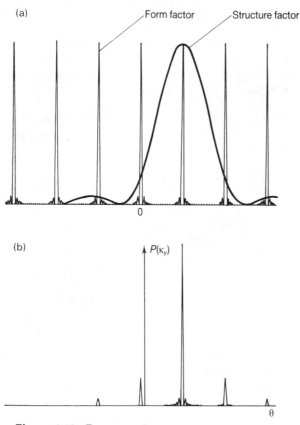

(a)

Form factor   Structure factor

0

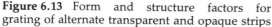

(b)

$P(x_y)$

θ

Figure 6.13 Form and structure factors for grating of alternate transparent and opaque strips

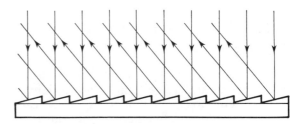

Figure 6.15 Blazed reflection grating

If the angle of reflection from the inclined surfaces is the same as the angle corresponding to one of the side maxima most of the light goes into the corresponding spectrum (Fig. 6.16). This condition is fulfilled strictly for only one wavelength but a very bright spectrum over a considerable wavelength range is obtained, and a blaze of light is seen when the grating is viewed at the correct angle relative to the source. Such gratings are called *blazed gratings*. It was not until 1911 that Wood and Trowbridge (using a diamond point prepared by Brackett) succeeded in making gratings that approximated to the form proposed by Rayleigh.

Gratings with up to 1200 lines mm^{-1} are readily available and even finer gratings are sometimes ruled. The ruled area on a medium sized grating is 10×10 cm^2, but gratings occupying 25×25 cm^2 are some-times made. Such a grating has 300 000 lines with a total length of 75 km. The demands on diamond technology to make a point that does not wear appreciably and on the machine which accurately maintains constant separation of the lines are very severe [4].

It is in fact possible to create a diffraction grating without depending on the stability of a ruling machine, for which the environmental control needs to be extremely close and which has a point that must inevitably wear in time. If it is arranged that two plane waves intersect at a small angle then provided that they are derived from the same source they will form interference fringes in the form of accurately parallel straight lines. A photographic plate placed in the region of the intersection will therefore record a set of parallel straight lines at a separation which can be adjusted by fixing the angle between the wave-fronts. With a photographic plate of high resolution, the spacing can be made to correspond to that needed for a diffraction grating, and the developed plate does in fact behave as a grating. A further possibility is to bleach the developed silver in the plate to produce transparent material of different refractive index—known as a *phase* grating. The process is an example of holography, which is dealt with in Chapter 13. Holographic gratings do not suffer from the imperfections common to ruled gratings, where spurious diffraction lines can occur as a result of imperfections and irregularities in the ruled lines.

So much for *far-field* diffraction, what about near-field effects? Can we calculate the intensity distribution in a plane close to the aperture?

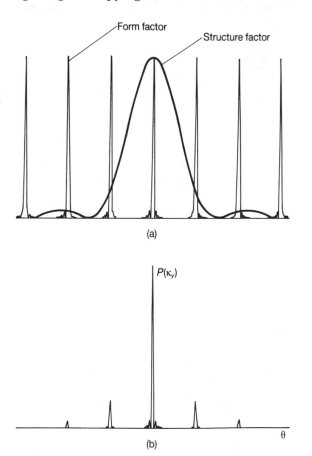

(a)

(b)

Figure 6.16 Form and structure for blazed grating

6.15 Near-field (Fresnel) diffraction by a circular aperture

Consider a plane wave advancing on a circular aperture. We wish to calculate the amplitude at a point Q_0 on the axis. We do this by imagining that the aperture is divided into zones, called Fresnel zones or Huygens zones, such that if A_n is a point on the outer

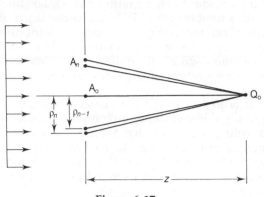

Figure 6.17

boundary of the nth zone (Fig. 6.17) then

$$A_n Q_0 - A_0 Q_0 = \frac{n\lambda}{2} \qquad \text{where } n \text{ is an integer}$$

(6.23)

(In general there will be an incomplete zone at the edge of the aperture.) If ρ_n is the radius of the outer boundary of the nth zone then

$$(z^2 + \rho_n^2)^{1/2} - z = \frac{n\lambda}{2} \qquad (6.24)$$

and

$$\rho_n^2 = n\lambda z + \frac{n^2\lambda^2}{4} \qquad (6.25)$$

The area of the nth zone is

$$S_n = \pi\left(\rho_n^2 - \rho_{n-1}^2\right) = \pi\lambda\left[z + \left(\frac{2n-1}{4}\right)\lambda\right] \quad (6.26)$$

and if r_n is the distance from Q_0 to the centre of the nth zone then

$$r_n = z + \left(n - \frac{1}{2}\right)\frac{\lambda}{2} = z + \left(\frac{2n-1}{4}\right)\lambda \quad (6.27)$$

so that

$$\frac{S_n}{r_n} = \pi\lambda \qquad (6.28)$$

(which is independent of n). When z is large compared with λ, both S_n and r_n increase only very slowly (and at the same rate) from zone to zone. Substituting S_n/r_n for dS/r in eqn (6.1c), the complex amplitude due to the nth zone is

$$E_{Qn} = i\pi E_0 F_n(\theta)\phi\left(\frac{r}{\lambda}\right) \qquad (6.29)$$

where $\phi(r/\lambda)$ is a factor that takes account of the variation of phase over the zone and $F_n(\theta)$ for the nth zone. For the case where $z \gg \lambda$, the value of $F(\theta)$ is equal to unity. To obtain $\phi(r/\lambda)$ for the first zone, we divide this zone into a set of very small subzones. The complex amplitude of one of these is represented by a vector whose magnitude and direction represent the area and phase of the subzone. If the subzones are of equal phase difference the magnitudes will all be equal (as eqn 6.26 shows). The vector diagram is shown in Fig. 6.18(a). Since the phase difference between the first and last subzones is π, the subzones are represented by a semicircle and the resultant by its diameter. Thus the resultant amplitude is $2/\pi$ times the amplitude that would be obtained if the phase were the same for light from all parts of the zone. Also the phase of the resultant is $\pi/2$ ahead of the phase of the first subzone so

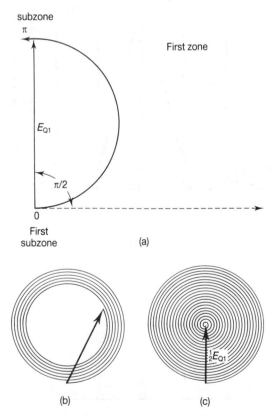

Figure 6.18 Vector diagrams for (a) first half-period zone, (b) several zones, (c) very large ($\rightarrow$infinite) number of zones

$\phi(r/\lambda) = 2i/\pi$ and, from (6.29), for $r \gg \lambda$,

$$E_{Q1} = 2E_0 \qquad (6.30)$$

The amplitude due to the first zone is twice the amplitude E_0 that would be obtained if the screen S_d were removed and the plane wave advanced without any obstruction. To find the resultant of several zones we continue the vector diagram as shown in Fig. 6.18(b). Each zone is represented by a semicircle but the diameters slowly become smaller as $F(\theta)$ decreases. The resultant amplitude is obtained by drawing a line from O to the final point of the curve. This shows that when the size of the aperture is increased from zero the resultant amplitude E_Q increases until the area of the aperture is equal to the area of one zone and then decreases to a minimum (which is small) when the aperture covers two zones. Maxima (corresponding to an odd number of completed zones) alternate with minima (when there is an even number). The maxima are less than $\frac{1}{2}E_{Q1} + \frac{1}{2}E_{Qn}$ and the minima greater than $\frac{1}{2}E_{Q1} - \frac{1}{2}E_{Qn}$. If $F(\theta)$ tends to zero as θ approaches $\pi/2$ then E_{Qn} approaches zero as n becomes very large[†] and, as we should expect, the amplitude due to a very large number of zones is $\frac{1}{2}E_{Q1} = E_0$, i.e. it is the same as the amplitude of the unobstructed wave (Fig. 6.18c).

If the aperture has a fixed radius ρ_f and a screen is brought inwards from an infinite distance to the right, maximum light (corresponding to an amplitude twice and an energy four times that of the unobstructed wave) will be obtained at a point on axis when z satisfies eqn (6.25) with $n = 1$, i.e. when the aperture constitutes one zone. At half this distance there will be two zones in the aperture and the amplitude will be a minimum; at a third there will be a maximum and so on. When there is a minimum the energy will have gone into bright rings around the axis.

The notion of Fresnel zones is useful for examining what happens on the axis of a system with a circular aperture. It is less useful for dealing with points off-axis. If,

from a given point on-axis, an aperture is 'seen' to contain very many half-period zones, then initially on moving the point away from the axis a large number of whole zones will still occupy the aperture and the resultant amplitude will remain large. At a large displacement from the axis, a large number of narrow arcs of the zone pattern will be seen. Since the resultants of the contributions from successive arcs are in opposite phase, the overall resultant amplitude will in this case be small.

6.16 Diffraction by a circular obstacle

Near-field diffraction by a truly circular obstacle leads to an unexpected result. Let us examine what happens when a plane wave falls normally on an opaque disc. On a ray picture, no light will penetrate the cylindrical region behind the obstacle. Suppose, however, we sit at the point behind the disc such that the disc diameter is also the diameter of the first half-period zone. The vector diagram representing this situation is similar to that of Fig. 6.18(c) but with the contribution from the first zone removed. Thus we peel off the first semicircular arc, leaving a resultant that is *practically the same as for the unobstructed wave*. For a slightly larger object—e.g. blocking out the first four zones—we peel off four half-loops and *still* have a large resultant. In fact, if we peel off any fraction of an outer zone we get a large resultant. Before anyone actually *looked* for a bright spot in the geometric shadow of a disc (or ball),

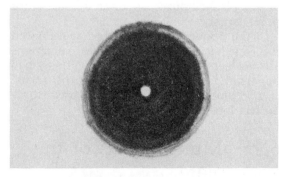

Figure 6.19 Bright spot in geometric shadow of circular aperture: the slightly irregular shape indicates that the object is not perfectly circular

---

[†] This is reasonable: the amplitude should be proportional to the projected area of the zone as θ increases. This area tends to zero as $\theta \to \pi/2$.

this prediction of the wave theory was used to show that the theory must be wrong! Fresnel and Arago showed that the bright spot *did* occur (but were, unknowingly, beaten to it by Lisle, who made the observation fifty years earlier). The spot is shown in Fig 6.19.

6.17   Fresnel zone plate

Suppose that we draw circles on a transparent plate with radii that are proportional to $1^{1/2}, 2^{1/2}, 3^{1/2}, ..., n^{1/2}$. These will correspond, for an incident plane wave, to half-period zones for a point Q_1 on the axis at some distance z_1 from the plate. Now let us blacken alternate zones. How will this affect the amplitude at Q_1? If we ignore for the moment the fact that the 'semicircles' of Fig. 6.18(c) diminish for successive zones, we see that for N zones, half of which are blacked out, we have $N/2$ semicircles, the resultants of which are all in phase. The resultant amplitude will be $N/2$ times that of the first zone, or N times the amplitude at Q_1 if there were no zone plate present. An N-fold increase in amplitude means an N^2-fold increase in intensity. Thus a bright point of light will occur—the plate will *focus* rather in the manner of a converging lens.

We can do even better if instead of blacking out alternate zones we cover them with a transparent layer which adds an optical thickness of one half-wavelength to that of the plate. This has the effect of reversing the directions of the semicircles on the left-hand side of the vector diagram (Fig. 6.18c), leading to a resultant amplitude double that of the 'blacked-out' case, or an intensity four times larger.

In the above discussion we have ignored the fact that the arcs of the vector diagrams are *not* semicircles. The effect of this small departure is that the amplitude at the 'focus' is $2/\pi$ times that which would arise for a perfect lens.

6.18   Multiple foci of the zone plate

Each of the zones for Q_1 will constitute three zones for a point Q_3 distant $z_1/3$ from the plate. The amplitudes of two of each trio of zones will cancel but the remaining areas will cooperate to give a concentration of light at Q_3 with energy one-ninth times that at Q_1. There will be still weaker concentrations at distances of $z_1/5$, $z_1/7$, etc. So far we have considered only the simplest situation where the image point is on-axis at infinity. It is easy to apply the same methods to light from a point on-axis at a distance z_0 to the left of the zone plate. It is found that the zone plate forms an image and the distances of object and image bear the same relationship as for a lens of focal length z_1 (Section 3.3). Arguments similar to the above apply to points slightly off-axis, so the zone plate will give an image of a small object near the axis and this image is magnified in the same ratio as would be obtained by a lens. A less bright image is also obtained with the circular obstruction described in Section 6.16.

Zone plates are not generally used in optics because lenses that concentrate the light into a single focus give clearer images. The Fresnel zone concept may, however, prove useful in connection with the X-ray region of the spectrum. Since the refractive index of all materials for X-rays is practically unity, no lenses are available. Finely focused electron beams can be used to make zone plates with very narrow zones, and this is one way in which it may be possible to make an X-ray microscope.

6.19   Fresnel integrals

The zone concept is limited to circular apertures and obstacles. Fresnel derived more general formulae for objects of any shape; the results are especially suitable for rectangular shapes including slits and a straight edge. Consider a diffraction screen S_d (Fig. 6.7) with one or more apertures and let (x, y, O) be the coordinates of a point P in one of the apertures. We wish to calculate the energy at a point Q (coordinate x_0, y_0, z_0) on the screen S_0. We assume that a plane wave is incident on S_d but the extension to an incident spherical wave is not difficult. Return to eqn (6.1c) and put

$$r^2 = z_0^2 + (x_0 - x)^2 + (y_0 - y)^2 \qquad (6.31)$$

Then

$$r = z_0 \left[1 + \frac{(x_0 - x)^2}{z_0} + \frac{(y_0 - y)^2}{z_0} \right]^{1/2}$$

and when z_0 is large compared with $x_0 - x$ and $y_0 - y$,

$$r = z_0 + \frac{1}{2} \frac{(x_0 - x)^2}{z_0} + \frac{1}{2} \frac{(y_0 - y)^2}{z_0} \quad (6.32)$$

Then, omitting the constants, and integrating over the aperture,

$$E_Q = \int \frac{\exp(-i\varkappa r)}{r} \, dS = \frac{\exp(-i\varkappa z_0)}{r}$$

$$\int \exp \left\{ -i\varkappa \left[\frac{(x_0 - x)^2}{2z_0} + \frac{(y_0 - y)^2}{2z_0} \right] \right\} dx_0 \, dy_0$$
$$(6.33)$$

The integral is to be taken over the whole area of the aperture. We have taken the $1/r$ factor outside the integral because its variation by a few wavelengths is of negligible importance.

Equation (6.33) is, in principle, applicable to any shape of aperture. If we consider a rectangular-shaped aperture the integral is the product of two integrals, one of which is

$$\int_{v_1}^{v_2} \exp(-\tfrac{1}{2} i\pi v^2) \, dv \quad (6.34)$$

where

$$\frac{\pi v}{2} = \left(\frac{\pi}{\lambda z_0} \right)^{1/2} (x_0 - x) = \left(\frac{\varkappa}{2z_0} \right)^{1/2} (x_0 - x) \quad (6.35)$$

The variation of illuminance in the x direction is

$$|E_Q|^2 = C^2 + S^2$$

where

$$C = \int_{v_1}^{v_2} \cos(\tfrac{1}{2} v^2) \, dv \quad \text{and} \quad S = \int_{v_1}^{v_2} \sin(\tfrac{1}{2} v^2) \, dv$$
$$(6.36)$$

C and S are called Fresnel's integrals. They can be integrated by series and tables of their values are available [3]. When $v_1 = -\infty$ and $v_2 = +\infty$, C^2 and S^2 each $= \tfrac{1}{4}$. For a vertical slit aperture, we put C^2 and S^2 equal to $\tfrac{1}{4}$ for the vertical direction and obtain values for the perpendicular direction from tables. Fig 6.20 shows the variation for slits of different widths. It is usual to employ a slit *source*; in this case, as with the Fraunhofer diffraction (Section 6.8), fringes are obtained.

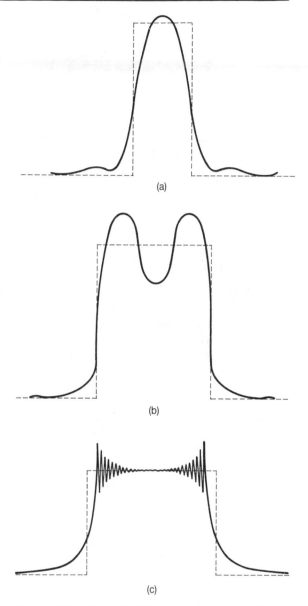

Figure 6.20 Fresnel diffraction intensity distributions for (a) very narrow, (b) narrow, (c) very wide slits

6.20 Cornu's spiral

Suppose a curve is plotted with C and S as abscissa and ordinates (limits of integration being 0 to v). The result is a spiral (Fig. 6.21) converging on the points $(+\tfrac{1}{2}, +\tfrac{1}{2})$ and $(-\tfrac{1}{2}, -\tfrac{1}{2})$. The marks on the curve correspond to the scale of v. The square of the length of

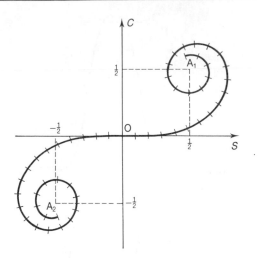

Figure 6.21 Cornu's spiral

a line joining two points on the spiral is proportional to the sum of the squares of the integrals (between the corresponding limits) and hence gives the energy at the corresponding point.

6.21 Diffraction by a straight edge

Suppose light from a slit illuminates a parallel straight edge. Then when O is far from the edge of the shadow and in the lighted region the limit of integration will be effectively ∞ on one side (so that A_1 is the representative point) and a point very near to A_2 on the other side. $(A_1A_2)^2$ represents the resultant, and this is constant so long as Q is far from the edge. As Q approaches the edge, one end of the line remains fixed on A_1 (because the limit on that side is ∞ and the other end moves round the spiral towards O; maxima and minima (i.e. fringes) are obtained when O is near the edge of the geometrical shadow. When Q reaches the edge the second point is at O and the energy has one-quarter of the value obtained when Q was far from the edge. When Q moves into the shadow the second point moves up the spiral gradually getting nearer to A_1 but without fluctuations. The light gradually

decreases to zero (Fig. 6.22) and there are no fringes in the shadow region.

Similarly we may use the spiral to obtain the diffraction pattern for a slit. We imagine a line of fixed length, corresponding to the slit width, overlaid on the spiral. When we are well away from the slit, the line is curled up in the neighbourhood of A_1 and the ends are close together, with a very small resultant. As we move across the field parallel to the plane of the slit the line unrolls, giving a resultant which oscillates until the line lies symmetrically about O. This variation is reversed as we proceed afar in the opposite direction.

6.22 Babinet's principle

If there is no diffraction screen in the apparatus shown in Fig. 6.7, the lenses will form an image of the source on S_0 with narrow fringes due to limitations of the beam by the edges of the lenses. Now consider two screens S_A and S_B which are such that apertures in S_A exactly coincide with opaque regions in S_B. If the amplitudes at a given point for (1) no screen at S_d, (2) S_A inserted, (3) S_B inserted are E_0, E_{0A} and E_{0B} respectively, then

$$E_0 = E_{0A} + E_{0B} \qquad (6.37)$$

in accordance with the principle of superposition (Section 4.6). If, at any point $E_0 = 0$ then $E_{0A}^2 = E_{0B}^2$ so that the two screens give the same energy at this point. This theorem is due to J. Babinet (1794–1872).

The theorem is correct as we have stated it, but it is often assumed that the diffraction patterns from the two screens will be nearly the same wherever E_0 is small compared with its value at the centre. This is not true. For example, a minute spot on the lens has very little effect but a screen containing a minute pinhole has a large one. The diffraction patterns are nearly the same only when E_0 is small compared with E_{0A} and E_{0B}. In practice the principle can be applied only when S_A and S_B each obstruct an appreciable fraction of the total area.

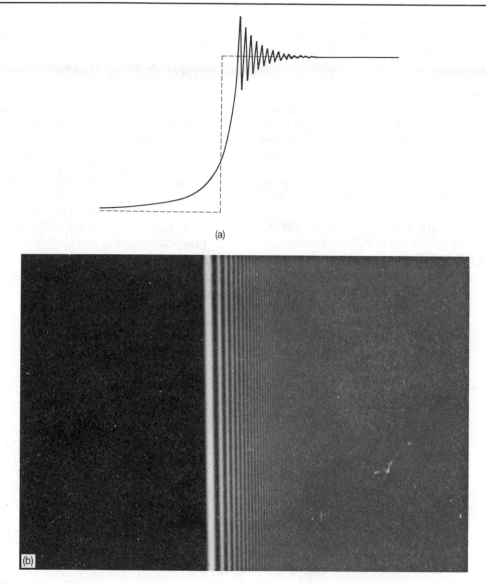

(a)

(b)

Figure 6.22 Intensity distribution near the shadow edge: (a) calculated, (b) experimental

References

1. A. Sommerfeld, *Optics Translation*, Academic Press, London and New York, 1954.
2. M. Born and E. Wolf, *Principles in Optics*, Pergamon Press, Oxford, London and New York, 1965.
3. Abramowitz and Stegun, *Handbook of Mathematical Functions*, Dover, 1965.
4. John Strong, *Procedures in Experimental Optics*, Dekker, 1988.

Problems

6.1 Verify the binomial expansion for r given in eqn (6.5).

6.2 Justify the neglect of the second-order terms in eqn (6.5) for the case where S_d, S_0 in Fig. 6.7 are each 10 mm square and $z_0 = 5000$ mm.

6.3 Fringes are formed from a slit 0.1 mm wide using a helium–neon laser

($\lambda = 633$ nm). Calculate the separation between the first minima on either side of the centre on a screen at a distance of 1 m.

6.4 You wish to produce a Fraunhofer diffraction pattern of a circular aperture. You have a laser ($\lambda = 633$ nm) and a lens of focal length 200 mm. You want the diameter of the first dark ring of the pattern to be 2 mm. What sized aperture would you use?

6.5 A blazed grating with 20 mm of ruling at 600 lines mm^{-1} is illuminated with light from a mercury lamp, where the mercury vapour temperature is 1000 K. Will the width $\Delta\lambda/\lambda$ of line observed be determined by the Doppler width of the mercury line or the properties of the grating? (The Doppler width of the line emitted by an atom moving with velocity v is given by $\Delta\nu_D/\nu = v/c$, where $\frac{1}{2}Mv^2 = \frac{1}{2}kT$ and M is the mass of the atom.)

6.6 You decide to make a zone plate by drawing circles, filling in alternate zones. The plate is to be produced photographically by projecting a reduced image on to a photographic plate which can barely resolve lines 0.002 mm apart. Your plate is to have its longest focal length 20 mm and to work with light of wavelength 500 nm. How many zones will you draw?

7

Polarised Light

7.1 Introduction

The phenomena discussed in previous chapters show clearly that any satisfactory description of light must recognise its essentially wave-like character. Interference and diffraction are general manifestations of any type of wave—sound waves, seismic waves, waves on surfaces, etc. We need to be able to specify the amplitude and phase of whatever is waving. We can immediately distinguish two types of wave motion, namely (1) longitudinal waves, and (2) transverse waves.

In sound waves the waving quantity is the displacement of the atoms of the material *in the direction of travel* of the wave. This is an example of a longitudinal wave. For surface waves on water the relevant quantity is the displacement of the water surface *perpendicular* to the direction of travel—an example of a transverse wave. To which class does light belong? In our references in previous chapters to the amplitude of light waves we ask (1) amplitude of *what*? and (2) in what direction is the amplitude measured?

We go back a couple of centuries and look at a simple experiment by Malus (1775–1812). Figure 7.1 represents the reflection of light from two pieces of glass. If the second ray is in the plane of the paper (as shown), i.e. if it is in the same plane as the first reflection, then strong reflection is

obtained. If, however, the second plate is rotated so that the beam is reflected out of the plane of the paper the reflection is much weaker. Brewster (1781–1868) showed that if the angle of incidence I on each plate is given by

$$\tan I = \mu \qquad (7.1)$$

where μ is the refractive index of the glass, then if the two reflections are in perpendicular planes, the reflection from the second plate is zero. The angle I is known as the polarising, or Brewster, angle. The results of Malus' and Brewster's experiments confirm that light is a *transverse* wave motion and that the vector characterising the light wave is perpendicular to the direction of propagation. Light reflected at the angle I has its oscillations confined to one direction only and is not reflected at this angle in a plane perpendicular to that of the initial reflection.

We shall see in Chapter 8 that light can be represented by oscillating electric and magnetic fields, both of whose directions are perpendicular to the direction of the light beam

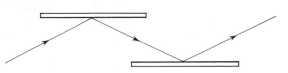

Figure 7.1 Malus experiment

(and to one another). It will emerge that the electric vector of the light reflected from a surface at the Brewster angle is perpendicular to the plane of incidence. We shall avoid using the term 'plane of polarisation'. Currently this is often used to refer to the plane of the electric vector, but historically it was used to denote the plane of what we now know to be the magnetic vector. To avoid confusion and ambiguity we shall refer to the 'plane of the electric vector'.

7.2   Types of polarisation

A beam of light in which the electric vector always lies in the same plane is termed *plane-polarised* light. If it were possible to see the electric vector then on looking at the oncoming beam the vector would be seen to be oscillating along a fixed line. It is possible, however, to produce a light beam in which the end of the electric vector would be seen to travel round a circle (*circularly polarised* light) or—the most general case—would describe an ellipse (*elliptically polarised* light). Since the light vector moves in the direction of light propagation, the true paths of the ends of the vectors will be circular or elliptical helices. The vector makes one revolution about the direction of propagation for a distance travelled of one wavelength.

7.3   Double refraction

Although a ray of light entering glass or water produces a single refracted ray, certain crystals behave in a more unusual fashion. Huygens had early observed that when a ray of ordinary ('unpolarised') light passed into a calcite ($CaCO_3$) crystal plate (Fig. 7.2a) *two* transmitted beams emerge. One, called the *ordinary* ray, obeys the usual laws of refraction. The other, called the *extraordinary* ray, does not. If the two beams are allowed to fall on a second calcite plate of the same thickness with the crystal axes orientated in the same way the separation is doubled (Fig. 7.2b). If the second plate is rotated through 180° about the direction of the beam the two emerging rays are brought together (Fig. 7.2c). In intermediate orientations of the crystal axes each beam splits into two and the directions of the separation are no longer in

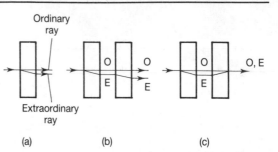

Figure 7.2   (a) Ray incident on calcite plate; (b) two plates, same orientation; (c) two plates, one rotated through π about the direction of incident ray

the plane of the paper. This property of double refraction of an incident beam is termed *birefringence*. In this experiment the passage through the first plate has produced two beams with different azimuthal properties. Study of the behaviour for various relative orientations of the two plates indicates that the two beams in Fig. 7.2(a) are each plane-polarised but in mutually perpendicular directions.

7.4   Methods of producing plane-polarised light

The fraction of light reflected from glass at the polarising angle amounts to only a few percent. Also the lateral separation produced by a calcite plate of moderate thickness is so small that the incident beam would need to be extremely narrow if the emerging beams were to be separated. Thus neither Malus' nor Huygens' experiments can provide us with an intense beam of polarised light. Combinations of crystals such as those by Rochon (Fig. 7.3a) and Wollaston (Fig. 7.3b) improved the situation. The arrangement devised by Nicol in 1828 was a considerable advance. He cut a calcite crystal in an appropriate plane and cemented the two halves together with Canada balsam (Fig. 7.4). We see later that the refractive index of the crystal for the ordinary ray is different from that for the extraordinary ray. In the Nicol prism it is arranged that the angles of incidence on the cement layer are such that the ordinary ray is totally reflected and the extraordinary ray goes through. A wide angle beam (24°) is transmitted in this way so the

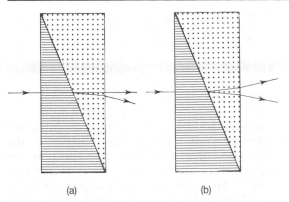

Figure 7.3 (a) Rochon prism; (b) Wollaston prism. The lines and dots (indicating ⊥ paper) show the directions along which the ordinary and extraordinary beams have the same velocity

cross-sectional area of the beam is limited only by the size of the prism.

The Glan–Thompson prism is an improved form of the Nicol prism with square ends. This prism gives very good linear polarisation over the visible spectrum. For work in the ultraviolet region a Foucault prism, with the cement replaced by an air gap, is used.

Very large pieces of optically clear calcite are rare and expensive and for beams of polarised light of large cross-section a different kind of device is needed. Edwin Land invented layers of artificial materials that have the convenient property of absorbing some 99.99% of light plane-polarised in one plane but of transmitting about 65% of light polarised in the perpendicular plane. Large sheets of this material can be produced inexpensively so that large area beams of plane-polarised light are easily produced. The degree of polarisation is not quite as good as that obtained from a Nicol or

Glan–Thompson prism, particularly at the ends of the visible spectrum, but sheet polarisers are good enough for many applications. A system that can *produce* plane-polarised light can also *detect* it since when suitably orientated it will extinguish a plane-polarised beam. When used in this way the device is referred to as an *analyser*.

7.5   Interference with polarised light

Fresnel and Arago carried out a series of experiments with polarised light and derived the following rules:

1. Two beams polarised in mutually perpendicular planes never produce interference fringes.
2. Two beams derived from unpolarised light (as in Fig. 7.2a) do not give fringes even when the plane of the light vector of one has been rotated (by devices described below in Section 7.8) so that the light vectors lie in the same plane.
3. If a beam of a plane-polarised light is split into two beams polarised in different planes (as in Fig. 7.3) and the plane of one beam is rotated so that the light vectors of both beams are in the same plane the interference fringes can be obtained.

The vector representing plane-polarised light may—as may any vector quantity—be resolved into perpendicular components. The above experiments show that the phase difference between the components is zero. They are said to be mutually coherent. The components of unpolarised light are not mutually coherent. In this context we use the term coherence to mean that the oscillations of the light vector at corresponding points of the two beams have a fixed phase difference which does not vary with time. We do *not* imply that either beam has a large coherent width or coherent length in the sense discussed in Chapter 6.

7.6   Wave surfaces in crystals

If at time $t = 0$ a small source of light starts to emit waves into a medium such as water or

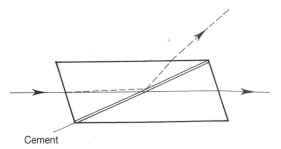

Cement

Figure 7.4   Nicol prism

glass the wave surface at a time t is a sphere of radius bt, where b is the wave velocity. Huygens proposed that in crystalline media the wave surface has two sheets, a sphere of radius bt and an ellipsoid of revolution (Fig. 7.5). The spherical surface is associated with the ordinary ray and the ellipsoid with the extraordinary ray. For the extraordinary ray the velocity depends on the direction relative to the crystal axes and for one direction $b_e = b_o$, where b_e is the velocity for the extraordinary ray and b_o that for the ordinary ray. This direction is called the *optic axis*. When the sphere includes the ellipsoid (Fig. 7.5a) the crystal is said to be *positive uniaxial*; when the ellipsoid is outside (Fig. 7.5b) the crystal is *negative uniaxial*. We shall see later (Chapter 11) that there are crystals with more complicated wave surfaces but for the present we consider only uniaxial crystals and not the more general *biaxial* types. If a crystal plate with plane parallel sides is produced the optical thickness for the ordinary ray will be $n_o e$, where $n_o = c/b_o$ and e is the thickness of the plate. Similarly, the optical thickness for the extraordinary ray is $n_e e / \cos \theta$ (Fig. 7.2a); n_e and θ depend on the relation between the direction of the normal to the plate and that of the crystal axis. Generally θ is small enough for $\cos \theta$ to be practically unity so that the phase difference between the two emerging rays is approximately $\varkappa e (n_o - n_e)$. (If $n_o < n_e$ then this is $\varkappa e (n_e - n_o)$.) What happens when such a plate is placed in a beam of plane-polarised light? We need first to establish the direction of the electric vector of the light emerging from our (sheet) polariser.

7.7   Where is the plane of polarisation?

Although a sheet of polariser produces plane-polarised light there is generally no indication on the sheet of the direction of the electric vector of the transmitted light. We can establish this direction by examining the light reflected from a glass surface at the Brewster angle. We know that the reflected light in this case has its electric vector perpendicular to the plane of incidence (Section 7.1). If we rotate the sheet until extinction is obtained then we know that the direction of the electric vector of light which would be *passed* by the sheet will be where the plane of incidence intersects the sheet (Fig. 7.6).

If we now place the marked polarising sheet before a crystal plate and rotate the latter about its normal we find that in one position only the ordinary ray is transmitted and that in the perpendicular orientation the extraordinary ray is transmitted. We mark on the crystal plate the directions of the electric vectors for the two cases: these are termed the *privileged directions* of the plate. The one for which the speed of light is higher (lower refractive index) is the 'fast' direction and the perpendicular direction is the 'slow' direction.

If a beam of plane-polarised light of intensity I_0 is incident on a birefringent crystal with the electric vector at an angle ϕ to one of the privileged directions, then the intensities of the two transmitted beams are

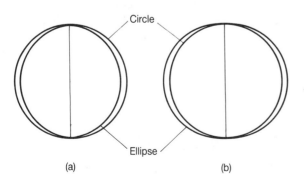

Figure 7.5   Wave surface in (a) positive and (b) negative uniaxial crystal

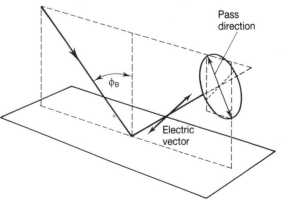

Figure 7.6   Direction of electric vector for light incident at the Brewster angle

$I_0 \cos^2\phi$ and $I_0 \sin^2\phi$ (neglecting any reflections at the crystal surfaces).

7.8 Quarter- and half-wave plates

Since the refractive indices of a plate for the ordinary and extraordinary rays differ there will be a phase difference between the waves emerging from the plate. When the phase difference is $\pi/2$ the plate is called a *quarter-wave plate*; when it is π we have a *half-wave plate*. For mica $n_e - n_o$ is only 0.0056 and a plate 16 μm thick will act as a quarter-wave plate. Large inexpensive quarter-wave plates can also be made from cellophane.

If plane-polarised light falls on a quarter-wave plate with the plane of the electric vector at $\pi/4$ to the privileged directions the emergent light consists of components of equal amplitude and with a phase difference of $\pi/2$ (Appendix 7.A). These combine to form circularly polarised light. If a plane-polarised beam falls on a half-wave plate so that the light vector makes an angle α with one of the privileged directions then the phase difference of π between the emerging beams is equivalent to reversing the amplitude of one beam while keeping the other constant. The emerging beams then combine to form a plane-polarised beam at an angle

$-\alpha$ to the privileged direction, i.e. the plane of the electric vector of the incident beam has been rotated by 2α (Fig. 7.7). This device thus enables the plane of the electric vector of plane-polarised light to be rotated through any angle.

7.9 Elliptically polarised light

Two coherent beams of light polarised parallel to the Ox and Oy axes respectively and having a phase difference of $\pi/2$ may be represented by

$$E_x = p \sin \omega t$$
and $$E_y = q \sin(\omega t - \pi/2) = q \cos \omega t \quad (7.2)$$

When $t = 0$ the resultant vector is of magnitude q and is directed along the positive Oy direction (Fig. 7.8a). When $t = \pi/2$ it is directed along the positive Ox direction with magnitude p (Fig. 7.8b). When $t = \pi$ it is directed along the $-y$ direction and the magnitude is again q (Fig. 7.8c). The vector rotates through 2π whenever ωt increases by 2π, i.e. in one period of the wave. The resultant of the two components is of magnitude $(E_x^2 + E_y^2)^{1/2}$ and eqn (7.2) gives

$$\frac{E_x^2}{p^2} + \frac{E_y^2}{q^2} = 1 \quad (7.3)$$

This is the equation of an ellipse and implies that the end of the resultant is always on the ellipse shown in Fig. 7.8. Thus two coherent beams of light polarised in mutually perpendicular planes with a phase difference of $\pi/2$ give *elliptically polarised* light. Since the rotation is clockwise to an observer who receives the beam this is called *right-handed* of *positive* elliptically polarised light. If the sign of the $\pi/2$ phase difference in eqn (7.2) is reversed then $E_y = -q \cos \omega t$ and the rotation is anti-clockwise: *negatively polarised* light.

In the special case where $p = q$, eqn (7.3) is the equation of a circle. The end of the resultant rotates round a circle and we have *circularly* polarised light. Again the rotation may be in either direction. What happens if the phase difference is neither zero nor $\pi/2$? Do we get some new, bizarre state of polarisation? No; the resulting polarisation is

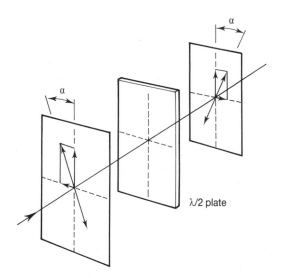

Figure 7.7 Rotation of plane of the electric vector by the $\lambda/2$ plate

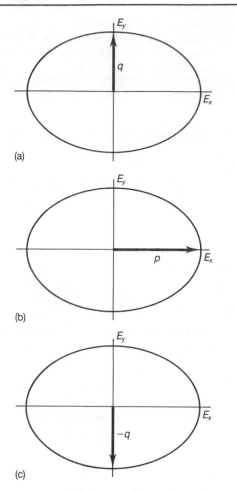

Figure 7.8 Right-handed elliptically polarised light: successive directions and magnitudes of electric vector at different times

still elliptical (Appendix 7.B), but with the axes of the ellipse inclined to the original polarisation directions. No new physics arises and we do not consider this further.

7.10   Stokes parameters

If a beam of polarised light, of any type, traverses a number of polarising devices, it can be algebraically extremely tedious to determine the final state of polarisation. A systematic way of characterising the state of polarisation was introduced by Stokes. Four parameters, determined from four simple measurements, serve to define the polarisation state. In addition, it is possible to

characterise polarisation devices by matrices. The combination of Stokes parameters and device matrices leads to a systematic and powerful method of polarisation analysis.

We make four measurements, considering for convenience a beam in a horizontal direction. The first determines the beam intensity with a polarisation-insensitive filter of transmittance of 50% to yield I_1. The second is a linear polariser orientated to pass plane-polarised light with the electric vector horizontal, I_2. The third entails rotating the polariser to an angle of $\pi/4$ with the horizontal, to give I_3. For the fourth we combine a quarter-wave plate and polariser orientated so that the combination passes only right-handed circularly polarised light, I_4.

The devices used for the latter three measurements also have a transmittance of 50%. The Stokes parameters are then defined by:

$$S_1 = I_1, \quad S_2 = I_2 - I_1, \quad S_3 = I_3 - I_1, \quad S_4 = I_4 - I_1$$
$$. \quad (7.4)$$

For a beam for which the amplitudes of the x and y plane-polarised components are p and q, and with a phase difference ε, the Stokes parameters are

$$\begin{aligned} S_1 &= \langle p^2 + q^2 \rangle \\ S_2 &= \langle p^2 - q^2 \rangle \\ S_3 &= \langle 2pq \cos \varepsilon \rangle \\ S_4 &= \langle 2pq \sin \varepsilon \rangle \end{aligned} \quad (7.5)$$

where the brackets indicate time averages. If the light is unpolarised, $p = q$ and so $S_2 = 0$. The time-averaged values of $\cos \varepsilon$ and $\sin \varepsilon$ are zero, so $S_3 = S_4 = 0$. If the light is polarised then ε is constant over the time of observation. From (7.5) we obtain

$$S_1^2 = S_2^2 + S_3^2 + S_4^2 \quad (7.6)$$

For unpolarised light S_1^2 is simply proportional to the intensity. Thus if for an unknown beam we find that

$$S_1^2 > S_2^2 + S_3^2 + S_4^2 \quad (7.7)$$

we know immediately that it is a mixture of unpolarised and polarised light.

The quantities S_2, S_3 and S_4 may be regarded as defining a vector (the Stokes vector). The magnitude is S_1 (eqn 7.6) and the direction is determined by the state of polarisation. Different points on the sphere of radius S_1 (the Poincaré sphere) correspond to different states of polarisation. Operation of different polarising devices may be described in terms of the movement of a point on the Poincaré sphere.

Since S_1 is simply a measure of the beam intensity we can normalise the values of S_2, S_3 and S_4. From eqn (7.5) we may characterise polarisation states for light travelling in the z direction, by four parameters as follows:

Plane-polarised light:

x direction $(a_y = 0)$: $\quad (1, 1, 0, 0)$
y direction $(a_x = 0)$: $\quad (1, -1, 0, 0)$

$+\dfrac{\pi}{4}$ to $\mathrm{O}x, \mathrm{O}y \left(a_x = a_y = \dfrac{a}{\sqrt{2}} \right)\ (1, 0, 1, 0)$

Circularly polarised light (right-handed)
$(a_x = a_y,\ \varepsilon = \pi/2)$ $\quad (1, 0, 0, 1)$

7.11 Matrix treatment of polarisation problems

We may represent the Stokes parameters by a column vector. A polarising device will result in light with a different Stokes vector. The operator that changes a column vector with n elements is an $n \times n$ matrix and a polarising device may be represented by a 4×4 matrix. Such are known as Mueller matrices. We give one or two examples.

If we have a light beam plane-polarised in the $+\pi/4$ direction and it strikes a polariser orientated to give light with its electric vector in the x-direction, we know that the Stokes vector $(1, 0, 1, 0)$ is changed into $\frac{1}{2}(1, 1, 0, 0)$ by the device. The factor $\frac{1}{2}$ arises because the intensity of the transmitted beam is half that of the incident . The matrix that will operate on a column vector $(1, 0, 1, 0)$ to give $\frac{1}{2}(1, 1, 0, 0)$ is

$$\frac{1}{2}\begin{pmatrix} 1 & 1 & 0 & 0 \\ 1 & 1 & 0 & 0 \\ 0 & 0 & 0 & 0 \\ 0 & 0 & 0 & 0 \end{pmatrix}$$

It may be easily verified that if this matrix operates on the Stokes vector $(1, 0, 0, 1)$, representing right-handed circularly polarised light, the resultant Stokes vector is

Table 7.1

Device	Mueller matrix
Linear polariser: x direction	$\frac{1}{2}\begin{pmatrix} 1 & 1 & 0 & 0 \\ 1 & 1 & 0 & 0 \\ 0 & 0 & 0 & 0 \\ 0 & 0 & 0 & 0 \end{pmatrix}$
Linear polariser: y direction	$\frac{1}{2}\begin{pmatrix} 1 & -1 & 0 & 0 \\ -1 & 1 & 0 & 0 \\ 0 & 0 & 0 & 0 \\ 0 & 0 & 0 & 0 \end{pmatrix}$
Linear polariser: $\pm \pi/4$ to $\mathrm{O}x$	$\frac{1}{2}\begin{pmatrix} 1 & 0 & \pm 1 & 0 \\ 0 & 0 & 0 & 0 \\ \pm 1 & 0 & 1 & 0 \\ 0 & 0 & 0 & 0 \end{pmatrix}$
Quarter-wave plate: fast axis vertical:	$\frac{1}{2}\begin{pmatrix} 1 & 0 & 0 & 0 \\ 0 & 1 & 0 & 0 \\ 0 & 0 & 0 & -1 \\ 0 & 0 & 1 & 0 \end{pmatrix}$
Quarter-wave plate: fast axis horizontal:	$\frac{1}{2}\begin{pmatrix} 1 & 0 & 0 & 0 \\ 0 & 1 & 0 & 0 \\ 0 & 0 & 0 & 1 \\ 0 & 0 & -1 & 0 \end{pmatrix}$

$\frac{1}{2}(1, 1, 0, 0)$—horizontally plane polarised light—as expected.

Matrices corresponding to one or two common polarising devices are given in Table 7.1.

If a light beam characterised by a Stokes vector S_i passes successively through N devices with Mueller matrices M_1, M_2, M_3, ..., M_N, then the Stokes vector S_f for the transmitted light is given by

$$S_f = M_N M_{N-1} \cdots M_4 M_3 M_2 M_1 S_i \quad (7.8)$$

7.12 Jones vectors and Jones matrices

An alternative way of expressing the four quantities needed to specify the state of polarisation is to use the amplitude and phase of the x and y components of the representative vector. In this approach unpolarised light cannot be represented. Polarising devices are associated with 2×2 matrices to operate on the two-element column vector. Equation (7.8) still applies, with the S's representing the Jones vector $(a_x\, \mathrm{e}^{\mathrm{i}\varepsilon_x},\ a_y\, \mathrm{e}^{\mathrm{i}\varepsilon_y})$ and the M's are 2×2 matrices. This system appears simpler, although in contrast to the Mueller matrices, the Jones matrices have elements that are complex numbers. Examples of the application of both Mueller and Jones matrices are given in ref. [1].

7.13 Analysis of polarised light

Ultimately, methods of analysing the state of polarisation of a light beam end up with the detection of plane-polarised light. We can recognise when light is plane-polarised by inserting a linear polariser and establishing that extinction occurs. Unfortunately both the *intensity* and the *rate of change of* intensity transmitted by a simple analyser tend to zero as the extinction position is approached, so that the accurate location of this position is not possible. This difficulty is overcome by a split-field device such as that illustrated in Fig. 7.9(a). This analyser consists of two

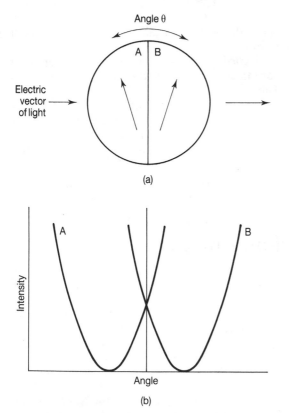

(a)

(b)

Figure 7.9 (a) Split-field analyser; (b) intensity distributions in the two halves of the field as a function of the angular setting

linear polarisers with their pass directions slightly inclined. When placed in a beam of plane-polarised light, equal intensities in the two halves of the field will be observed when the direction of the electric vector bisects the pass directions of the two components. However, the intensity is not zero at the balance point and the intensity in one half decreases while that of the other increases as the analyser is rotated through the balance position (Fig. 7.9b). Under this condition the equal-intensity position can be detected with great precision, either visually or photoelectrically.

7.14 Circularly and elliptically polarised light

Since this type of light may be represented as two oscillations in perpendicular directions with a phase difference of $\pi/2$ between them (Section 7.9), then on passing the light through a suitably orientated quarter-wave plate (which introduces a phase difference of $\pm \pi/2$) the light emerging will be plane-polarised. This can then be detected by a device such as that described in Section 7.13. Extinction of the emerging beam will occur *only* when the privileged directions of the quarter-wave plate are parallel to the axes of the ellipse, in the case of elliptically polarised light, so that the balance position needs to be reached by successive adjustment of the orientations of the quarter-wave plate and the analyser. The approximate directions of the axes can be found initially from the settings for minima and maxima of the light transmitted by a linear analyser alone.

7.15 Mixtures

Suppose now that our light beam consists of a mixture of different types of polarisation. How can analysis be effected in this case? We shall not deal with all cases but give an illustration. Consider two beams, one of which is elliptically polarised and the other a mixture of unpolarised and linearly polarised light. If we rotate a linear analyser in the path of each beam we shall, on rotation observe a maximum and (non-zero) minimum. How do we distinguish between them? We insert a quarter-wave plate before the analyser with its privileged directions parallel to the analyser orientations at maximum and minimum. If the light is elliptically polarised then the analyser will be able to extinguish the light from the plate. For a mixture of unpolarised and linearly polarised light extinction will not occur [2].

7.16 Ellipsometers

Instruments for determining the characteristics of elliptically polarised light ('ellipsometers') have found applications in many fields of physics. The optical constants (Chapter 9) of materials may be obtained from measurements of the polarisation of reflected light. When these are measured over a wide range of wavelengths ('spectroscopic ellipsometry') invaluable information can be obtained on the crystalline order of the sample. The optical constants of alloys reflect the composition so that ellipsometry offers a non-destructive method of analysis. The adsorption of less than one monolayer of gas on a clean surface is readily detectable by ellipsometry, giving the field of surface studies an immensely powerful tool. One important aspect is that surfaces immersed in liquids may be studied, thus assisting the exploration of interfaces in electrolytic experiments. An idea of the extent of the use of the method may be gleaned from ref. [2] (which lists 315 references!).

7.17 Compensators

The quarter- and half-wave plates discussed above introduce phase differences of $\pi/2$ and π respectively. Suppose one needs to measure phase differences other than these values? For such it would be convenient to have a device that can introduce a known, continuously variable phase shift into a beam. The Babinet compensator is one of many devices that will do just this (Fig. 7.10a). Made of birefringent wedges with the optic axes perpendicular, the phase difference introduced by the system at a given point may be varied continuously by moving the wedges past one another. In the Soleil compensator (Fig. 7.10b) the phase difference introduced by the wedge in the

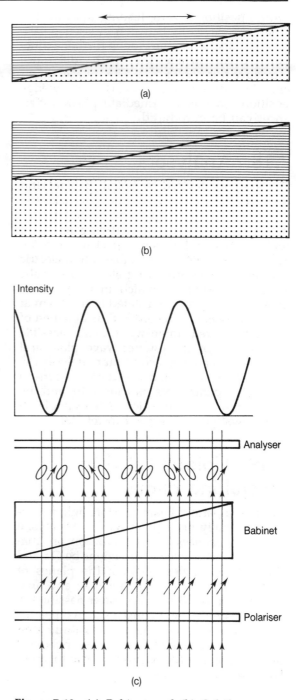

Figure 7.10 (a) Babinet and (b) Soleil compensators. In each case one wedge can be moved in a horizontal direction. Lines in upper wedges represent the direction of the optic axis. (c) Fringe formation in the Babinet compensator, in which the lower wedge has its optic axis perpendicular to the plane of the paper

central position is zero, leading to extinction when the device is between crossed polarisers. On displacing the wedge further extinctions occur when the path difference introduced is $2\pi, 4\pi, \ldots$. The phase difference introduced varies linearly with wedge position, so that intermediate phase differences can be introduced.

7.18   Analysers

The systems devised by Soleil, Babinet and their successors were designed to make the best use of the properties of the eye as detector. Today interest often extends to regions outside the visible spectrum and the light is detected by suitable photoelectric devices, usually producing electrical signals. An example is that in which an analyser and quarter-wave plate are rotated in the beam at different speeds. A record is thus obtained of the transmitted intensity for all possible orientations of the quarter-wave plate and analyser. An attached computer analyses the intensity variation and calculates therefrom the Stokes parameters (Section 7.10) and the polarisation characteristics. An example of such a system is described in ref. [3].

7.19   Colours of thin crystalline plates

Suppose a parallel beam of white light is incident normally on a polariser, a thin crystal plate of thickness t and finally on an analyser. Let the privileged directions be Ox and Oy and the angles between the planes of transmission and Ox for polariser and analyser be α and β (Fig. 7.11). After passing the crystal plate the light consists of two beams, one polarised along Ox of amplitude proportional to $\cos\alpha$ and the other polarised along Oy of amplitude proportional to $\sin\alpha$. There is a phase difference

$$\delta = \frac{2\pi t(n_e - n_o)}{\lambda} \tag{7.9}$$

After passing the analyser the beams are polarised in the same plane and are coherent so that the complex amplitude of the

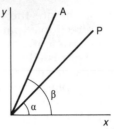

Figure 7.11   Thin crystal slice: Ox and Oy are privileged directions. OP is the plane of the polariser and OA that of the analyser. All planes are perpendicular to the plane of the paper

resultant is

$$R = E_0(\cos\alpha \cos\beta + \sin\alpha \sin\beta \exp i\delta) \tag{7.10}$$

$$W = E_0^2(\cos^2\alpha \cos^2\beta + \sin^2\alpha \sin^2\beta + \tfrac{1}{2}\sin 2\alpha \sin 2\beta \cos\delta) \tag{7.11a}$$

After putting $\cos\delta = 1 - 2\sin^2\tfrac{1}{2}\delta$ and rearranging terms,

$$W = E_0^2[\cos^2(\alpha-\beta) - \sin^2 2\alpha \sin^2 2\beta \sin^2 \tfrac{1}{2}\delta] \tag{7.11b}$$

The first term represents the light that would pass if the crystal plate were absent. It is therefore white light. The second term is zero for some wavelengths and large for others according to the value of δ. If the light is passed into a spectroscope a channelled spectrum of light and dark bands is seen. If the crystal plate is thick and cut so that $n_e - n_o$ is not very small there are many bands in the visible spectrum and the light emerging from the system appears white. If the product $t(n_e - n_o)$ is fairly small there are only one or two broad bands in the spectrum and the light appears coloured. The colour is strongest if $\alpha = +\pi/4$ and $\beta = -\pi/4$ so that the white term is zero and $\sin 2\alpha = -\sin 2\beta = 1$.

A piece of mica shows this effect very well. If it has been made of uniform thickness the whole plate has one colour. If it has been irregularly split so that areas of different thickness are present adjacent areas often show highly contrasting colours. If white light is present the colour term is changed from positive to negative by rotating the analyser and then each colour is replaced by a complementary colour.

7.20 The Lyot filter

In the experiment described above the number of maxima between two wavelengths such as 400 and 600 nm increases with t. If a succession of plates $P_1, P_2, \ldots$, is arranged with polarizers p between them such that each plate is twice the thickness of its predecessor (Fig. 7.12) the number of maxima doubles at each stage. It can be arranged that two well-separated maxima are passed by the first plate and associated polarisers. Over this range the second plate alone would pass four bands but two of these coincide with the minima of the first plate and are chosen to make these minima zero. Thus only two bands pass the combination and these are half the widths of those passed by the first plate. This process may continue until only two very narrow bands pass the whole assembly. An auxiliary glass filter may be used to select one of these. Lyot [4] made filters of this type using seven plates for observation of the solar corona when the Sun is *not* eclipsed. He was careful to have high-quality calcite polarisers matched with the glass so that the ordinary ray is undeviated. The scattered light is reduced to a very low value.

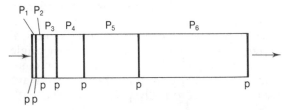
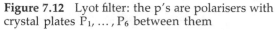

Figure 7.12 Lyot filter: the p's are polarisers with crystal plates $P_1, \ldots, P_6$ between them

7.21 Savart–Françon plate

It is sometimes desirable to produce a copy of a wave-front sheared sideways without introducing a phase difference. A pair of crystal plates designed by Savart (1791–1841) achieved this object for normal incidence. Figure 7.13 shows a greatly improved version due to Françon. The axes for the plates are shown in the figure and a half-wave plate with its privileged directions at $45°$ to the privileged directions of the plates is interposed [5]. When rays are incident at

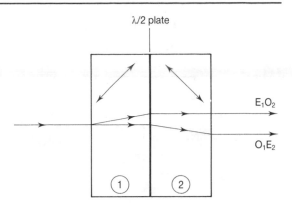

Figure 7.13 Savart-Françon plate

an angle of α in the plane of the paper, there is a path difference proportional to $\sin \alpha$ and with convergent light fringes are obtained. These fringes form a sensitive method of detecting a small proportion of plane-polarised light.

7.22 Interference microscopy

In the normal microscope detail in the image is seen as a result of the variation in the intensity of light transmitted by the object, leading to a corresponding variation of intensity in the image. In many situations, however, there are variations in *phase* of the transmitted light but without any variations in intensity. (This is a characteristic of many biological specimens.) If an incident wave of amplitude E_0 emerges as $E_0 e^{i\phi}$, where ϕ is the phase change introduced by the object, the corresponding intensity is $|E_0 e^{i\phi}|^2 = E_0^2$, the same as the incident. Thus a region of the specimen that produces only a phase change will not be visible.

Using the properties of polarised light discussed above, such phase-change regions may be made visible. Consider the arrangement of Fig. 7.14. C_1 and C_2 are birefringent plates mounted as shown in Fig. 7.2(c). With no obstruction in either beam the wavefront of the incident light will be reproduced after passage through the plates. If, however, there is a phase object in *one* of the sheared beams, then the *intensity* of the emerging light will change as a result of interference between beams A and B of Fig. 7.14.

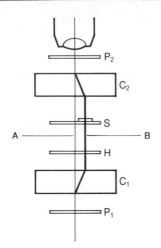

Figure 7.14 Development of the Lebedev interference microscope. P_1, P_2—polarisers, H—$\lambda/2$ plate, C_1, C_2 — birefringent crystals

7.23 Rotatory polarisation

When a beam of plane-polarised light is propagated in quartz along the optic axis the plane of the electric vector rotates as the wave advances. This effect is called *rotatory polarisation* or *optical activity*. In some specimens of quartz the rotation is clockwise to an observer looking in the direction in which the wave is advancing. This is called *right-handed rotatory polarisation*. Substances that show right-handed rotation are labelled *dextro-rotatory* whereas those showing left-handed rotation are *laevo-rotatory*. The rotation for a piece of quartz 1 mm thick is $\pm 21.7°$.

Some liquids (including solutions of sugar) also show this effect. For solutions of moderate strength the rate of rotation is proportional to the concentration. The *specific rotation* is the rotation for a column of length 10 cm for a concentration (of sugar) of one gramme per millilitre. The specific rotation for cane sugar is $66.7°$. The determination of the strength of sugar solutions by measurement of the optical rotation is called *saccharimetry*. The angle of rotation can be measured visually to within about 5 seconds of arc. Photoelectric devices can manage somewhat better.

7.24 Allogyric birefringence

Just as a beam of circularly polarised light may be regarded as the resultant of two coherent beams of light plane-polarised in perpendicular directions so a beam of plane-polarised light can be regarded as the resultant of two beams of circularly polarised light whose vectors rotate in opposite directions (Fig. 7.15). In a non-active medium these rotate at the same rate and the light is propagated with a constant plane of polarisation. However, in some substances the velocity of light for right-handed circularly polarised light is different from that for left-handed circularly polarised light. In such cases the plane of the electric vector of the light will rotate as the wave advances.

If we accept this representation of optical activity we should expect to find different indices of refraction for right- and left-handed circularly polarised light in optically active substances. This *allogyric birefringence* as it is called has been found experimentally by measuring angles of incidence and refraction.

For propagation in the direction of the optic axis quartz shows only optical activity. A plate cut with the optic axis in the surface shows ordinary birefringence with the usual pair of privileged directions and no optical activity. For a plate cut at an arbitrary angle to the optic axis the theory is rather complicated. It is found that there are always two kinds of elliptically polarised light (with

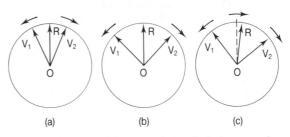

(a)   (b)   (c)

Figure 7.15 (a) Plane polarised light as the resultant of two vectors OV_1 and OV_2 which rotate in opposite directions, (b) direction of resultant fixed when rotations have the same speed, (c) resultant rotating when OV_1 and OV_2 have different speeds

similar semi-axes but with oppositely directed rotations) which are transmitted unchanged but with different velocities.

7.25 Dispersion

Both ordinary birefringence and rotatory polarisation show dispersion, i.e. n_e and n_o are different functions of wavelength so that $n_e - n_o$ changes with wavelength. It would be convenient for some purposes to have a material with $n_e - n_o$ proportional to λ, thus making δ (eqn 7.10) constant, but no substance that is available in good large pieces has this property. The dispersion of $n_e - n_o$ is put to good use in the phase-matching of frequency-doubling crystals, as described in Section 15.3.

7.26 Photoelasticity

The optical properties of a material depend ultimately on the way in which the constituent atoms are arranged, which can be altered by the application of mechanical stress. An initially isotropic material can become anisotropic, or the anisotropic constants of an initially anisotropic material can be changed. It should be possible therefore to deduce information on the stresses on a body by the study of its optical properties. This is not only possible but provides us with a very powerful tool in the study of stress distribution in loaded structures.

Consider an isotropic, unstrained sample placed between crossed polarisers and illuminated with white light. No light is transmitted by the analyser since the sample does not change the state of polarisation of the light passing through it. If the sample is now stressed, the resultant strain gives rise to birefringence: the value of $|n_e - n_o|$ depends on the applied stress. If the phase difference introduced (eqn 7.9) is a multiple of 2π, no light is transmitted. For other values of δ, the light emerging from the sample is generally elliptically polarised, with ellipticity depending on $|n_e - n_o|$ and on the wavelength. Thus coloured regions are observed between the dark ($2m\pi$) regions. If the stress in one part of the sample is very high then a large concen-

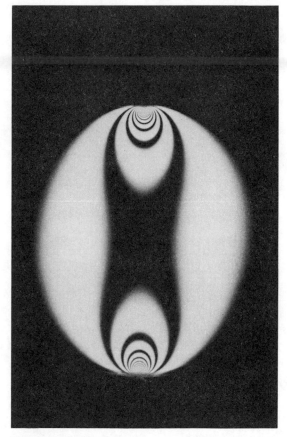

Figure 7.16 Study of stress distribution by photoelasticity. Reproduced by permission of Sharples Stress Engineering Ltd

tration of fringes is seen. Thus in order to investigate the stress distribution in a loaded structure, a model is made in a transparent plastic and the fringes viewed when the model is placed between crossed polarisers and stress applied. An example of such a model is shown in Fig. 7.16.

Appendix 7.A Plane- and circularly polarised light

At any point in the path of a circularly polarised wave the light vector rotates with time. The end of the vector at a given time will have coordinates $x = a \cos \phi$ and $y = \pm a \sin \phi$, where the '+' indicates anticlockwise and the '−' clockwise rotation. As ϕ increases, the end of the vector travels round a circular path, with the equation $x^2 + y^2 = a^2$.

The equations $x = a \cos \phi$ and $y = \pm a \sin \phi$ respectively represent plane polarised oscillations along the x and y directions, differing in phase one from another by $\pi/2$. Thus circularly polarised light may be regarded as the resultant of two orthogonal plane-polarised oscillations of equal amplitude and with a phase difference of $\pi/2$. Conversely, a plane-polarised beam may be regarded as the resultant of two oppositely directed circularly polarised beams.

Appendix 7.B   Elliptical polarisation—the most general case

If coherent perpendicular oscillating components have amplitudes p and q and if the phase difference between them is ε, they may be represented by

$$x = p \cos \omega t \quad \text{and} \quad y = q \cos(\omega t - \varepsilon) \quad (7.12)$$

Expanding the cosine in the second term we have

$$y = q(\cos \omega t \cos \varepsilon + \sin \omega t \sin \varepsilon)$$

$$= q\left(\frac{x}{p} \cos \varepsilon + \sqrt{1 - \frac{x^2}{p^2}} \sin \varepsilon\right) \quad (7.13)$$

whence

$$q^2 x^2 - (2pq \cos \varepsilon)xy + p^2 y^2 = p^2 \sin^2\varepsilon \quad (7.14)$$

The equation $ax^2 + 2hxy + by^2 = 1$ represents an ellipse if $ab - h^2 > 0$. The corresponding condition for eqn (7.14) is $\sin^2\varepsilon > 0$, which is clearly the case for all values of p, q and ε.

References

1. A. Gerrard and J. M. Burch, *Introduction to Matrix Methods in Optics*, John Wiley and Sons, London and New York, 1975.
2. D. E. Aspnes, *J. Opt. Soc. Am.*, **65**, 1274 (1975).
3. R. M. Azzam and N. M. Bashara, *Ellipsometry and Polarised Light*, North Holland Publishing Company, Amsterdam, New York and Oxford, 1977.
4. B. Lyot, *Ann d'Astrophys*, **7**, 1 (1944).
5. M. Françon and S. Mallick, *Polarisation Interferometers*, Wiley–Interscience, London and New York, 1971.

Problems

7.1 A germanium plate is sometimes used as a polariser for the infrared, in which region the refractive index is about 4. What size of plate would be needed for a reflection polariser to produce a beam 20 mm in diameter?

7.2 What types of polarised light are represented by the following?

(a) $E_x = E_0 \cos(\omega t - \varkappa z)$,
$$E_y = E_1 \cos(\omega t - \varkappa z)$$
(b) $E_x = E_0 \sin(\omega t - \varkappa z)$,
$$E_y = E_0 \cos(\omega t - \varkappa z)$$
(c) $E_x = E_0 \sin(\omega t - \varkappa z)$,
$$E_y = E_2 \sin(\omega t - \varkappa z + \pi/2)$$

7.3 Use the Mueller matrices given in Section 7.11 to operate on the Stokes vectors given for plane- and circularly polarised light. Confirm that the resultant vector correctly describes what would be expected.

7.4 A split-field analyser (Section 7.13) consists of linear polarisers with their pass directions inclined at $5°$. If the light level is such that the eye can detect a difference of intensity of 2%, with what accuracy will the device locate the direction of polarisation?

7.5 What is observed if the analyser in Fig. 7.10(c) is rotated?

7.6 If the polariser in Fig. 7.10(c) is removed and circularly polarised light directed on to the compensator, what would be observed when (a) the analyser and compensator are rotated as one unit, (b) the analyser alone is rotated?

7.7 A mica sheet 0.213 mm thick is placed between crossed polarisers. White light is incident on the sandwich and the transmitted light is dispersed by a spectrograph. Bright bands are seen at wavelengths 494, 535, 584, 642 mm. Calculate the birefringence of the mica. Comment.

7.8 The tube of a saccharimeter, originally 10.00 cm long, was broken. When repaired its length was 9.82 cm. The rotation measured for an unknown sugar solution was $8.24 \pm 0.01°$. Calculate the sugar concentration.

Electromagnetic Theory of Dielectric Media

8.1 The nature of light waves

The results discussed in Chapter 4 leave no doubt that any satisfactory theory of light must incorporate a wave-like description, involving quantities varying periodically in space and with time. The work of James Clerk Maxwell, in the period 1860–1865, showed that the periodically varying quantities involved are electric and magnetic fields. Maxwell summarised the known results of electricity and magnetism (Coulomb's, Ampère's and Faraday's laws) in a concise and elegant way through a set of four equations (Maxwell's equations) and showed that as a consequence of these relationships, electrical and magnetic disturbances should propagate in space with a velocity that could be calculated from basic electrical and magnetic measurements. The velocity so calculated turned out to agree closely with the measured velocity of light. Further developments of the theory enabled the behaviour of light in dielectrics and metals to be predicted, as well as accounting for the phenomena of polarisation, reflection and refraction. Electromagnetic theory has been found to give a satisfactory description for propagation in homogeneous and inhomogeneous media, in isotropic and anisotropic materials, and over a wavelength range covering some eighteen orders of magnitude.

8.2 The basis of electromagnetic theory

The electric field E at a distance r from a point charge q is given by

$$E = \frac{q}{4\pi\varepsilon_0 r^3} r \qquad (8.1)$$

If q is in coulombs and r is in metres, the electric field is in volts per metre where $\varepsilon_0 = 8.854 \times 10^{-12}$ farads per metre. The flux through an area dS at a distance r from the charge is

$$E \cdot ds = \frac{q}{4\pi\varepsilon_0 r^3} r \cdot dS = \frac{q}{4\pi\varepsilon_0} d\Omega \qquad (8.2)$$

where $d\Omega$ is the solid angle subtended at q by dS. Now consider an extended charged region and an arbitrary surface surrounding the charge. By the principle of superposition, the total flux through the surrounding surface is

$$\int E \cdot dS = \frac{1}{4\pi\varepsilon_0} \int \rho \, d\Omega \qquad (8.3)$$

where ρ is the charge density—a function of position. Since $\int_s E \cdot dS = \int \nabla \cdot E \, dV$, then

$$\nabla \cdot E = \frac{\rho}{\varepsilon_0} \qquad (8.4)$$

Charges in motion constitute electric currents and give rise to magnetic forces. The force between two current loops (Fig. 8.1) may be written

$$F_{ab} = \frac{\mu_0}{4\pi} I_a I_b \oint_a \oint_b \frac{dl_b \times (dl_a \times r)}{r^2} \qquad (8.5)$$

$$= I_b \oint_b dl_b \times B \qquad (8.6)$$

where

$$B = \frac{\mu_0}{4\pi} I_a \oint_a \frac{dl_a \times r}{r^2} \qquad (8.7)$$

serving to define the *magnetic induction*; μ_0 is known as the *permeability* of *free space*.

A measurement of the ratio of electric to magnetic forces gives a relation between μ_0 and ε_0, in the form

$$(\mu_0\varepsilon_0)^{-1/2} = 2.99792 \times 10^8 \text{ m s}^{-1} \qquad (8.8)$$

One of the quantities $\mu_0\varepsilon_0$ may therefore be fixed and the measurement used to

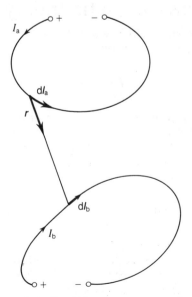

Figure 8.1 Force between current loops

determine the other. We set $\mu_0 = 4\pi \times 10^{-7} \text{ m kg C}^{-2}$ in consequence of which ε_0 has the value $8.854 \times 10^{-12} \text{ F m}^{-1}$.

In the event of the existence of magnetic monopoles (which are predicted theoretically but for which experimental evidence so far is limited, uncertain or non-existent, depending on one's level of scepticism!) a magnetic equivalent of eqn (8.4) would exist. In the absence of free magnetic charges, we have

$$\nabla \cdot B = 0 \qquad (8.9)$$

8.3 Electric and magnetic properties of materials

When material media are exposed to electric and magnetic fields, the forces exerted on the atoms or molecules result in changes in the distributions of charges. An electric field E produces a dipole moment per unit volume (electric polarisation) P and a magnetic induction B produces a magnetic moment per unit volume M. Although in isotropic materials P and E are collinear (as also are M and B), this is not the case for anisotropic media. It proves convenient to define vectors D and H such that

$$D = \varepsilon_0 E + P \qquad (8.10)$$

and

$$H = \mu_0^{-1} B - M \qquad (8.11)$$

For isotropic dielectric materials D, E and P are collinear, and for isotropic magnetic materials H, B and M are collinear. The initial part of the curve of P versus E for a dielectric is found to be linear, so for this region we have

$$P = \chi\varepsilon_0 E \qquad (8.12)$$

Figure 8.2 illustrates the behaviour observed in typical ferroelectric materials. (For a further consideration of the behaviour of materials in high fields, see Chapter 15.)

The *electric susceptibility* of the dielectric is χ. From (8.10) and (8.12) we have

$$D = \varepsilon_0(1 + \chi)E = \varepsilon_r\varepsilon_0 E = \varepsilon E \qquad (8.13)$$

where ε is the *permittivity*† of the dielectric and $\varepsilon_r = \varepsilon/\varepsilon_0$ the *relative permittivity*.

---

† Still, despite its vagueness, referred to as the 'dielectric constant'.

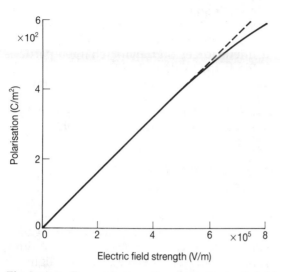

Figure 8.2 Dependence of polarisation of field strength

Corresponding expressions exist to describe the magnetic properties of materials. $B = \mu H = \mu_r\mu_0 H$ where μ_r is the *relative permeability*.

8.4 Conductivity

Although electric currents consist of flows of discrete charges—e.g. electrons or ions—it is nevertheless convenient to envisage currents as a continuous flow of charge and thus to define a charge density ρ that varies continuously from point to point. Thus in a region where a charge density ρ is moving with velocity v, we define the current density J as ρv. The magnitude of J represents the amount of positive charge crossing unit area of surface (normal to the direction of v) per unit time. When the flow of current is the result of applying an electric field E, the value of J is often linearly proportional to E, allowing an electrical conductivity σ to be defined as

$$J = \sigma E \qquad (8.14)$$

In isotropic materials, σ is a scalar quantity: in crystalline materials, there may be a variation of conductivity with direction such that σ needs to be represented by a tensor (see Section 11.2).

8.5 The laws of electromagnetism

Faraday's law of induction, relating the e.m.f. induced in a circuit by a changing magnetic flux, may be written

$$\oint_C E \cdot dl = -\frac{d\Phi}{dt} \qquad (8.15)$$

where Φ is the magnetic flux linking the circuit in question. If B is the magnetic induction, then

$$\Phi = \int_S B \cdot dS \qquad (8.16)$$

where dS is an area element and S is any surface bounded by the path of integration. The line integral in (8.15) may be written, using Stokes' theorem, as a surface integral to give

$$\int_S (\nabla \times E) \cdot dS = -\frac{d}{dt} \int_S B \cdot dS \qquad (8.17)$$

so that

$$\nabla \times E = -\frac{\partial B}{\partial t} \qquad (8.18)$$

Ampère's law gives the line integral of B around a current I as

$$\oint_C B \cdot dl = \mu_0 I \qquad (8.19)$$

which for a volume distribution of current may be written

$$\oint_C B \cdot dl = \mu_0 \int_S J \cdot dS \qquad (8.20)$$

Again using Stokes' theorem, this gives

$$\nabla \times B = \mu_0 J \qquad (8.21)$$

In the case of dielectric media a further term—the *displacement current*—has to be added to the right-hand side of (8.21) to allow for the effect of the displacement of charges within the medium. The complete expression is

$$\nabla \times B = \mu_0 J + \mu_0 \dot{D} \qquad (8.22)$$

8.6 Maxwell's equations

The results of Sections 8.2 to 8.5 may be summarised as follows:

$$D = \varepsilon E, \qquad \mu H = B \quad \text{and} \quad J = \sigma E = \rho v \qquad (8.23)$$

together with

$$\nabla \cdot D = \varepsilon(\nabla \cdot E) = \rho \qquad (8.24\text{a})$$

$$\nabla \cdot B = 0 \qquad (8.24\text{b})$$

$$\nabla \times E = -\mu \dot{H} \qquad (8.24\text{c})$$

$$\nabla \times H = J + \varepsilon \dot{E} \qquad (8.24\text{d})$$

8.7   Electromagnetic waves

With the help of the relation

$$\nabla \times (\nabla \times \quad) = \nabla(\nabla \cdot \quad) - \nabla^2 \quad (8.25)$$

E and H may be separately eliminated from eqns (8.24) to yield

$$\nabla \times \nabla \times E = \nabla(\nabla \cdot E) - \nabla^2 E = -\mu(\nabla \times \dot{H}) \qquad (8.26)$$

so that, with (8.24d),

$$\nabla^2 E - \nabla\left(\frac{\rho}{\varepsilon}\right) = \mu(\dot{J} + \varepsilon\ddot{E}) = \mu\sigma\dot{E} + \mu\varepsilon\ddot{E} \quad (8.27)$$

Similarly,

$$\nabla^2 H = \mu\sigma\dot{H} + \mu\varepsilon\ddot{H} \qquad (8.28)$$

8.8   Propagation in vacuum

In this case, $\rho = 0$, $\sigma = 0$, $\mu = \mu_0$ and $\varepsilon = \varepsilon_0$ so that (8.26) and (8.27) reduce to

$$\nabla^2 E = \mu_0\varepsilon_0 \frac{\partial^2 E}{\partial t^2} \qquad (8.29\text{a})$$

$$\nabla^2 H = \mu_0\varepsilon_0 \frac{\partial^2 H}{\partial t^2} \qquad (8.29\text{b})$$

which may be compared with eqn (4.42) in Appendix 4.A. These forms of wave equation indicate wave propagation with a velocity equal to $(\mu_0\varepsilon_0)^{-1/2}$. As noted in Section 8.2, the experimental value of $(\mu_0\varepsilon_0)^{-1/2}$ from a comparison of electric and magnetic forces on moving charged particles yielded the value $2.99792 \times 10^8 \, \text{m s}^{-1}$ for $(\mu_0\varepsilon_0)^{-1/2}$, which agrees closely with the measured velocity of light in vacuum. This result therefore strongly suggests that light is an electromagnetic phenomenon.

8.9   Propagation in a perfect dielectric

With $\rho = \sigma = 0$, the factor $\mu_0\varepsilon_0$ in (8.29) is replaced by $\mu\varepsilon$, indicating wave propagation with velocity $b = (\mu\varepsilon)^{-1/2}$. Writing $(\mu_0\varepsilon_0)^{-1/2} = c$ for the velocity of light in vacuum, we may write the refractive index n of the medium as

$$n = \frac{c}{b} = \left(\frac{\mu\varepsilon}{\mu_0\varepsilon_0}\right)^{1/2} \div \left(\frac{\varepsilon}{\varepsilon_0}\right)^{1/2} = \varepsilon_\text{r}^{1/2} \quad (8.30)$$

since μ is found to be equal to μ_0 at optical frequencies. If the value of the permittivity of a dielectric were the same at optical frequencies as at low frequencies, then we should expect the refractive index to be equal to $\varepsilon_\text{r}^{1/2}$. A selection of values is shown in Table 8.1.

The extent to which the relation $n = \varepsilon_\text{r}^{1/2}$ holds depends on the mechanism by which the dielectric becomes polarised. In high-permittivity dielectrics, this often arises through the alignment of polar molecules. This process can occur for steady fields and for alternating fields of low frequency. However, the inertia of a polar group of atoms is such that it cannot rotate quickly enough to follow the rapid ($\sim 10^{14} \, \text{s}^{-1}$) reversals of the electric field of a light wave. Such dipoles

Table 8.1

Material	$\varepsilon^{1/2}$	n_D
Helium	1.000034	1.000035
Hydrogen	1.000136	1.000132
Quartz	2.08 ($\perp$ optic axis)	1.553 (ordinary ray)
Quartz	2.07 ($\parallel$ optic axis)	1.544 (extraordinary ray)
Water	8.97	1.333

Permittivity for frequencies below 10^8 Hz.
Refractive indices for sodium D-line ($f = 5 \times 10^{14}$ Hz).
Room temperature.

thus play no part in contributing to the high-frequency permittivity. In cases where the dominant contribution to the permittivity is made by outer electrons of the atoms, the low-frequency and optical-frequency permittivities will be the same and relation (8.30) will hold.

8.10 Properties of electromagnetic waves

Thus far we have established only that electric and magnetic field characteristics can propagate with a velocity $(\mu_0 \varepsilon_0)^{1/2}$ in vacuum or $(\mu \varepsilon)^{-1/2}$ in a dielectric. Many questions remain. What are the directions of the E and H vectors in a propagating wave? Can electric and magnetic disturbances propagate separately? If not, are the magnitudes of the E and H vectors related? Are they in phase, or are arbitrary phase relations possible? Questions such as these are readily answered with the help of Maxwell's equations. We first establish the following.

8.10.1 The waves are transverse

First consider a plane wave travelling in the z direction. The components of the E and H fields can be represented by expressions with the form

$$E_x = E_{0x} \exp i(\omega t - \varkappa z) \qquad (8.31)$$

Since the waves considered are plane, the values of E_x, E_y, E_z, H_x, H_y and H_z are the same over all planes perpendicular to Oz:

$$\frac{\partial E_x}{\partial x} = 0, \quad \frac{\partial E_y}{\partial y} = 0, \quad \frac{\partial E_z}{\partial x} = 0 \quad \text{and} \quad \frac{\partial E_z}{\partial y} = 0 \qquad (8.32)$$

with corresponding expressions involving H_x, H_y, H_z. For a region with no free charges ($\rho = 0$), (8.24a) therefore implies that

$$\frac{\partial E_z}{\partial z} = 0 \qquad (8.33)$$

For a wave to propagate in the z direction, a variation of amplitude with z is essential. The formal solution to (8.33) allows E_z to be a function of x and y, but this is excluded by the fact that we are considering plane-wave solutions. Equation (8.33) has $E_z =$ constant as a solution, but this would not form part of a *propagating* field for which, from (8.32), the fluctuating field components lie in planes perpendicular to Oz. Thus the waves are transverse. This argument applies equally to the magnetic field components.

We next show the following.

8.10.2 The electric and magnetic vectors are mutually perpendicular

For a plane-polarised wave with an initially arbitrary direction of E in a plane perpendicular to Oz, we may rotate axes about Oz so that the x axis lies along the direction of E. Thus $E_y = 0$.

The x component of (8.24c) is

$$\frac{\partial E_z}{\partial y} - \frac{\partial E_y}{\partial z} = -\mu \frac{\partial H_x}{\partial t} \qquad (8.34)$$

E_z is independent of y and E_y is zero so that the left-hand side of (8.34) vanishes. For the fluctuating field, this indicates that $H_x = 0$. Thus with $E_y = 0$, $E_x \neq 0$ we have $H_x = 0$, $H_y \neq 0$, showing that E and H are perpendicular to one another *and* to the direction of the wave-normal (Fig. 8.3). (This result does not apply for certain crystalline materials, as discussed in Section 11.2.)

We also establish the following.

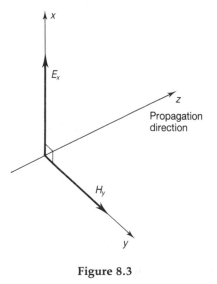

Figure 8.3

8.10.3   The electric and magnetic vectors oscillate in phase

Together with (8.31) for E_x, we write

$$H_y = H_{0y} \exp i(\omega t - \varkappa z) \qquad (8.35)$$

E_{0x} and H_{0y} may both be complex. If however, the ratio E_{0x}/H_{0y} is real and positive, the vectors E_x and H_y are in phase. If the ratio is negative, the phase difference between E_x and H_y is π. A complex ratio E_{0x}/H_{0y} implies a phase difference which is neither zero nor π.

From (8.24c) we have

$$-\frac{\partial H_y}{\partial t} = \frac{\partial E_x}{\partial z} - \frac{\partial E_z}{\partial x} \qquad (8.36)$$

from which, using (8.35) and (8.31),

$$\mu \omega H_{0y} = \varkappa E_{0x} \qquad (8.37)$$

so that

$$\frac{E_{0x}}{H_{0y}} = \frac{\mu \omega}{\varkappa} = \mu b = \left(\frac{\mu}{\varepsilon}\right)^{1/2} \qquad (8.38)$$

which is real and positive: thus the E and H vectors oscillate in phase. Moreover, for a given value of H_{0y}, the value of E_{0x} is fixed by (8.38).

The spatial variation and directions of E and H in an electromagnetic wave at a given instant are shown in Fig. 8.4. The whole disturbance moves along the z direction with velocity b.

The ratio of E_{0x} to H_{0y} is termed the 'wave impedance', Z. Its dimensions are those of (volts per metre)/(amperes per metre)—i.e. of resistance. For free space, the value Z_0 is

given by

$$Z_0 = \left(\frac{E_{0x}}{H_{0y}}\right)_{\text{free space}} = \mu_0 c = 377 \text{ ohms} \qquad (8.39)$$

The electric field E, magnetic field H and the direction of propagation form a right-handed system (Fig. 8.3).

In general, for a wave propagating in the general direction $\varkappa$ with its magnetic vector in a (perpendicular) direction H, the vector E is perpendicular to both and of magnitude $Z_0 |H|$. The direction of E is then given by

$$\varkappa \times E = \varkappa Z H \qquad (8.40)$$

8.11   Polarisation of light

In 1808, Malus observed that when a light beam had been reflected from a glass surface, its behaviour on reflection from a second glass surface depended on the angle between the planes of incidence for reflection at the first and second surfaces (Section 7.1). This indicated that light waves were transverse: for a longitudinal wave all planes containing the ray direction are equivalent. Electromagnetic theory is not only consistent with a wave phenomenon exhibiting polarisation characteristics: it provides the means for suitably labelling 'planes of polarisation'. Conveniently *either* the plane containing the magnetic vector *or* that containing the electric vector could be used. (Malus's definition corresponds to that of the magnetic vector.) The risk of confusion is minimised if the term 'plane of polarisation' is avoided. In many situations, the strength of interactions due to the electric field of a light wave is enormously larger than that for the magnetic vector and it is convenient to specify, as an indication of polarisation, the plane containing the electric vector. It must be emphasised, however, that *both* electric *and* magnetic disturbances are always present, with directions and magnitudes as given in Section 8.10.

The wave described in Section 8.10 has its electric vector E_x in the x direction and related magnetic vector B_y in the y direction. There can also exist a wave travelling in the z direction with the electric vector E_y in the y direction, in which case the associated magnetic vector B_x is in the x direction; the

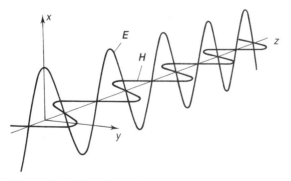

Figure 8.4 Electric and magnetic wave vectors. The pattern moves along the direction of propagation with velocity c

directions of E, H and K again form a right-handed set. The two waves are independent: E_x and E_y may take any values, but when these are fixed, the associated values of the magnetic fields are determined (eqn 8.38).

8.12 Energy of the electromagnetic field

Since energy is required to establish electric and magnetic fields in initially free space, an energy density (energy per unit volume) is associated with any region of space through which electromagnetic waves are passing. If the electric and magnetic fields are E, H in an infinitesimal volume $d\tau$, then the associated energy dW is given by

$$dW = \tfrac{1}{2}(D \cdot E + B \cdot H)d\tau \qquad (8.41)$$

For an isotropic medium, for which $D = \varepsilon E$ and $B = \mu H$ and in which waves of amplitudes E_0, H_0 exist, the total energy in a volume V is

$$W = \tfrac{1}{2} \int (\varepsilon \bar{E}^2 + \mu \bar{H}^2)d\tau \qquad (8.42)$$

where $\bar{E}^2 = \tfrac{1}{2}E_0^2$ and $\bar{H}^2 = \tfrac{1}{2}H_0^2$. Since $E_0 = (\mu/\varepsilon)^{1/2}H_0$ (8.38) this may be written in terms of either field strength:

$$W = \tfrac{1}{4} \int_V (\varepsilon E_0^2 + \mu H_0^2)d\tau$$
$$= \tfrac{1}{2} \int_V \varepsilon E_0^2 \, d\tau = \tfrac{1}{2} \int_V \mu H_0^2 \, d\tau \qquad (8.43)$$

8.13 Power flow: the Poynting vector

Since electromagnetic waves travel with velocity $(\mu\varepsilon)^{-1/2}$ and establish an energy density in the region through which they travel, we may determine the rate of flow of energy associated with such waves. From (8.41) we see that the time rate of change of energy density is

$$\frac{\partial W}{\partial t} = \frac{1}{2} \frac{\partial}{\partial t} \int_V (D \cdot E + B \cdot H)d\tau \qquad (8.44)$$

$$= \varepsilon \int_V E \cdot \frac{\partial E}{\partial t} \, d\tau + \mu \int_V H \cdot \frac{\partial H}{\partial t} \, d\tau \qquad (8.45)$$

and for a current-free region, using (8.24c)

and (8.24d), we have

$$\frac{\partial W}{\partial t} = \int_V (E \cdot \nabla \times H - H \cdot \nabla \times E)d\tau \qquad (8.46)$$

$$= - \int_V \nabla \cdot (E \times H)d\tau \qquad (8.47)$$

$$= - \int_\Sigma (E \times H) \cdot d\boldsymbol{\sigma} \qquad (8.48)$$

Thus the flux of the vector $S \equiv E \times H$ over the surface Σ surrounding the volume V is equal to the rate of decrease of the energy density in that volume. The vector $E \times H$ represents the rate of flow of energy per unit area in the direction of the wave. S is known as the Poynting vector, after J. H. Poynting (1852–1914).

8.14 Reflection and refraction of light at an interface

Thus far we have considered the propagation of electromagnetic waves in an infinite medium, either of free space or a perfect dielectric. We now consider the behaviour when radiation meets an interface between distinct media. We know from experiment that at the boundary between media with different refractive indices, part of the incident radiation is reflected and part transmitted. Electromagnetic theory enables us to calculate the proportions reflected and transmitted and to appreciate how these quantities depend on the state of polarisation of the radiation. For static electric and magnetic fields, the laws of electromagnetism lead to the result that the tangential components of E and H are continuous across a discontinuity between two media. In an electromagnetic wave, the values of E and H are varying with time but the condition for continuity of the tangential components of the field vectors applies at all times. This leads to fixed relationships between the amplitudes of the field components in the incident, reflected and transmitted waves. Thus for given amplitudes of the electric vectors of the incident wave, the values of E and H for the reflected and transmitted waves may be determined. The ratios of the reflected/incident and transmitted/incident fluxes may be calculated. It should be noted that although the area of cross-section of a reflected beam is the same as that of the incident beam, this is not the case for the transmitted and incident beams. The Poynting vector gives the power flow *per unit area*. In fact the results given below apply strictly only to infinitely wide beams, but give accurate results for beams whose widths are large compared with the wavelength.

The theory enables us to determine how the reflection and transmission at a plane boundary varies with the refractive indices of the media and with the angle of incidence. In addition, the relationships between the angles of reflection, refraction and incidence are given.

8.15   Boundary conditions

We choose a coordinate system in which the interface between medium 1 and medium 2 is perpendicular to the z direction, the x and y axes lie in the surface and the plane of incidence is $y = 0$ (Fig. 8.5). Continuity of the tangential field components requires that

$$E_{1x} = E_{2x} \qquad E_{1y} = E_{2y}$$
$$H_{1x} = H_{2x} \qquad H_{1y} = H_{2y} \tag{8.49}$$

From eqns (8.24c) and (8.24d), noting that $J = 0$ in a perfect dielectric, we see that (8.49)

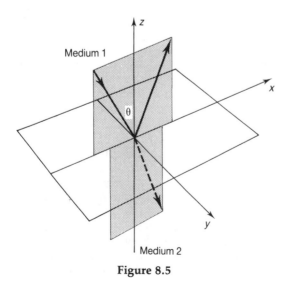

Figure 8.5

implies that

$$\varepsilon_1 E_{1z} = \varepsilon_2 E_{2z} \qquad \mu_1 H_{1z} = \mu_2 H_{2z} \tag{8.50}$$

We consider a plane wave whose normal is at an angle of incidence θ_1 in medium 1, noting that although this implies a wave of infinite lateral extent, the theory will be expected to apply in cases where diffraction effects are of small consequence. This is the case for a beam of lateral extent very large compared with the wavelength.

Consider the case where the electric vector of the incident wave is perpendicular to the plane of incidence and hence is in the y direction. General expressions for the incident, reflected and refracted waves may be written as:

Incident:
$$E_{1y} = A_{1y} \exp i[\omega_1 t - \varkappa_1(\mathbf{l}_1 x + \mathbf{m}_1 y + \mathbf{n}_1 z)] \tag{8.51}$$

Reflected:
$$E'_{1y} = A'_{1y} \exp i[\omega'_1 t - \varkappa'_1(\mathbf{l}'_1 x + \mathbf{m}'_1 y + \mathbf{n}'_1 z)] \tag{8.52}$$

Transmitted:
$$E_{2y} = A_{2y} \exp i[\omega_2 t - \varkappa_2(\mathbf{l}_2 x + \mathbf{m}_2 y + \mathbf{n}_2 z)] \tag{8.53}$$

We assume A_{1y} to be real and positive. If A'_{1y} is real and positive, this implies that the incident wave suffers no change of phase on reflection. A negative value of A'_{1y} implies a phase change of π on reflection. A complex value for A'_{1y} implies a phase change other than zero or π. It will be seen later that for other than total reflection at a perfect dielectric interface the phase change on reflection is always either zero or π; under conditions of total reflection, other values of phase change obtain.

From the choice of coordinate system, the value of the direction cosines of the incident wave are

$$\mathbf{l}_1 = \sin \theta_1, \qquad \mathbf{m}_1 = 0, \qquad \mathbf{n}_1 = -\cos \theta_1 \tag{8.54}$$

The tangential components in eqn (8.40) represent the total fields. For the case of Fig. 8.5, the total electric field in medium 1 is the sum of the fields of the incident and reflected waves. Thus at $z = 0$ and for all values of x, y and t,

$$E_{1y} + E'_{1y} = E_{2y} \tag{8.55}$$

which is satisfied only if the coefficients of x, y and t in the exponentials of eqns (8.51) to (8.53) are equal.

1. Equating coefficients of t,

$$\omega_1 = \omega_1' = \omega_2 \equiv \omega \qquad (8.56)$$

2. Equating coefficients of x, noting that $l_1' = \sin \theta_1'$ and $l_2 = \sin \theta_2$ and that $x_1' = x$ (since incident and reflected waves are in the same medium),

$$\sin \theta_1' = \sin \theta_1 \qquad (8.57)$$

and

$$x_1 \sin \theta_1 = x_2 \sin \theta_2 \qquad (8.58a)$$

Since $xb = \omega$, where b is the wave velocity, then

$$\frac{x_2}{x_1} = \frac{b_1}{b_2} = \frac{n_2}{n_1}$$

and

$$n_1 \sin \theta_1 = n_2 \sin \theta_2 \qquad (8.58b)$$

3. Equating coefficients of y,

$$0 = \mathbf{m}_1' = \mathbf{m}_2 \qquad (8.59)$$

so that the incident beam, surface normal and reflected and transmitted beams all lie in the same plane.

The foregoing results summarise the experimental laws of reflection and refraction at a surface separating two isotropic media. If the case had been considered of an incident beam with the electric vector *in* the plane of incidence, then similar considerations would be made applied to the associated magnetic vector, which by the orthogonality of the E and H vectors would lie in the y direction. The same results would then be obtained.

8.16   Reflected and transmitted amplitudes

The foregoing results would be obtained for any form of transverse wave motion, for which continuity of the tangential components of the vector quantity representing the wave occurs. Confirmation that electromagnetic theory is the appropriate description of light waves is obtained by further consideration, in which the proportions of

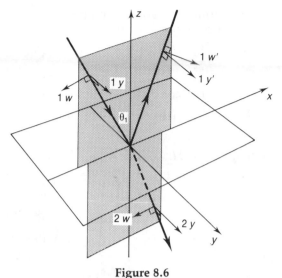

Figure 8.6

light flux reflected and transmitted are calculated. We treat separately the cases where the electric vector of the incident wave is parallel and perpendicular to the plane of incidence. We introduce sets of local axes in the three beams (Fig. 8.6). The direction of propagation and the normal to the plane of incidence (y direction) provide two axes and the third axis (w) is chosen so that w, y and the direction of propagation form a right-handed set. The incident beam may be specified by writing

$$E_{1w} = A_{1w} \exp i[\omega t - x_1(x \sin \theta_1 - z \cos \theta_1)] \qquad (8.60)$$

$$E_{1y} = A_{1y} \exp i[\omega t - x_1(x \sin \theta_1 - z \cos \theta_1)] \qquad (8.61)$$

with corresponding expressions for E_{1w}', E_{2w} and E_{1y}', E_{2y}, noting that for the reflected beam the value of $\mathbf{n}_1'$ (eqn 8.52) is $+\cos \theta_1$. To determine the reflected and transmitted amplitudes we need the values of A_{1w}'/A_{1w}, A_{2w}/A_{1w} and A_{1y}'/A_{1y}, A_{2y}/A_{1y}.

From eqn (8.38), and taking account of the directions of the w axes, we have

$$\mu_1^{1/2}H_{1w} = -\varepsilon_1^{1/2}E_{1y} \quad \text{and} \quad \mu_1^{1/2}H_{1y} = \varepsilon_1^{1/2}E_{1w} \qquad (8.62)$$

with corresponding expressions for the reflected and transmitted beams. Field vectors in the y direction are already parallel to the boundary. For w components, the tangential components are in the x direction.

$$E_{1x} = -E_{1w} \cos \theta_1 \qquad (8.63a)$$

$$E'_{1x} = +E'_{1w} \cos \theta_1 \qquad (8.63b)$$

$$E_{2x} = -E_{2w} \cos \theta_2 \qquad (8.63c)$$

$$\mu_1^{1/2} H_{1x} = -\mu_1^{1/2} H_{1w} \cos \theta_1 = +\varepsilon_1^{1/2} E_{1y} \cos \theta_1 \qquad (8.64a)$$

$$\mu_1^{1/2} H'_{1x} = +\mu_1^{1/2} H'_{1w} \cos \theta_1 = -\varepsilon_1^{1/2} E'_{1y} \cos \theta_1 \qquad (8.64b)$$

$$\mu_2^{1/2} H_{2x} = -\mu_2^{1/2} H_{2w} \cos \theta_2 = +\varepsilon_2^{1/2} E_{2y} \cos \theta_2 \qquad (8.64c)$$

$$\mu_1^{1/2} H_{1y} = \varepsilon_1^{1/2} E_{1w} \qquad (8.65a)$$

$$\mu_1^{1/2} H'_{1y} = \varepsilon_1^{1/2} E'_{1w} \qquad (8.65b)$$

$$\mu_2^{1/2} H_{2y} = \varepsilon_2^{1/2} E_{2w} \qquad (8.65c)$$

Each of the above equations contains an exponential factor on each side. At the boundary $z = 0$ these factors are equal.

From the boundary condition (8.55) we have

$$A_{1y} + A'_{1y} = A_{2y} \qquad (8.66)$$

and from the corresponding boundary condition for H_x, using (8.64),

$$(\mu_2 \varepsilon_1)^{1/2} (A_{1y} - A'_{1y}) \cos \theta_1 = (\mu_1 \varepsilon_2)^{1/2} A_{2y} \cos \theta_2 \qquad (8.67)$$

For optical frequencies, $\mu_1 \doteqdot \mu_2 \doteqdot \mu_0$, so (8.67) gives

$$n_1 (A_{1y} - A'_{1y}) \cos \theta_1 = n_2 A_{2y} \cos \theta_2 \qquad (8.68)$$

whence

$$\frac{A'_{1y}}{A_{1y}} = -\frac{n_2 \cos \theta_2 - n_1 \cos \theta_1}{n_2 \cos \theta_2 + n_1 \cos \theta_1} = -\frac{\sin(\theta_1 - \theta_2)}{\sin(\theta_1 + \theta_2)} \qquad (8.69)$$

(using Snell's law) and

$$\frac{A_{2y}}{A_{1y}} = \frac{2 \sin \theta_2 \cos \theta_1}{\sin(\theta_1 + \theta_2)} \qquad (8.70)$$

Similarly, from consideration of the E_x components we obtain

$$(A_{1w} - A'_{1w}) \cos \theta_1 = A_{2w} \cos \theta_2 \qquad (8.71)$$

and, from (8.65),

$$(\mu_2 \varepsilon_1)^{1/2} (A_{1w} + A'_{1w}) = (\mu_1 \varepsilon_2)^{1/2} A_{2w} \qquad (8.72)$$

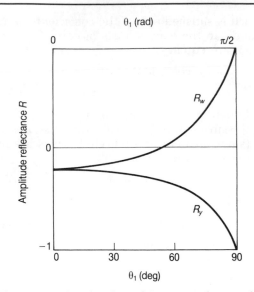

Figure 8.7 Amplitude reflectances of w and y components for different angles of incidence

which for optical frequencies is

$$n_1 (A_{1w} + A'_{1w}) = n_2 A_{2w} \qquad (8.73)$$

and these relations yield the results

$$\frac{A'_{1w}}{A_{1w}} = \frac{n_2 \cos \theta_1 - n_1 \cos \theta_2}{n_2 \cos \theta_1 + n_1 \cos \theta_2} = \frac{\sin 2\theta_1 - \sin 2\theta_2}{\sin 2\theta_1 + \sin 2\theta_2} \qquad (8.74)$$

or

$$\frac{A'_{1w}}{A_{1w}} = \frac{\tan (\theta_1 - \theta_2)}{\tan (\theta_1 + \theta_2)} \qquad (8.75)$$

and

$$\frac{A_{2w}}{A_{1w}} = \frac{2 \sin \theta_2 \cos \theta_1}{\sin(\theta_1 + \theta_2) \cos(\theta_1 - \theta_2)} \qquad (8.76)$$

The variation with angle of incidence of the y and w amplitude reflection coefficients is shown in Fig. 8.7. Note that whereas the sign of the y coefficient does not change, that of the w coefficient changes for a particular

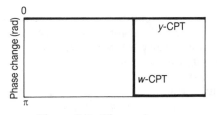

Figure 8.8 Phase changes

angle of incidence. Figure 8.8 shows the phase changes on reflection for the y and w components for light incident from air on to glass ($n = 1.52$).

8.17 Normal incidence

When θ_1, and therefore θ_2, is zero, the transmission and reflection ratios become

$$\frac{A_{2y}}{A_{1y}} = \frac{A_{2w}}{A_{1w}} = \frac{2n_1}{n_1 + n_2} \qquad (8.77)$$

$$\frac{A'_{1w}}{A_{1w}} = \frac{n_2 - n_1}{n_2 + n_1} \qquad (8.78a)$$

$$\frac{A'_{1y}}{A_{1y}} = \frac{n_1 - n_2}{n_1 + n_2} \qquad (8.78b)$$

Equations (8.78) need a comment. At normal incidence there is no distinction between w and y components, since all directions about the surface normal are equivalent. The apparent discrepancy in eqn (8.78) arises from the choice of directions of the w axes. Whereas the positive directions of the $1y$ and $1y'$ axes are the same, those of the $1w$ and $1w'$ tend, as $\theta \to 0$, towards opposite directions. For $n_1 < n_2$ (e.g. air to glass) the ratio A_{1y}/A'_{1y} is negative, indicating a phase change of π on reflection. Equation (8.78a) gives a positive ratio but due to the choice of w axis directions, this implies that the electric vectors in the reflected and incident beams are in opposite directions, i.e. their phases differ by π.

8.18 Energy reflection and transmission coefficients

The power flow per unit area normal to the direction of a wave is given by the magnitude of the Poynting vector. For a wave with the electric vector in the w direction, the corresponding magnetic vector is in the y direction. The magnitude of the Poynting vector is, from (8.60) and (8.65a), $(\varepsilon_1^{1/2}/\mu_1^{1/2})A_{1w}^2$. Correspondingly, for the reflected beam the value is $(\varepsilon_1^{1/2}/\mu_1^{1/2})(A'_{1w})^2$ so that the ratio of reflected to incident flux density is given simply by $(A'_{1w})^2/A_{1w}^2$. The same argument applies to the wave with the electric vector in the y direction. Thus the *energy reflectances*

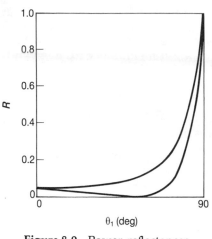

Figure 8.9 Power reflectances

R_w, R_y are given by

$$R_y = \left(\frac{A'_{1y}}{A_{1y}}\right)^2 = \frac{\sin^2(\theta_1 - \theta_2)}{\sin^2(\theta_1 + \theta_2)} \qquad (8.79a)$$

$$R_w = \left(\frac{A'_{1w}}{A_{1w}}\right)^2 = \frac{\tan^2(\theta_1 - \theta_2)}{\tan^2(\theta_1 + \theta_2)} \qquad (8.79b)$$

Curves of R_y and R_w as functions of angle of incidence are shown in Fig. 8.9 for reflection in air from a dielectric with a refractive index of 1.52.† From (8.79b) it is seen that when $\theta_1 + \theta_2 = \pi/2$, the w component of reflectance is zero. The angle of incidence at which this occurs is referred to as the Brewster Angle. At this angle, light reflected from the surface is plane-polarised with the electric vector perpendicular to the plane of incidence.

The case of the transmitted beam is more complicated. The ratio of magnetic to electric vector in the transmitted beam is different from that for the incident beam, since the transmission medium is characterised by ε_2, μ_2 and the incident beam by ε_1, μ_1. In addition there is a change of area of cross-section due to refraction at the surface. The flux arriving at unit area of the surface comes from an area $\cos\theta_1$ of the incident beam and is delivered into an area $\cos\theta_2$ in the refracted beam. Thus the incident flux

---

† The y and w components, in the notation used here, are sometimes referred to as the s and p components (where s stands for senkrecht = perpendicular and p for parallel, indicating the direction of the electric vector in relation to the plane of incidence).

striking a unit area of surface for the w component is $(\varepsilon_1^{1/2}/\mu_1^{1/2})A_{1w}^2 \cos\theta_1$. The flux in the emerging beam is $(\varepsilon_2^{1/2}/\mu_2^{1/2})/A_{2w}^2 \cos\theta_2$ so that the transmittance T_w is given by

$$T_w = \frac{(\varepsilon_1^{1/2}/\mu_1^{1/2})\cos\theta_1}{(\varepsilon_2^{1/2}/\mu_2^{1/2})\cos\theta_2}\frac{A_{2w}^2}{A_{1w}^2}$$

$$= \frac{n_1 \cos\theta_1}{n_2 \cos\theta_2}\frac{2\sin\theta_2\cos\theta_1}{\sin(\theta_1+\theta_2)\cos(\theta_1-\theta_2)} \quad (8.80a)$$

by (8.76) and assuming optical frequencies ($\varepsilon_i^{1/2} = n_i$ and $\mu_1 = \mu_2 = 1$). For the y component,

$$T_y = \frac{n_2 \cos\theta_1}{n_1 \cos\theta_2}\frac{2\sin\theta_2\cos\theta_1}{\sin(\theta_1-\theta_2)} \quad (8.80b)$$

8.19 Degree of polarisation

For a mixture of plane-polarised and unpolarised light, the degree of polarisation is defined as

$$Q = \frac{A_{max}^2 - A_{min}^2}{A_{max}^2 + A_{min}^2} \quad (8.81)$$

where A_{max} is the maximum amplitude of the electric vector for different orientations around the direction of propagation and A_{min} is the minimum amplitude. In terms of measured irradiances I, this may be written

$$Q = \frac{I_{max} - I_{min}}{I_{max} + I_{min}} \quad (8.82)$$

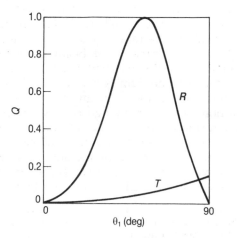

Figure 8.10 Degree of polarisation Q of reflected and transmitted light as a function of angle of incidence

The variation with angle of incidence of the y and w reflection coefficients shows that for unpolarised incident light, both reflected and transmitted beams will show some degree of polarisation, which is readily calculated from (8.81) using the reflection coefficient (8.69) and (8.75). The result is shown in Fig. 8.10. Complete polarisation ($Q = 1$) of the reflected light occurs at the Brewster angle: at no angle is the transmitted light completely polarised.

The case of reflection and transmission by a parallel-sided film is dealt with in Appendix 8.A.

8.20 Total reflection

If the refractive index n_1 of the incident medium is greater than n_2, then for some values of θ_1 the quantity $n_1 \sin\theta_1/n_2$ ($\equiv \sin\theta_2$, by 8.58b) will exceed unity, so that no real value of θ_2 exists. Let us calculate the reflection coefficients for this condition. We note that $\cos\theta_2$ is a pure imaginary, since

$$\cos\theta_2 = (1 - \sin^2\theta_2)^{1/2} = \frac{i}{n_2}\left(n_1^2\sin^2\theta_1 - n_2^2\right)^{1/2} \quad (8.83)$$

From eqn (8.68), we obtain

$$\frac{A_{1y}'}{A_{1y}} = \frac{n_1\cos\theta_1 - i(n_1^2\sin^2\theta_1 - n_2^2)^{1/2}}{n_1\cos\theta_1 + i(n_1^2\sin^2\theta_1 - n_2^2)^{1/2}} \quad (8.84)$$

and from eqn (8.74),

$$\frac{A_{1w}'}{A_{1w}} = \frac{n_2^2\cos\theta_1 - in_1(n_1^2\sin^2\theta_1 - n_2^2)^{1/2}}{n_2^2\cos\theta_1 + in_1(n_1^2\sin^2\theta_1 - n_2^2)^{1/2}} \quad (8.85)$$

both of which are of the form $(a - ib)/(a + ib)$ which has a modulus of unity. Thus in each case the amplitude of the reflected wave is equal to that of the incident wave. The reflection is total, for both y and w components. The limiting angle of incidence beyond which total reflection occurs (the critical angle, θ_c) is given by

$$\theta_c = \arcsin\left(\frac{n_2}{n_1}\right) \quad (8.86)$$

There is a phase shift on total reflection, which differs for the y and w components. Since the moduli of (8.84) and (8.85) are unity, we may write

$$\frac{A_{1y}'}{A_y} = e^{-2i\delta_y} \quad \text{and} \quad \frac{A_{1w}'}{A_{1w}} = e^{-2i\delta_w} \quad (8.87)$$

from which we find that

$$\delta_y = \text{arc tan}\left[\frac{(n_1^2 \sin^2\theta_1 - n_2^2)^{1/2}}{n_1 \cos \theta_1}\right] \quad (8.88)$$

and

$$\delta_w = \text{arc tan}\left[\frac{n_1(n_1^2 \sin^2\theta_1 - n_2^2)^{1/2}}{n_2^2 \cos \theta_1}\right] \quad (8.89)$$

Thus the electric vectors of the reflected beams, for incident waves given by (8.60) and (8.61), have the form

$$E_{1y}' = A_{1y} \exp i[\omega t - \varkappa_1(x \sin \theta_1 + z \cos \theta_1) - 2 \delta_y] \quad (8.90)$$

$$E_{1w}' = A_{1w} \exp i[\omega t - \varkappa_1(x \sin \theta_1 + z \cos \theta_1) - 2 \delta_w] \quad (8.91)$$

The phase difference $\Delta \equiv 2\delta_w - 2\delta_y$ between the totally reflected w and y components is given by

$$\tan\frac{\Delta}{2} = \tan(\delta_w - \delta_y) = \frac{\cos \theta_1(n_1^2 \sin^2\theta_1 - n_2^2)^{1/2}}{n_1 \sin^2\theta_1}$$
$$(8.92)$$

8.21 Total reflection: the region beyond the interface

Since the reflectances given by (8.84) and (8.85) are identically unity, this would appear to imply that there *is* no electromagnetic disturbance in the second medium. In fact, we may obtain A_{2y} and A_{2w} from (8.67) and (8.68) and (8.72) and (8.73) and, using the value of $\cos \theta_2$ from (8.83), obtain

$$A_{2y} = A_{1y}(1 + e^{-2i\delta_y}) = 2A_{1y}e^{-i\delta_y} \cos \delta_y \quad (8.93a)$$

and

$$A_{2w} = A_{1w}(1 + e^{-2i\delta_w}) = 2A_{1w}e^{-i\delta_w} \cos \delta_w$$
$$(8.93b)$$

so that, for the y component,

$$E_{2y} = 2A_{1y} \cos \delta_y \exp\left[-\frac{\varkappa_2 z}{n_2}(n_1^2 \sin^2\theta_1 - n_2^2)^{1/2}\right.$$
$$\left. + i(\omega t - \varkappa_1 x \sin \theta_1 - \delta_y)\right] \quad (8.94)$$

with a corresponding expression for E_{2w}. There appears to be a paradox. How, if *all* the incident energy is reflected at the interface, can there be any disturbance in the second medium? The answer is revealed if the values of the magnetic field vectors in the second medium are calculated. They are found to be $\pi/2$ out of phase with the electric vectors. Consideration of the rate of energy flow shows that the average value of the Poynting vector over one period is zero in the second medium, so the existence of electric and magnetic fields in the second medium is not in conflict with the unit reflectance at the interface.

It should be noted that this argument applies only for plane waves, of infinite lateral extent. In fact, the results apply reasonably well for beams of practical widths—vastly larger than the wavelength.

The wave in the second medium is described as evanescent. From (8.94) it is seen that the wave is periodic in the co-ordinate parallel to the interface and falls exponentially with distance into the second medium. For angles away from the critical angle, the field decays to a very low value at distances of the order of a few wavelengths.

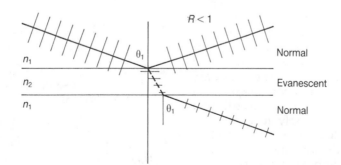

Figure 8.11 Normal and envanescent waves on reflection at a film at greater than the critical angle

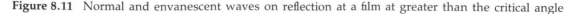

If, however, a second interface is placed close to the first, as shown in Fig. 8.11, then a normal wave may propagate into the third medium. If the refractive indices of the first and third media are equal, then the refraction angle in the third medium is equal to the angle of incidence in the first. Under this condition, the reflection is not total and is described as *frustrated total reflection*. This phenomenon is used in coupling radiation into the thin-film structures used in integrated optical devices (see Chapter 14).

8.22   Is it right?

Although the predictions of electromagnetic theory as outlined above have received substantial experimental verification, small discrepancies are evident when measurements of very high precision are made. These are not such as to invalidate the theory but are generally due to the fact that the experimental arrangements do not exactly satisfy the assumptions of the theory. Thus it cannot generally be assumed that the refractive index of a glass specimen is perfectly constant right to the surface, since the action of polishing produces a layer of slightly different refractive index in the near neighbourhood of the surface.

Appendix 8.A   Optical multilayers

Many important optical devices employ thin, parallel-sided films, so we need to be able to determine how they behave. For a single thin film, it is possible to calculate the reflectance and transmittance by summing the amplitudes of waves reflected and transmitted at the two film boundaries. For a stack of many films this procedure becomes excessively tedious. It is in fact unnecessary. In each film, the multiply-reflected and transmitted waves add up to a resultant. The boundary conditions—continuity at each interface of the tangential components of electric and magnetic field vectors—apply to the vector sums of all the reflected and transmitted waves. Thus we may imagine the summations made before imposing the boundary conditions, without the need to consider the

(very large number of) reflections and transmissions of individual beams. For simplicity, we consider the reflection and transmission of a stack of films for light at normal incidence. The case of an arbitrary angle of incidence introduces only additional algebraic complexity.

Let us consider a film of refractive index n_p surrounded perhaps by many others on both sides. A plane wave of the form $E_0 \exp i(\omega t - \varkappa x)$ is incident on the system, where the x axis is normal to the film plane. The resultant wave in the pth film is made up of waves travelling in the $+x$ direction, with resultant amplitude E_p^+, and of waves in the $-x$ direction, of resultant amplitude E_p^-. Corresponding quantities $E_{p\pm 1}^+$ and $E_{p\pm 1}^-$ exist in the adjacent films. Thus the resultant electric field vector at any point x within the film may be written

$$E_p(x) = E_p^+ e^{-i\varkappa_p x} + E_p^- e^{i\varkappa_p x}$$
$$= (E_p^+ + E_p^-)\cos \varkappa_p x - i(E_p^+ - E_p^-)\sin \varkappa_p x$$
$$(8.95)$$

From Section 8.16 the corresponding magnetic field vector is given by

$$H_p(x) = -in_p E_p^+ e^{-i\varkappa_p x} + in_p E_p^- e^{i\varkappa_p x}$$
$$= -in_p(E_p^+ - E_p^-)\cos \varkappa_p x$$
$$- n_p(E_p^+ + E_p^-)\sin \varkappa_p x \quad (8.96)$$

Take the origin as the $(p-1)/p$ interface and let the film thickness be d_p. Boundary conditions require that $E_{p-1}(0) = E_p(0)$, $H_{p-1}(0) = H_p(0)$ and $E_p(d_p) = E_{p+1}(d_p)$, $H_p(d_p) = H_{p+1}(d_p)$. Imposing these conditions leads to the relations

$$E_p = (E_p^+ + E_p^-)\cos \varkappa_p d_p - i(E_p^+ - E_p^-)\sin \varkappa_p d_p$$
$$(8.97)$$

$$H_p = -n_p(E_p^+ - E_p^-)\cos \varkappa_p d_p$$
$$- n_p(E_p^+ + E_p^-)\sin \varkappa_p d_p \quad (8.98)$$

and since $E_{p-1}(0) = E_p^+ + E_p^-$, $H_{p-1}(0) = in_p(E_p^+ - E_p^-)$ these equations may be written in matrix form as

$$\begin{pmatrix} E_p \\ H_p \end{pmatrix} = \begin{pmatrix} \cos \varkappa_p d_p & \dfrac{i}{n_p}\sin \varkappa_p d_p \\ in_p \sin \varkappa_p d_p & \cos \varkappa_p d_p \end{pmatrix} \begin{pmatrix} E_{p-1} \\ H_{p-1} \end{pmatrix}$$
$$(8.99)$$

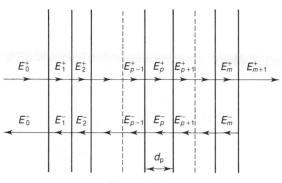

Figure 8.12

Thus the effect of the presence of the film on the field components E_{p-1}, H_{p-1} at the entrance to the film is to create at the exit plane components E_p, H_p, which are obtained from the column vector (E_{p-1}/H_{p-1}) through multiplication by the matrix

$$M_p \equiv \begin{pmatrix} \cos \varkappa_p d_p & \dfrac{i}{n_p} \sin \varkappa_p d_p \\ i n_p \sin \varkappa_p d_p & \cos \varkappa_p d_p \end{pmatrix} \quad (8.100)$$

The effect of many films is therefore calculated by forming the product of the set of 2×2 matrices representing each film of the stack. For the system of Fig. 8.12, we may therefore calculate the fields at the final $m/(m+1)$ interface in terms of that at the $0/1$ interface. In the final, $(m+1)$th, medium there is no negative-going wave, so that $E_{m+1} = E_{m+1}^+$. At the first boundary, $E_1 = E_0^+ + E_0^-$. E_{m+1} is related to E_1 through the product of the film matrices. From this relationship, the ratio E_{m+1}^+/E_0^+—the amplitude transmission coefficient—and E_0^-/E_0^+—the amplitude reflection coefficient—are easily obtained.

Problems

8.1 Calculate the electric and magnetic fields due to the solar radiation of 1.3 kW m^{-1} at the top of the Earth's atmosphere,

with the (bizarre) assumption that the Sun's light is monochromatic. (Or does it matter?)

8.2 The beam from a 5 mW helium-neon laser is of Gaussian form $E = E_0 \, e^{-r^2/r_0^2}$ with $r_0 = 0.3$ mm. Calculate the value of E_0.

8.3 Light is incident normally in air on a thick glass plate of refractive index n. Taking account of multiple reflections, calculate the reflected and transmitted powers for unit incident power. Satisfy yourself that energy conservation has not been violated. Why a *thick* plate?

8.4 For a collimated beam incident at exactly the Brewster angle, the degree of polarisation of the reflected light is unity. If the angle of incidence on a germanium plate $(n = 4.0)$ is 1° above the Brewster angle, what degree of polarisation would result?

8.5 Equation (8.94) can be written as

$$E_{2y} = A_{2y} \cos \delta_y \, e^{-z/z_0} \exp i(\omega t - \varkappa_1 x \sin \theta - \delta_y)$$

where z_0 represents the depth at which the amplitude E_{2y} falls to $1/e$ of its value at the surface. For $n_1 = 1.52$ and $n_2 = 1.33$, calculate z_0 for angles $(\theta_{\text{critical}} + 2^\circ)$ and $(\theta_{\text{critical}} + 5^\circ)$ for light of wavelength 488 nm.

8.6 Determine the direction and rate of flow of energy across the interface between two media over one period when light is incident in the denser medium at greater than the critical angle.

8.7 Design a device for producing circularly polarised light by totally reflecting an initially plane-polarised beam twice at the opposite faces of a block of glass. The light should enter and leave two of the faces of the block normally. (The ingenious device that you have invented was first produced by Fresnel.) What are the angles of the block for a glass with a refractive index of 1.50?

9

Electromagnetic Theory of Absorptive Materials

9.1 Preamble

In the previous chapter we considered the propagation of electromagnetic waves in 'perfect' dielectric media which are completely characterised by a refractive index. Such media do not exist. The electric and magnetic fields of the light wave interact with the charges in the constituent atoms with the result that energy from an incident beam of radiation is either absorbed or redirected. Nevertheless, some materials absorb extremely little. Radiation of certain wavelengths in the solar spectrum pass through the Earth's atmosphere (equivalent to 8 km of air at STP) with practically no loss. Also the type of glass used for optical fibre communications (see Chapter 14) transmits more than 99% of the power of an incident beam over a distance of a kilometre, again for certain wavelengths. At longer and shorter wavelengths, however, significant absorption occurs, even in this material. Thus absorption is the general characteristic of real materials and for some (e.g. metals) this can be extremely high. Very little light of visible wavelengths will penetrate to a depth of the order of one micrometre in a typical metal.

In this chapter we examine the modifica-

tions needed to the theory given in the previous chapter in order for it to describe correctly the behaviour of absorbing materials. We consider first materials that absorb but do not scatter. The power in a collimated beam traversing such a material will either be absorbed or transmitted in the beam: it will *not* emerge in other directions. For such a material, the power flux per unit area (irradiance) is found experimentally to fall exponentially with distance. This relation (Lambert's law) can be written, for a beam travelling in the z direction,

$$I(z) = I_0 e^{-2\alpha z} \qquad (9.1)$$

which implies that the wave amplitude $A(z)$ follows the relation

$$A(z) = A_0 e^{-\alpha z} \qquad (9.2)$$

from which

$$\alpha = -\frac{1}{A} \frac{dA}{dz} \qquad (9.3)$$

showing that α is the fractional change in amplitude per unit path in the medium. In place of eqn (8.31), we need to write

$$E_x = E_{0x} \exp\left[-\alpha z + i(\omega t - \varkappa z)\right] \qquad (9.4)$$

where $\varkappa$ is the spatial frequency in the

medium. If the refractive index is n, then $\varkappa = n\omega/c$. Writing $k = \alpha c/\omega$, eqn (9.4) may be written as

$$E_x = E_{0x} \exp i\omega \left[t - \frac{(n-ik)z}{c} \right] \quad (9.5)$$

showing that the behaviour of an absorbing medium is correctly represented if the refractive index n is replaced by a complex quantity $\mathbf{n} \equiv n - ik$,[†] where k is termed the *extinction coefficient*. Note, however, that when light is incident on an absorbing medium at non-normal incidence, the ratio of the sines of the angles of incidence and refraction is *not* given by the real part of $\mathbf{n}$. From eqn (9.1) we see that the flux in a beam falls to $1/e$ of its initial value after a distance of $l = (2\alpha)^{-1}$. In terms of k, this is equal to $c/2k\omega$. For light of wavelength 500 nm ($\omega = 3.77 \times 10^{15} \, \mathrm{s}^{-1}$) and a material with $k = 1$, the value of l is 4×10^{-8} m or 40 nm.

9.2 Light incident on an absorbing medium

For light incident in air[‡] at an angle θ_1 and with its electric vector perpendicular to the plane of incidence, the results of Section 8.16 give, for the incident and reflected waves,

$$E_{1y} = A_{1y} \exp i\omega \left(t - \frac{x \sin \theta_1 - z \cos \theta_1}{c} \right) \quad (9.6)$$

$$E'_{1y} = A'_{1y} \exp i\omega \left(t - \frac{x \sin \theta_1 + z \cos \theta_1}{c} \right) \quad (9.7)$$

A corresponding expression for E_{2y} may be written

$$E_{2y} = A_{2y} \exp i\omega \left[t - \frac{\mathbf{n}(x \cos \theta_2 - z \cos \theta_2)}{c} \right] \quad (9.8)$$

where, from eqn (9.58b), we require

$$\sin \theta_1 = \mathbf{n} \sin \theta_2 \quad (9.9)$$

where θ_1 is real. Since $\mathbf{n}$ is complex, then so is $\sin \theta_2$, showing that θ_2 is not simply an angle. This is a manifestation of the fact that a plane wave in an absorbing medium generally differs from that in a transparent

---

† This is sometimes written $n(1 - ik)$.
‡ Strictly vacuum

medium, in which the amplitude of the wave at all points of the wave-front is constant —planes of constant phase are parallel to planes of constant amplitude. This is not true for a beam refracted into an absorbing medium (except for normal incidence). Planes of constant phase are perpendicular to the beam direction but the wave amplitude varies across the beam since different parts of the beam have traversed different distances in the absorbing medium. On planes parallel to the surface in the second medium, the field vectors vary sinusoidally with x and with time, but the maximum value of the amplitude of the wave is constant. We refer to these as 'planes of constant amplitude' (Fig. 9.1).

Expressions for A'_{1y}/A_{1y} and A'_{1w}/A_{1w} can be obtained as was done for transparent media in Section 8.16, but these are extremely cumbersome, as is readily seen when the (complex) forms of $\mathbf{n}$ and $\cos \theta_2$ are substituted for n_2 and $\cos \theta_2$ in eqns (8.69) and (8.74). The forms of the curves of R_y and R_w versus angle of incidence have some similarity (Fig. 9.2a) with those for the case of a transparent medium (Fig. 9.2b), but the minimum of the R_w curve is not zero. For normal incidence, this complexity disappears

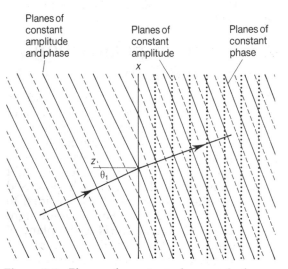

Figure 9.1 Planes of constant phase and of constant amplitude on transmission of a plane wave into an absorbing medium

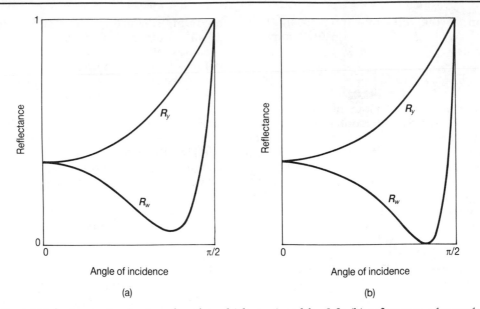

Figure 9.2   (a) Reflectances in air at surface for which $n = 4$ and $k = 0.3$; (b) reflectances for surface with $n = 4$ and $k = 0$

and we obtain

$$R = R_y = R_w = \left(\frac{n-1}{n+1}\right)\left(\frac{n^*-1}{n^*+1}\right)$$

$$= \frac{(n-1)^2 + k^2}{(n+1)^2 + k^2} \qquad (9.10)$$

For many metals and for visible wavelengths, values of k can be much larger than unity, with the result that R can be very high. Thus for $\lambda = 700$ nm, the reflectance of silver is 98.7%.

9.3   State of polarisation of reflected light

The complex values of n and θ_2 lead in general to complex values of A'_{1y}/A_{1y} and A'_{1w}/A_{1w}, indicating changes of phase on reflection which are neither zero nor π, as is so for reflection (other than total) at transparent media. Plane-polarised incident light is generally reflected as elliptically polarised light with the axes of the ellipse inclined to the plane of incidence. For incident light with the electric vector at $\pi/4$ to the plane of incidence, $A_{1y} = A_{1w}$ and the ratio of the reflected amplitudes may be written as

$$\frac{A'_{1w}}{A_{1w}} \equiv a\, e^{i\delta} = -\frac{\cos(\theta_1 + \theta_2)}{\cos(\theta_1 - \theta_2)} \qquad (9.11)$$

where (8.69) and (8.75) are used. For one particular angle of incidence, the value of δ is equal to $\pi/2$. The angle is known as the *principal angle of incidence* Θ_1. For this angle, eqn (9.11) may be rearranged to give

$$\frac{1+ia}{1-ia} = \frac{\tan\Theta_1 \sin\Theta_1}{(n^2 - \sin^2\Theta_1)^{1/2}} \qquad (9.12)$$

where θ_2 in eqn (9.11) has been eliminated using eqn (9.9).

Under this condition, the axes of the ellipse lie parallel and perpendicular to the plane of incidence. Insertion of a compensator (e.g. a quarter-wave plate, suitably orientated) converts the elliptically polarised reflected light to plane-polarised. The angle between the electric vector and the plane of incidence is termed the *principal azimuth* (Φ) and depends on the ratio of the A'_{1w} and A'_{1y} reflected amplitudes (Fig. 9.3):

$$\tan \Phi = a \qquad (9.13)$$

Thus measurements of Θ_1 and Φ enable $n(= n - ik)$ to be evaluated from eqn (9.12), somewhat tediously. For many metals, it is found that $n^2 + k^2 \gg \sin^2\Theta_1$, in which case eqn (9.12) simplifies to

$$n\, e^{i2\Phi} = \tan\Theta_1 \sin\Theta_1 \qquad (9.14)$$

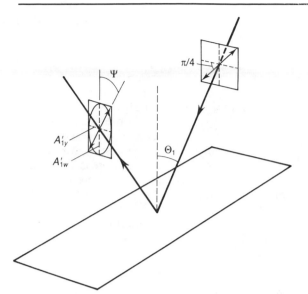

Figure 9.3 Reflectance at principal incidence Θ_1

from which we obtain

$$n = \tan\,\Theta_1\,\sin\,\Theta_1\,\cos\,2\Phi \qquad (9.15)$$

and

$$k = n\,\tan\,2\Phi \qquad (9.16)$$

Measurements of n and k as described above may be used to calculate the reflectance of a surface, with the help of eqn (9.10), and the result may then be compared with the measured value. The agreement is often not very impressive, a consequence of the fact that surface films (e.g. oxides) or disordered surfaces affect the two types of measurement in different ways. The theory has assumed a perfectly homogeneous material with a mathematically plane surface. Freshly cleaved crystal surfaces in a high vacuum may approach this ideal; polished, air-exposed surfaces generally do not.

9.4 Optical constants of conducting materials

When an electromagnetic wave traverses a material containing free charges the latter are accelerated by the electric field, are scattered by atoms in the lattice and hence remove energy from the beam. Such absorption should therefore be related to the electrical conductivity of the material. This was fore-

shadowed in the last chapter by eqn (8.27). For an uncharged specimen, we have for the amplitude of the electric field strength

$$\nabla^2 E = \mu\sigma\dot{E} + \mu\varepsilon\ddot{E} \qquad (9.17)$$

The plane wave in the z direction, represented by eqn (9.5), is a solution of eqn (9.17) provided

$$(n - \mathrm{i}k)^2 = c^2\mu\varepsilon - \frac{\mathrm{i}c^2\sigma\mu}{\omega} \qquad (9.18)$$

We note that $c^2 = (\mu_0\varepsilon_0)^{-1}$ and that, for optical frequencies, $\mu = \mu_0$. Equation (9.18) then yields

$$n^2 - k^2 = \frac{\varepsilon}{\varepsilon_0} \qquad (9.19)$$

$$2nk = \frac{\sigma}{\varepsilon_0\omega} \qquad (9.20)$$

If we measure n and k using the method indicated in Section 9.3, and the values of ε and σ by electrical methods, we can test the validity of eqns (9.19) and (9.20). For light of visible wavelengths there is total disagreement. The value of k is generally larger than n; eqn (9.19) would imply a negative permittivity! Moreover, the variation of n and k with frequency ω predicted by eqn (9.20) is not observed. We return to this point in Section 9.5, but note in passing that notwithstanding the total failure of the above theory for visible wavelengths, good agreement is obtained for the infrared ($\lambda \gtrsim 5\,\mu\mathrm{m}$) and radio-frequency regions of the spectrum. For such frequencies, the value of $\sigma/\varepsilon_0\omega$ for metals is large compared with unity. Under this condition, the value of R of eqn (9.10), evaluated using n and k from (9.19) and (9.20), is found to be

$$R = 1 - 4\left(\frac{\pi\varepsilon_0 c}{\sigma\lambda}\right)^{1/2} \qquad (9.21)$$

Experimental measurements on many metals, at wavelengths in the 5–25 μm region, agree well with this prediction.

9.5 Why the theory fails for the visible region

The reason for the failure of the theory in the visible region is easy to see. The conductivity σ describes the behaviour of a metal to which a steady electrical field is applied. Electrons are accelerated by the field until they are scattered by the ions in the lattice, which occurs after a certain mean time τ. If the period of an incident electromagnetic wave is large compared with τ, the number of electron collisions per second will be much the same as in the steady-field case. When the period of the wave is less than τ, the conditions are different and the theory invoking the electrical conductivity (the 'd.c.' value) is not expected to apply. In fact, the value of $2nk\varepsilon_0\omega$ (see eqn 9.20) is referred to as the *optical conductivity*.

9.6   Optical behaviour of non-conductors (dielectrics)

From the preceding sections it would appear that materials for which $\sigma \approx 0$, such as dielectrics, must necessarily be non-absorbing (eqns 9.19 and 9.20). This is not so: although dielectrics (gases, liquids or solids) may show very little absorption over wide ranges of wavelength, there are always some regions where very high absorption occurs. Moreover, the values of n and k vary with frequency in a highly characteristic fashion. We shall now see that electromagnetic theory is well suited to account for this behaviour and that it predicts frequency dependences of n and k that have the correct characteristics. Problems *do* arise but we should not be surprised: there are many areas of physics in which classical models break down and need to be replaced by quantum descriptions.

Non-conductors contain no free carriers (electrons or holes) but their constituent atoms have bound electrons which will respond to the electric field of an incident electromagnetic wave. We envisage the electron as undergoing forced oscillations due to the wave. Since the electrons are surrounded by, and interact with, other electrons and atoms, we assume—reasonably—that the oscillating electron is subjected to damping forces.

The displacement of the electrons by the field (and the very small opposite displacement of the ion cores, which we ignore) creates electric dipoles which result in an overall polarisation P. The electrons are subjected not only to the applied field E but to the resultant polarisation. The local field in an isotropic material is given by

$$E_{\text{loc}} = E + \frac{P}{3\varepsilon_0} \qquad (9.22)$$

(see ref. [1]). For isotropic materials, P and E are parallel, so that amplitudes are used in eqn (9.22). In some circumstances, e.g. for gases at other than very high pressures, the correction term $P/3\varepsilon_0$ can be ignored. For liquids and solids, its retention is essential.

If a field E is applied to a system with $\mathcal{N}$ atoms per unit volume, each with a bound electron with natural oscillation frequency ω_0 and damping constant γ, then the equation of motion of the oscillating electron is†

$$m(\ddot{r} + \gamma\dot{r} + \omega_0^2 r) = \left(E + \frac{P}{3\varepsilon_0}\right)e \qquad (9.23)$$

and since the polarisation $P = \mathcal{N}er$, we have

$$\ddot{P} + \gamma\dot{P} + \omega_0^2 P = \left(E + \frac{P}{3\varepsilon_0}\right)\frac{\mathcal{N}e^2}{m} \qquad (9.24)$$

The relation between P, E and the complex index n follows from eqns (8.10), (8.13) and (8.30), where n is replaced by n and the permittivity ε_r is complex

$$P = (n^2 - 1)\varepsilon_0 E \equiv \xi E \qquad (9.25)$$

ξ is complex, indicating a phase difference between the oscillations of E and P. Equation (9.24) is rapidly solved for ξ, so giving the variation of $n^2 - 1$ with the applied frequency ω.

9.7   Dispersion in gases at low pressure

For air at STP and for visible wavelengths, the absorption is small and $n \approx n = 1.00029$. From eqn (9.25), we have $P/3\varepsilon_0 = 1.93 \times 10^{-4}E$. Thus the local field correction term in eqn (9.24) may be ignored. For an applied field of the form $E = E_0 e^{i\omega t}$, the solution for P/E, and hence $n^2 - 1$, is obtained as

$$n^2 - 1 = \frac{\mathcal{N}e^2}{m\varepsilon_0}\frac{1}{\omega_0^2 - \omega^2 + i\gamma\omega} \qquad (9.26)$$

---

† For an isotropic material, the vectors P and E may be replaced by their magnitudes P and E. These, however, are generally complex, since P and E will have differing phases.

Atoms and molecules generally contain many electrons. If each contains f_s electrons with resonant frequency ω_s and damping constant γ_s and there are N atoms per unit volume, then

$$n^2 - 1 = \frac{Ne^2}{m\varepsilon_0} \sum_s \frac{f_s}{\omega_s^2 - \omega^2 + i\gamma_s\omega} \quad (9.27)$$

At low pressures, gases are almost transparent except in narrow regions close to absorption lines, at which the frequency corresponds to a resonant frequency of one of the electrons. Except in these regions, the damping term $i\gamma_s\omega$ in eqn (9.27) may be ignored. Thus n^2 is real $(= n^2)$ and

$$n^2 - 1 = \frac{Ne^2}{m\varepsilon_0} \sum_s \frac{f_s}{\omega_s^2 - \omega^2} \quad (9.28)$$

Expressed in terms of wavelengths, this relation is referred to as the Sellmeier dispersion formula.

For most gases at low pressure (e.g. less than one atmosphere), n is less than 1.001, so that $n^2 - 1 \approx 2(n - 1)$. Equation (9.28) then simplifies to

$$n - 1 = \frac{Ne^2}{2m\varepsilon_0} \sum_s \frac{f_s}{\omega_s^2 - \omega^2} \quad (9.29)$$

Note that $n - 1 \propto N$, the number of atoms per unit volume and so is proportional to the density of the gas, a result confirmed by experiment. The divergence of the right-hand side as $\omega \to \omega_s$ serves to remind us that the relation is valid *only* for frequencies that are not near to the electron resonant frequencies ω_s.

In the neighbourhood of absorption lines, it is found to be possible to fit experimental curves of $(n - 1)$ versus ω to expressions of the form of eqn (9.27), where ω_s is given by the position of absorption lines and γ_s by their width. The f_s values have to be determined by curve-fitting and, contrary to what might be expected, the values required to ensure agreement with experiment are *not* integers and are generally less than unity. This suggests that the dipole moment created by the applied field is less by a factor f_s than that expected on the basis of the simple classical model. The latter is known as the *oscillator strength*. Fractional oscillator strengths are entirely comprehensible in quantum mechanical terms and their values can be calculated in some cases.

9.8 Dispersion in condensed matter

When atoms or molecules are packed together at densities of the order of those of solids, the influence of the polarisation on the local field strength can no longer be ignored. Equation (9.24) is solved for a sinusoidal varying field $E = E_0 \exp(i\omega t)$, using eqn (9.25) to give

$$\frac{n^2 - 1}{n^2 + 2} = \frac{\mathcal{N}e^2}{3m\varepsilon_0} \frac{1}{\omega_0^2 - \omega^2 + i\gamma\omega} \quad (9.30)$$

which, for the case of N atoms per unit volume and f_s electrons with ω_s, γ_s, becomes

$$\frac{n^2 - 1}{n^2 + 2} = \frac{Ne^2}{3m\varepsilon_0} \sum_s \frac{f_s}{\omega_s^2 - \omega^2 + i\gamma_s\omega} \quad (9.31)$$

In contrast to the case of low-pressure gases, the absorption regions in liquids and solids are broad—a consequence of the disturbance to electron behaviour by surrounding atoms or molecules. In some cases, e.g. where there are electrons in unfilled shells—the $3d$ electrons of the transition metals—screening by the outermost electrons occurs and sharp absorption lines result. For such cases we may calculate the dispersion relations for n and k in the neighbourhood of absorption lines in the following way. We assume that the effect of all other absorption lines is to contribute an amount n_0 to the real part of n and that they make no contribution to k. We also write

$$(\omega_s')^2 \equiv \omega_s^2 - \frac{Nf_s e^2}{3m\varepsilon_0} \quad (9.32)$$

as a result of which eqn (9.31) becomes

$$n^2 - 1 = n_0^2 - 1 + \frac{Nf_s e^2}{m\varepsilon_0} \frac{1}{(\omega_s')^2 - \omega^2 + i\gamma_s\omega} \quad (9.33)$$

For values of ω close to ω_s', we have $(\omega_s')^2 - \omega^2 \doteq 2\omega_s'(\omega_s' - \omega)$ so that

$$n^2 = n_0^2 + \frac{Ne^2 f_s}{m\varepsilon_0\omega_s'} \frac{1}{2(\omega_s' - \omega) + i\gamma_s} \quad (9.34)$$

whence

$$n^2 - k^2 = n_0^2 + \frac{2Ne^2 f_s}{m\varepsilon_0\omega_s'} \frac{\omega_s' - \omega}{4(\omega_s' - \omega)^2 + \gamma_s^2} \quad (9.35)$$

and

$$nk = \frac{Ne^2 f_s}{2m\varepsilon_0\omega_s'} \frac{f_s\gamma_s}{4(\omega_s' - \omega)^2 + \gamma_s^2} \quad (9.36)$$

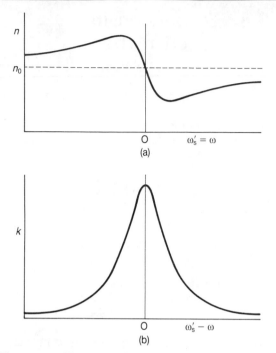

Figure 9.4 Variation of n and k in neighbourhood of absorption line

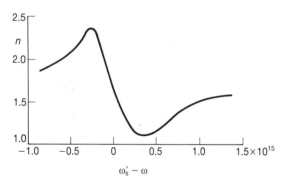

Figure 9.5 Dispersion curve for n for cyanin

The forms of the curves of n and k versus $\omega_s' - \omega$ are shown in Fig. 9.4. In some cases, generally similar curves are obtained for solids. Thus Fig. 9.5 shows the dispersion of n for cyanin; note the general similarity to Fig. 9.4(a).

9.9   Measurement of oscillator strength

Returning to the case of low-pressure gases, we note that since the value of n differs very

little from unity, we may put $n = 1$ in eqn (9.36) and so obtain k. From section 9.3, we note that $2\alpha = 2k\omega/c$ and for values of ω close to ω_s', eqn (9.36) becomes

$$2\alpha = \frac{2k\omega}{c} \doteq \frac{2Ne^2}{2m\varepsilon_0}\frac{f_s\gamma_s}{4(\omega_s' - \omega)^2 + \gamma_s^2} \quad (9.37)$$

The total absorption over the line is thus

$$\int_0^\infty 2\alpha \ d\omega = \frac{\pi Ne^2 f_s}{2m\varepsilon_0 c} \quad (9.38)$$

in which the only unknown is f_s. Thus measurement of the absorption integrated over the whole absorption line yields f_s.

The value of f_s may also be obtained by methods that effectively explore the variation of n in the neighbourhood of an absorption line. In the 'hook' method of Roschdestvenski, fringes from a Jamin interferometer are projected on to the slit of a spectrometer, so arranged that the spectrum displays fringes whose shape resembles the n versus $\omega_s' - \omega$ curve. Measurement of the frequency interval between the turning points yields the value of f_s.

9.10   Optical behaviour of metals

In Section 9.8, the effects on the optical constants of solids of bound electrons in the constituent atoms are considered. For conducting materials, we need to include consideration of the free (conduction) electrons. We make two reasonable assumptions, viz.(1) that such electrons have zero resonant frequency and (2) that the effect of the ions being in a conducting 'sea' of electrons removes the need for the local field correction (eqn 9.22). For this case eqn (9.33) is replaced by

$$n^2 - 1 = \frac{Ne^2}{m\varepsilon_0}\left(\frac{fe}{i\gamma_e\omega - \omega^2} + \sum_s \frac{f_s}{\omega_s^2 - \omega^2 + i\gamma_s\omega}\right)$$
$$(9.39)$$

where f_e is the number of free electrons per atom and γ_e is the damping term which allows for collisions with the lattice ions. In terms of band theory, the terms in the summation correspond to excitation from the valence band to the conduction band while the f_e, γ_e term covers excitation from occu-

·pied to unoccupied conduction band states.

In many cases the bound electron contribution in some frequency regions is small compared with the free electron term. Also at large enough frequencies, the $i\gamma_e\omega$ term may be neglected compared with ω^2, so that (9.39) becomes

$$n^2 = 1 - \frac{Ne^2 f_e}{m\varepsilon_0 \omega^2}$$

$$= 1 - \frac{\omega_p^2}{\omega^2} \qquad (9.40)$$

where

$$\omega_p^2 = \frac{Ne^2 f_e}{m\varepsilon_0} \qquad (9.41)$$

Equation (9.40) indicates that for $\omega < \omega_p$, the value of n is a pure imaginary, in which case (eqn 9.10) the reflectance of the medium at normal incidence should be unity. For $\omega \gtrsim \omega_p$, n is real and small compared with unity, indicating a low reflectance. Thus the classical treatment of metallic reflection predicts an abrupt change in reflectance from a low value to unity as ω increases through a critical frequency $\omega_p = (Ne^2 f_e/m\varepsilon_0)^{1/2}$. This is the behaviour observed for the alkali metals. From the known electronic structure of the alkali metals we expect f_e to be unity. N may be determined from the known crystal structure so that ω_p may be calculated and compared with experimental values. The results are shown in Table 9.1. In the light of the rather brutal assumptions made, the agreement is very good.

The frequencies in Table 9.1 correspond to the ultraviolet region, from 155 to 340 nm.

Table 9.1

Metal	ω_p (expt)	ω_p (calc)
Lithium	1.22×10^{16}	1.22×10^{16}
Sodium	8.98×10^{15}	9.02×10^{15}
Potassium	5.98×10^{15}	6.57×10^{15}
Rubidium	5.54×10^{15}	5.85×10^{15}

9.11 Non-local effects

The discussion of metals in the previous section assumes that the free electrons in the metal are accelerated by the electric field of the light wave, that energy is transferred by collisions with the atoms of the lattice and that the dissipation process can be described by an optical constant k. Although the theory successfully accounts for many of the observed optical properties of metals, there are regions of the spectrum where failure can occur.

The density of free electrons in a metal is such that the collective behaviour of the electrons is important: they are, after all, charged particles and so interact one with another. The collective motion of the whole electron 'gas' in the solid may be described in terms of quantised oscillations known as plasmons. The important point to note is that, in contrast to electromagnetic wave effects as described above, which are transverse, the collective electron oscillations can be longitudinal.

Consider, then, the effect of incident radiation on a metal surface, dealing separately with s and p components. The s component of the oscillating electric field has no component in the plane of incidence. It cannot therefore create a longitudinal wave of electron motion along the metal surface so that, for the s component of an incident wave, the equations in Section 9.10 apply. In contrast, the electric field for p polarisation does have a component along the metal surface in the plane of incidence and is therefore capable of generating surface plasmon waves in the electron gas. However, this now means that the boundary conditions previously adopted no longer apply. Since the plasmon waves generated along the surface have an associated momentum, then this must be taken into account in specifying the boundary conditions at the surface.

Such effects become important only at frequencies close to the surface plasmon frequency ω_s, related to the value ω_p (eqn 9.41) by $\omega_s = \omega_p/\sqrt{2}$. In fact, optical measurements on metals in such frequency regions are invaluable for gaining an understanding of the electron behaviour in metals.

Reference

1. P. Lorrain, D. R. Corson and F. Lorrain *Electromagnetic Fields and Waves*, Freeman and Company, New York, 1988.

Problems

9.1 What is the maximum tolerable value of the absorption coefficient 2α (eqn 9.1) (a) for the glass of a window which should absorb less than 5% of the daylight (take $\lambda \sim 550$ nm) (b) for an optical communication fibre, for which the loss should not exceed 50% over a distance of 20 km. For this case, the wavelength is 1.3 μm.

9.2 The value of the extinction coefficient for platinum ($\lambda = 600$ nm) is 2.8. For what thickness of specimen will the incident flux be reduced to half?

9.3 Calculate the phase change on reflection at normal incidence when light is incident from air on palladium, for which $n = 1.81$ and $k = 2.64$.

9.4 The value of $1 - R$ (eqn 9.21) for copper at 450 K is found to be 1.17×10^{-2} for a wavelength of 25.5 μm. Its resistivity at this temperature is 2.84×10^{-8} Ω m. Does this support the argument of Section 9.4?

9.5 A laser pulse of mean power 100 watts and duration 0.1 second is incident normally on an isolated platinum foil 3 mm square and 1 mm thick. Calculate the temperature rise of the foil ($n = 1.3$, $k = 2.8$, density of Pt = 21 370 kg, specific heat = 138 J kg^{-1}).

10

Scattering of Light

10.1 How does scattering arise?

In the previous chapters we have considered the way in which the velocity and amplitude of a plane light wave depends on the properties of the ambient medium. For a plane wave in an absorbing medium, the amplitude falls exponentially with distance, but there is *no change* in the form of the wavefront. A plane wave retains its planeness and a spherical wave its sphericity. There is no change in the direction of the energy flow (the Poynting vector).

This approach is satisfactory provided the actual effect of the radiation on the medium is reasonably represented by an average value, implying that in a volume with linear dimensions small compared with the wavelength, there is a sufficiently large number of atoms to display average behaviour. In this case, if we apply an electric field E to a certain volume of a solid, we observe a polarisation P which is accurately proportional to E. If, however, we consider the same volume of a gas at very low pressure, the observed value of P will vary with time because the number of atoms (N) in the specified volume will fluctuate with time. The fractional fluctuation is $1/N^{1/2}$ and is very small only for very large values of N.

Each atom may be thought of as scattering a spherical wave. The phase of the scattered wave is determined by that of the incident wave. The resultant of the individually scattered waves *in the forward direction* is large since all the individually scattered waves are in phase and constitute a transmitted wave. In all other directions, and for a large concentration of atoms, the phases of the waves arriving from individual atoms vary randomly, covering all values with equal probability. Their mean resultant is therefore zero. In the case of a low density of scatterers, the fluctuating numbers of scatterers in a given volume means that the resultant amplitude will show a significant fluctuation about zero, leading to a finite amount of scattered power. We return to this point in Section 10.4.

10.2 Scattering cross-section

We need a quantitative measure to indicate the extent of the scattering from an incident beam.

Imagine the scatterer placed in a uniform beam of N_0 quanta per unit area and that the projected area of the scattering particle seen from the incident beam is A. Then the number of quanta N striking the scatterer is

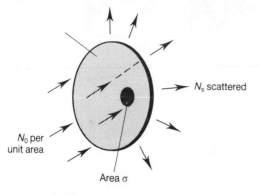

Figure 10.1 Definition of scattering cross-section

$N_0 A$. Suppose the total number of quanta scattered is N_s. This will be proportional to N_0, so we may write

$$N_s = \sigma N_0 \qquad (10.1)$$

so that

$$\sigma = \frac{N_s}{N_0} = \frac{N_s}{N} A \qquad (10.2)$$

N_s/N is the fraction of the number of incident particles scattered (Fig. 10.1); σ is termed the total scattering cross-section; the number of scattered quanta, N_2, is σ/A times the total number N striking the target and σ/A is the probability that an incident quantum will be scattered.

The cross-section σ has the dimensions of an area. For Raman scattering (Section 10.7) its value is of the order of 10^{-32} m^2. Taking the value of A for a molecule as 10^{-18} m^2, we see that the probability of an incident photon being scattered is $10^{-32}/10^{-18} = 10^{-14}$. It is no surprise that the arrival of the laser caused a great boost to the field of Raman spectroscopy.

10.3   Elastic and inelastic scattering

On the model discussed in Section 10.1, the scattered wave is thought of as arising from radiation from the oscillating dipole which is created when electrons in the atom or molecule are caused to oscillate by the incident electric field of the electromagnetic wave. The frequency of the scattered radiation is thus inevitably equal to that of the incident wave: the scattering is termed *elastic*.

Other possibilities arise, however. The result of irradiation of the atom or molecule could be that some energy transfer occurs (in either direction) between the radiation and the atom. The scattered radiation may then have a higher or lower frequency than the incident radiation. Since the relevant energy levels of atoms and molecules are quantised, this means that for irradiation with a monochromatic beam with frequency ω, the scattered radiation may have discrete frequencies above or below ω. Such scattering is *inelastic* and was first observed by Raman in 1928 (Section 10.7). Similarly, when scattering by a solid occurs, energy exchange may take place between the radiation and the (quantised) energy of the lattice vibrations leading to Brillouin scattering (Sections 10.8 and 10.9).

10.4   Rayleigh scattering

If we consider a collection of atoms or molecules at a sufficiently low concentration (where they are too far apart to interact appreciably with one another), in a unidirectional incident beam of radiation, then the waves scattered in all but the forward direction will be incoherent. Their resultant will therefore be the sum of the intensities of the scattering from individual particles. For monochromatic incident radiation with frequency ω and maximum electric field strength E_0, the maximum induced dipole moment is given by $p_0 \equiv \alpha E_0$, where α is the polarisability. From the result derived in Appendix 10.A, the mean scattered power $\bar{S}_N(r, \theta)$ from a collection of N particles in a direction $(\pi/2 - \theta)$ from the beam direction is

$$\bar{S}_N(r, \theta) = \frac{\omega^4 \alpha^2 E_0^2 \sin^2\theta}{32\pi^2 \varepsilon_0 r c^3} \qquad (10.3)$$

Since $\omega = 2\pi c/\lambda$, this yields the result that the scattered power depends on λ^{-4}, a long-known characteristic of Rayleigh scattering.

The blue colour of the sky is correctly associated with Rayleigh scattering, since the ratio of scattered blue ($\lambda_B \sim 420$ nm) to red ($\lambda_R \sim 650$ nm) light is $(\lambda_R/\lambda_B)^4 = 5.7$. However, the story is not quite so simple. If blue light is scattered predominantly from the light leaving a sunlit mountain, then the mountain should appear reddened when seen from a large distance. In fact, on a clear day, mountains appear their usual colour, even at distances of tens of kilometres.

At sea level, the average separation between air molecules is about 3 nm, a distance very small compared with light wavelengths. The individual scattered waves, in other than the forward direction, from large numbers of atoms will combine *coherently*. Since the *positions* of the scatterers are randomly distributed in space, the phases of the arriving waves will take all values, and the resultant of the combination of large numbers of amplitudes with random positive and negative phases will be nearly zero. Thus the sideways scattering of light from a given number of atoms of air at atmospheric pressure is in fact much less than that from the same number at a much lower pressure. The incoherence of the scattered waves in this case leads to a simple summation of scattered powers from the individual scatterers.

Thus the paradox of the not-red mountain is resolved: the observed Rayleigh scattering originates mainly from the upper, low-pressure part of the atmosphere. From satellites or space laboratories, no blue sky is seen. The sky appears an inky black.

With the help of a polariser, you can readily confirm that light from a blue sky has a high degree of linear polarisation, and yet the light arriving from the Sun is unpolarised. The electric field of the solar radiation will indeed create oscillating dipoles with all orientations perpendicular to its direction (Fig. 10.2). However, those

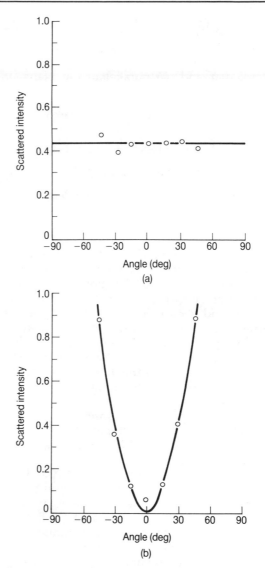

Figure 10.3 Rayleigh scattering by argon gas: electric vector (a) perpendicular to the plane of observation and (b) parallel to the plane of observation

whose axes lie along the viewing direction EM radiate no power in this direction. The scattered radiation arriving at E arises only from the components of the induced dipoles perpendicular to the plane containing E, M and the Sun.

Until the arrival of the laser, laboratory measurements of Rayleigh scattering from low-pressure gas was difficult: the scattered intensity in a direction perpendicular to the

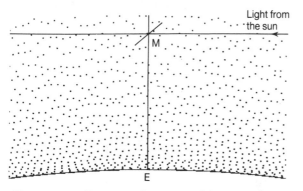

Figure 10.2 Scattering by the Earth's atmosphere

beam for conventional sources was beyond the limits of measurement. Figure 10.3 shows the results for argon gas for an incident beam with the electric vector (a) perpendicular and (b) parallel to the plane of observation. The full lines are theoretical curves.

10.5   Rayleigh scattering in quantum mechanical terms

The discussion above treats Rayleigh scattering in purely classical terms. How do we interpret this phenomenon in quantum mechanical language? We need first to note that the quantum mechanical description of a radiation field involves representation as a collection of quantised harmonic oscillators. Interaction between a (quantised) atomic system and a radiation field is described in terms of an atom and radiation field gaining or losing quanta. Thus in considering absorption by an atom, the radiation field loses a quantum and the atom gains one. In Rayleigh scattering, the atom initially gains and then loses a quantum, returning to its initial state. We do not observe the intermediate state (i.e. after the atom has gained a quantum but before it hands it back to the radiation field).

If the initial state of the atom is $\langle a |$ and the intermediate state $| b \rangle$, then the number of transitions per unit time from state a to state b depends on the matrix element $\langle a | R | b \rangle$,† where R is the dipole moment induced by the field. It should be noted that the frequency ω_{ab} corresponding to the energy difference between states a and b ($\hbar \omega_{ab} = E_b - E_a$) is not necessarily equal to the frequency ω of the incident wave. The fact that energy is not conserved in a transition to an *unobservable* intermediate state is irrelevant. Conservation must, however, apply between initial and *final* states.

The transition to the intermediate state corresponds to the atom acquiring an oscillating dipole moment. It will therefore radiate

according to the discussion of Appendix 10.A, with the indicated angular dependence. The direction associated with the field oscillator excited by the return of the quantum to the field may therefore be different from that of the one originally taken. It is possible, therefore, to calculate the number of quanta scattered per second into a solid angle $d\Omega$ in a direction θ to the electric field of the incident electromagnetic wave. For a collection of N atoms at a sufficiently low density that the scattered waves combine incoherently, the ratio k of scattered to incident quanta is given by

$$k = \frac{16\pi^2 e^4 \, | \langle a | R | b \rangle |^2 N \, \sin^2\theta}{9\hbar^2 c^4} \frac{\omega^4 \omega_{ab}^2 \, d\Omega}{(\omega^2 - \omega_{ab}^2)^2}$$

(10.4)

We note that the classically predicted $\sin^2\theta$ angular dependence is preserved. The worrying infinity which occurs at $\omega = \omega_{ab}$ arises because we have not taken any form of damping into account. If, however, the frequency of the incident wave *is* equal to that corresponding to transitions between states a and b, then the scattering is large. The effect is then known as *resonant* scattering.

10.6   Mie scattering

One feature of Rayleigh scattering is its symmetry in the forward and backward directions. The scattered flux in a direction making an angle β to the beam direction is equal to that at $\pi - \beta$. Where scattering occurs from, for example, droplets of moisture in the atmosphere, this symmetry does not occur. The relevant parameter is the size of the scattering particle in relation to the wavelength of the radiation. For Rayleigh scattering to occur, the particle size must be very small compared to the wavelength. With increasing particle size there is an increase of forward scattering and a generally more complex behaviour, as shown in Fig. 10.4. Scattering both by dielectric spheres and by highly conducting spheres with a wide range of diameters was exhaustively investigated by Mie, after whom this form of scattering is named.

---

† In terms of wave-functions,

$$< a | R | b > = e \int \psi_a^* r \psi_b \, d\tau$$

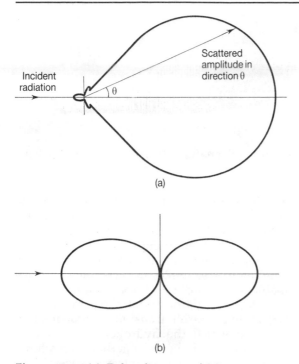

(a)

(b)

Figure 10.4 (a) Polar diagram of Mie scattering for a dielectric sphere; (b) corresponding distribution for radius ≪ wavelength

10.7 Raman scattering

In the discussion in Section 10.4 it was assumed that the effect of an electromagnetic wave incident on an atom or molecule was to produce an oscillating dipole moment with frequency equal to that of the incident wave. This certainly can occur and gives rise to the Rayleigh scattering discussed in that section. However, real life is somewhat more involved in the case of molecules. Individual atoms in a molecule may perform vibratory motions, or the molecule as a whole may undergo rotations and these may be excited by the incident electromagnetic wave. The behaviour of the molecule may be characterised in terms of a set of normal modes and associated frequencies of (assumed simple harmonic) oscillation. If one of the modes is excited, then an oscillatory polarisation results. For a system with n normal frequencies, the polarisability α will be made up of a component such as gave rise to Rayleigh scattering, together with an oscillatory one arising from internal oscillations of

the atoms in the molecule. We could expect, therefore, a polarisability of the form

$$\alpha = \alpha_0 + \sum_n \alpha_n \cos \omega_n t \qquad (10.5)$$

where ω_n is the angular frequency of the nth mode.

It is clear that α cannot be a scalar quantity. The direction of oscillation of a given atom on application of a plane-polarised electromagnetic wave will *not* generally be that of the electric vector of the field. The motion of the atom will be governed by the directions of the force field due to all the other atoms in the molecule. The polarisability α will therefore be a tensor.

What frequencies will be radiated when a molecule is exposed to an electromagnetic wave of frequency ω? With a polarisability of the form (10.5), the resultant polarisation P for a field $E = E_0 \cos \omega t$ will be

$$P = \alpha_0 E_0 \cos \omega t + \sum_n \alpha_n E_0 \cos \omega t \cos \omega_n t$$

$$= \alpha_0 E_0 \cos \omega t + \frac{E_0}{2} \sum_n \alpha_n [\cos(\omega - \omega_n)t$$

$$+ \cos(\omega + \omega_n)t] \qquad (10.6)$$

The rate of radiation from an oscillating dipole is proportional to the square of the amplitude of the dipole moment; thus in addition to radiation scattered with frequency ω (Rayleigh scattering), frequencies $\omega - \omega_n$ and $\omega + \omega_n$ will be observed, respectively known as Stokes and anti-Stokes radiation (Fig. 10.5).

In order to determine the amount of radiation scattered in the Raman process, values of the components of the polarisability tensor are needed. These may in some cases be determined with the help of quantum mechanics, in terms of which we need to take a somewhat modified view from that used so far.

We note first that the energy (and therefore frequency) associated with different modes of oscillation or vibration is quantised. The mode frequencies are related to the differences between the energies of the eigenstates of the system corresponding to different vibrational or rotational states. If the latter are represented by $\langle j|$, then the matrix element representing a transition from state

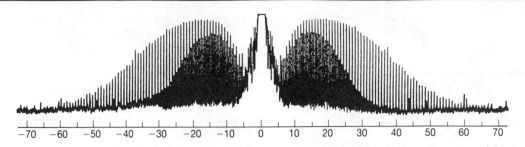

Figure 10.5   Stokes and anti-Stokes lines in a Raman spectrum. Reproduced by permission of Springer-Verlag from Demtröder, *Laser Spectroscopy*

a to state b is given by

$$|R_{ab}| = 2\langle a|R|b\rangle \cos \omega_{ab}t$$
$$+ \left[\sum_j \left| \frac{\omega_{aj}}{\omega_{aj}+\omega} - \frac{\omega_{jb}}{\omega_{jb}-\omega} \right| \right.$$
$$\langle a|R|j\rangle\langle j|R|b\rangle \sin(\omega+\omega_{ab})t$$
$$+ \sum_j \left| \frac{\omega_{bj}}{\omega_{bj}+\omega} - \frac{\omega_{ja}}{\omega_{ja}-\omega} \right|$$
$$\left. \langle a|R|j\rangle\langle j|R|b\rangle \sin(\omega-\omega_{ab})t \right]$$

$$(10.7)$$

per incident quantum of electromagnetic field. The second and third terms in (10.7) represent Raman radiation at frequencies $\omega - \omega_{ab}$ (Stokes) and $\omega + \omega_{ab}$ (anti-Stokes). The energy radiated per second per incident quantum is given by

$$W = \frac{4(\omega \pm \omega_{ab})^4}{3c^3} |R_{ab}|^2 \qquad (10.8)$$

As in the case of Rayleigh scattering, the matrix element R_{ab} for Raman scattering appears to diverge when $\omega = \omega_{jb}$ or $\omega = \omega_{ja}$ (eqn 10.7). When damping is taken into account, this divergence disappears. Nevertheless, the magnitude of Raman-scattered radiation does increase enormously for incident frequencies close to ω_{jb}, ω_{ja}, an effect known as *resonance* Raman scattering.

Since the intensities of the Raman-scattered lines depend on the configuration of the scattering molecules, Raman spectroscopy is an invaluable tool for the elucidation of molecular structure. Such information can also be deduced from infrared absorption spectroscopy. Both forms of spectroscopy have associated with them selection rules,

which determine which of all possible lines actually appear. The selection rules applicable to Raman spectra differ from those for infrared spectroscopy so that the two techniques effectively complement one another. For example, in molecules such as methane, where the carbon atom is surrounded by hydrogens at the corners of a regular tetrahedron, transitions involving the mode in which the hydrogen atoms move outwards and inwards together ('breathing mode') do not occur in infrared spectra but do give rise to Raman lines.

10.8   Scattering in solids

In the foregoing sections we have been concerned with light scattering by atoms or molecules. Account has been taken of interactions between the radiation field and the atom or molecule: it has been assumed that there are no interactions *between* atoms or molecules.

In the solid state we deal with the opposite extreme. Atoms are locked in close proximity to their neighbours and the strong interactions that result lead to numerous collective effects which strongly influence the scattering of electromagnetic radiation. In addition to direct electromagnetic effects, such as have been discussed so far, there are effects resulting from the propagation of acoustic (longitudinal) waves. In fact, the observations of light scattered from solids have enabled a thorough and often highly detailed picture to emerge on the behaviour of assemblies of atoms in crystals.

If a solid crystal consisted of a perfectly regular array of (stationary) atoms, then it would exhibit *no* Rayleigh scattering. For a macroscopic sample, the coherent superposition of the very large numbers of scattered waves would be zero except in the forward direction. The effect of forward scattering is simply to give the solid a refractive index. In fact, Rayleigh scattering from 'perfect' crystals *is* observable. There is no contradiction. Atoms are not stationary, so that the local density fluctuates and the conditions that led to a complete cancellation of scattered waves do not exist. Such local density fluctuations are of thermal origin and are described as 'non-propagating'.

In contrast, the collective motions of the atoms in a crystal, described in terms of quantised lattice vibrations (phonons), also give rise to scattering. In the interaction between the light wave and the (propagating) phonon, momentum conservation requires that the frequency of the scattered radiation be different from that of the incident wave. This is termed Brillouin scattering and is discussed below in Section 10.9.

The discussion of Section 10.7 on Raman scattering applies equally to scattering from solids, except that instead of considering excitation between energy levels of an isolated molecule, we need to consider excitations in the solid that contribute to the susceptibility of the sample, but do not give rise to density fluctuations. Examples of such excitations are optical phonons, magnetic spin waves and transitions between electronic or magnetic energy levels. Where interaction with a single 'lattice quantum' is involved the process is termed 'first-order Raman scattering'. It is possible, however, for *two* lattice excitations to be involved, giving rise to harmonics and to sum and difference bands. This is termed 'second-order Raman scattering'.

The three forms of scattering so far described are characterised by somewhat different magnitudes of the frequency shifts involved, as shown in Table 10.1.

10.9   Brillouin scattering

The thermal motions of atoms in a crystal may be discussed in terms of superpositions of plane acoustic waves propagating throughout the crystal. Consider what would happen to an incident light wave passing through a crystal supporting a single progressive acoustic wave. This will create periodic variations of density and hence of refractive index, and so will turn the crystal into a diffraction grating—a 'two-dimensional' one in the sense that, instead of an array of parallel *lines*, we have a (moving) array of parallel *planes*. Diffraction at such a grating will occur only if Bragg's law is obeyed. If the acoustic wave were stationary (e.g. a standing wave) then the frequency of the light in the diffracted beam would be equal to that of the incident beam. However, for a progressive wave, which produces a *moving* grating, we need to allow for the Doppler effect. If the incident wave frequency is ω_i and its velocity $v_i (= c/n)$, where n is the refractive index, then the frequency at which wave crests arrive at a grating, moving as shown in Fig. 10.6, is given by

$$\omega_i' = \omega_i \left(1 + \frac{n v_s}{c} \sin \frac{\theta}{2} \right) \qquad (10.9)$$

where v_s is the acoustic wave velocity. Thus the grating believes that the incident light has a frequency shifted from ω_i by $(\omega_i v_s n \sin \theta/2)/c$. The moving grating has a velocity component $v_s \sin \theta/2$ in the direction of the diffracted beam, so that the frequency of this beam suffers a further shift of $(\omega_i v_s n$

Table 10.1

Type	Frequency shift $\Delta\omega$ (GHz)	Wave number shift $(cm^{-1})^a$
Rayleigh	0	0
Brillouin	20–200	0.1–1
Raman	10^3–10^6	5–5000

^{a} The unit cm^{-1} is frequently used by spectroscopists.

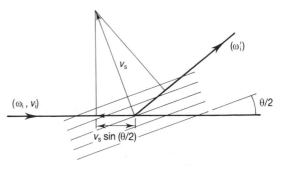

Figure 10.6   Brillouin scattering

$\sin\,\theta/2)/c$. We observe therefore scattered light with a frequency ω_i'' where

$$\omega_i'' = \omega_i \left(1 \pm \frac{2nv_s \sin\,\theta/2}{c}\right) \qquad (10.10)$$

(The negative sign would refer to an oppositely directed acoustic wave in Fig. 10.6. As in the case of Raman scattering, we describe the Brillouin-scattered light as Stokes and anti-Stokes.)

Typical sound velocities in solids are of the order 1000 m s^{-1}. From eqn (10.10) it is clear that the shift observed in Brillouin scattering is small, requiring very high optical resolution for study. The Fabry–Perot interferometer is usually employed. Figure 10.7 shows the experimental arrangement for observing Brillouin scattering, together with

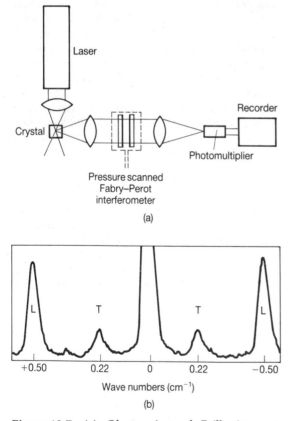

(a)

(b)

Figure 10.7 (a) Observation of Brillouin scattering with a pressure-scanned interferometer. (b) Resulting spectrum for triglycine sulphate. L and T respectively show the longitudinal and transverse modes. Reproduced by permission of H. Z. Cummins

a spectrum of the scattered light from such an experiment (see also Fig. 20.18).

Since the acoustic properties of a solid depend on its crystal structure, then in the event of a phase transition occurring when the temperature is changed, we should expect to see an abrupt change in the Brillouin scattering exhibited. This is indeed the case, and the study of Brillouin scattering provides an invaluable tool for the study of phase transitions.

10.10 Raman–Nath and Bragg scattering

The underlying basis of these forms of scattering is essentially that described in Section 10.9 under Brillouin scattering. Diffraction occurs at gratings produced by acoustic waves. Instead of the waves arising from the normal lattice vibrations in the material, they are produced artificially, with a transducer.

10.10.1 Raman–Nath scattering

For Raman–Nath scattering, the thickness of the scattering medium in the direction of the light beam is small. The acoustic wave produces modulations of density along the specimen and in effect produces a linear diffraction grating (Fig. 10.8). Diffraction then occurs into orders $\pm m$ in accordance with the condition

$$m\lambda_L = \lambda_s \sin\,\theta_m \qquad (10.11)$$

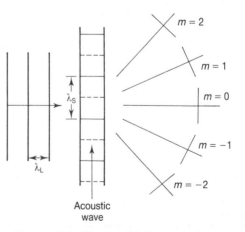

Figure 10.8   Raman–Nath scattering

where λ_s is the wavelength of the acoustic wave in the medium. The *amount* of light diffracted depends on the amplitude of the acoustic wave. Thus modulation of the acoustic wave amplitude produces a corresponding modulation in the diffracted intensity.

10.10.2 Bragg scattering

If we now envisage a scattering medium of large dimension in the incident beam direction, it would behave as a continuous superposition of 'line' gratings—in other words it would resemble the planar grating mentioned in Section 10.9. The diffraction condition is in this situation the Bragg condition, whereby for certain directions θ_B relative to the diffracting planes, constructive interference occurs and a large diffracted intensity results (Fig. 10.9). The path differences between successive wave-fronts is $2\lambda_s \sin \theta_B$ so that the diffraction condition is

$$2\lambda_s \sin \theta_B = m\lambda_L \qquad (10.12)$$

As discussed in Section 10.9, the Doppler effect on scattering from the moving grating produced by the acoustic wave means that the frequency of the diffracted light is shifted. For the $\pm m$th-order diffracted beams, the light frequency is $\omega_L \pm m\omega_s$. We thus have a direct method of producing frequency-modulated light beams, by varying the frequency of the impressed acoustic wave. The maximum modulation frequency attainable is determined by the time t_m the acoustic wave takes to travel a distance equal to the beam width W, namely W/v_s. For a 0.3 mm diameter (laser) beam with $v_s \sim 3000 \text{ m s}^{-1}$, $t_m = 10^{-7}$ s, frequency modulation up to ~ 10 MHz is indicated.

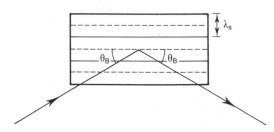

Figure 10.9 Bragg scattering

10.11 Scattering by free electrons

In the foregoing sections we saw that the amount of scattering from free atoms or molecules or from solids is a function of the frequency of the radiation. What happens if the radiation encounters a *free* electron, where there are no restoring forces and no resonant frequency? The electric field of the light wave will accelerate the electron and there will in consequence be a Lorentz force on the moving electron due to the magnetic field of the light wave. Unless the radiation intensity is so high as to give the electron a relativistic velocity, this interaction will be small. The magnitude of the magnetic force is of order v/c times that due to the corresponding electric force. In these circumstances the scattered power is independent of the frequency (Appendix 10.B).

Appendix 10.A Radiation from oscillating dipole

Application of an oscillating electric field to an atom causes a shift in the positions of the centres of gravity of the positive and negative charges, thus creating an oscillating dipole moment $p(t) = p_0 e^{i\omega t}$. Such an oscillating dipole radiates electromagnetic waves. We need to determine the form of the waves and the rate at which the dipole radiates energy. Our interest is in the field at distances large compared with the size of the atom; thus the charge separation of the dipole is regarded as infinitesimal, but the dipole moment is finite.

We need to take account of the finite velocity with which electromagnetic disturbances propagate: the state of affairs at a distance r from the dipole at time t depends on the condition of the dipole at time $t - r/c$. We write $[p]$ for the value of p at time $t - r/c$ and note that the vector potential A, defined by $B = \nabla \times A$, may be written in terms of p, and taking the time retardation into account, as

$$A(r, t) = -\frac{\mu_0}{4\pi r} \frac{\text{d}}{\text{d}t} [p] \qquad (10.13)$$

Since the current element with which A is

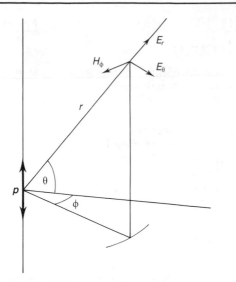

Figure 10.10 Polar coordinate system: non-vanishing components of E and H shown

associated corresponds to $\mathrm{d}p/\mathrm{d}t$ at the origin so that

$$H = \frac{B}{\mu_0} = \frac{\nabla \times A}{\mu_0} = -\nabla \times \left(\frac{[p']}{4\pi r}\right) \quad (10.14)$$

where

$$[p'] \equiv \frac{\mathrm{d}p(t - r/c)}{\mathrm{d}(t - r/c)}$$

Thus

$$H = -\left\{\frac{1}{r}\,\nabla \times [p'] + [p'] \times \nabla\,\left(\frac{1}{r}\right)\right\} \quad (10.15)$$

Using polar coordinates (Fig. 10.10), with p along $\theta = 0$, evaluation of H reveals that $H_r = H_\theta = 0$ and

$$H_\phi = -[p]\,\frac{\omega^2 \sin \theta}{4\pi rc}\,\left(1 - \frac{ic}{\omega r}\right) \quad (10.16)$$

The second term in (10.16) may be written as $i\lambda/2\pi r$, so that for distances $\gg \lambda$, it may be ignored, leaving

$$H_\phi = -\frac{\omega^2 \sin \theta}{4\pi rc}\,[p] \quad (10.17)$$

From Maxwell's equations for a current-free region,

$$\dot{E} = \frac{1}{4\pi\varepsilon_0}\,(\nabla \times H) \quad (10.18)$$

from which E_r, E_θ and E_ϕ may be found from

(10.18) to give

$$E_r = \frac{[p]\,\cos \theta}{4\pi\varepsilon_0 r^2}\,\left(\frac{1}{r} + \frac{i\omega}{c}\right)$$

$$E_\theta = \frac{[p]\,\sin \theta}{4\pi\varepsilon_0 r}\,\left(\frac{1}{r^2} + \frac{i\omega}{rc} - \frac{\omega^2}{c^2}\right)$$

$$E_\phi = 0 \quad (10.19)$$

For the case $r \gg \lambda$, the only significant term is

$$E_\theta = -\frac{[p]\omega^2 \sin \theta}{4\pi\varepsilon_0 rc^2} \quad (10.20)$$

At distances $\gg \lambda$ from the dipole, the electric and magnetic vectors are perpendicular to one another and to the direction of r. This indicates a power flow at a distance r in a direction away from the dipole and with magnitude (given by the Poynting vector, eqn 8.48)

$$S(r, \theta, t) = \frac{[p]^2\omega^4 \sin^2\theta}{(4\pi)^2\varepsilon_0 r^2 c^3} \quad (10.21)$$

and time-averaged value

$$\bar{S}(r, \theta) = \frac{p_0^2\omega^4 \sin^2\theta}{32\pi^2\varepsilon_0 r^2 c^3} \quad (10.22)$$

A polar diagram of the radiated power is shown in Fig. 10.4(b). There is no radiation along the dipole axis: the maximum value of $p_0^2\omega^4/32\pi^2\varepsilon_0 r^2 c^3$ occurs in the plane $\theta = \pi/2$.

The total energy radiated per second is obtained by integrating (10.22) over a sphere of radius r. Since the mean value of $\sin^2\theta$ over a sphere is $\frac{2}{3}$, the result for the total radiated power P is

$$P = \frac{p_0^2\omega^4}{12\pi\varepsilon_0 c^3} \quad (10.23)$$

Appendix 10.B Scattering by free electrons

If a free electron is in an oscillating electric field $E = E_0 \exp(i\omega t)$, then it is subjected to an electrostatic force eE. The resulting motion creates an oscillating current and therefore a magnetic field. Also the accelerating charge radiates energy, giving rise to radiation damping. If these two effects are small, the resulting equation of motion may be written as

$$mx'' = eE = eE_0\,\mathrm{e}^{i\omega t} \quad (10.24)$$

a solution of which is

$$x = -\frac{eE_0}{m\omega^2}\, e^{i\omega t} \qquad (10.25)$$

where the origins are chosen so that $x = 0$ at $t = 0$. The electron displaced from the origin by a distance x creates an electric moment ex, equivalent to a dipole moment. The total power radiated from the oscillating electron is thus (from 10.25)

$$P_e = \frac{e^4 E_0^2}{12\pi\varepsilon_0 m^2 c^3} \qquad (10.26)$$

The power flow in the electromagnetic wave (Poynting vector) is $\frac{1}{2} c\varepsilon_0 E_0^2 \equiv P_0$ and the total power scattered by one free electron in unit volume is

$$\frac{P_e}{P_0} = \frac{e^4}{6\pi\varepsilon_0^2 m^2 c^4} = \frac{8\pi}{3}\, r_0^2 \qquad (10.27)$$

where r_0, the classical electron radius, is given by

$$r_0 = \frac{e^2}{4\pi\varepsilon_0 m c^2} \qquad (10.28)$$

Note that subject to the conditions that magnetic effects and radiation resistance can be neglected, the fraction of the incident electromagnetic power scattered by a free electron is independent of frequency.

Problems

10.1 Consider the extent to which the assumption $P \propto E$ of Section 10.1 applies (a) to air at sea level and (b) to the atmosphere at an altitude of 100 km for visible light.

10.2 A monolayer of atoms of diameter 0.3 nm covers a surface 3 mm square. An argon laser beam ($\lambda = 488$ nm) of power 4 watts strikes the surface normally. If the Raman scattering cross-section is 2.4×10^{-32} m^2, calculate the number of Raman quanta produced per second.

10.3 Show that the frequency shift in Brillouin scattering given by eqn (10.10) is equal to the frequency ω_s of the acoustic wave.

11

Electro- and Magneto-optics

11.1 General considerations

Since atoms consist of positively charged nuclei and electrons, with associated spin and orbital magnetic moments, then collections of atoms will be affected by electric and magnetic fields: they can be polarised and magnetised.

Since light is an electromagnetic disturbance, then the associated electric and magnetic fields will interact with the charges in a collection of atoms. The velocities of light waves will decrease on entering a transparent medium from vacuum; the amplitude will decrease as the wave propagates in an absorbing medium.

The corollary to the above is that the behaviour of a light wave in a medium is expected to change if an external electric or magnetic field is applied to the medium. These effects—electro-optic and magneto-optic effects—are discussed in this chapter. Far from being merely academic manifestations of physical phenomena, they are of immense importance in the rapidly emerging technology of opto-electronics.

11.2 Light propagation in crystals

We saw in Chapter 7 that the propagation of light in certain crystalline materials could be accounted for by assuming that the wave surface of light radiating from a point in a birefringement medium consisted of two sheets, one spherical and the other ellipsoidal. A consequence of this can be seen in Fig. 11.1, showing the traces of the two wave surfaces when a light wave enters a crystal where the optic axis of the crystal is parallel to the surface and in the plane of incidence. For the ordinary ray (spherical wave surface), the directions of the ray and wave-normal are identical. In contrast these directions are *not* the same for the ellipsoidal surface. The extraordinary wave moves crab-wise—it looks in one direction while moving in another. This applies to all directions in the crystal *except* along and perpendicular to the optic axis.

In fact the most general behaviour of crystals is somewhat more complicated. The two-sheeted wave surface is as shown in Fig 11.2 and consists of interpenetrating sheets, neither of which is spherical. Unlike the uniaxial crystal described above, it possesses *two* optic axes, along which the velocities of the two waves are identical. With the axes as

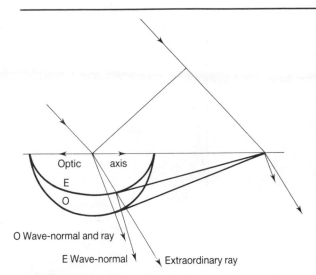

Figure 11.1 Shape of wave-fronts in plane of incidence for the case where the optic axis is parallel to the surface and the plane of incidence. E is the extraordinary and O the ordinary wave surface

shown in Fig. 11.2, the wave-normals of the two sheets are parallel along the directions of the coordinate axes. Three distinct velocities (and hence three refractive indices) are required to specify the surface of Fig. 11.2. Such crystals are termed *biaxial*.

This behaviour is accounted for by electromagnetic theory provided that the susceptibility (Section 8.3), which is a scalar quantity

in the case of isotropic media, has the form of a tensor. The physics is as follows. In an isotropic medium, an electron is free to move in any direction equally easily from its equilibrium position. If an electric field E is applied, the displacement is in the direction of E and hence so also is the direction of the polarisation P. Since $\mathbf{P} = \varepsilon_0 \chi \mathbf{E}$, then χ is a scalar. In an anisotropic medium, the atomic configuration is such that the restoring forces on the displaced electron vary with direction and so the electron does *not* necessarily move in the direction of the applied force (Fig. 11.3). The resultant polarisation P is inclined to that of E; each component of P depends, with different proportionality constants, on each component of E, which we can write as

$$P_x = \varepsilon_0 (\chi_{11} E_x + \chi_{12} E_y + \chi_{13} E_z)$$
$$P_y = \varepsilon_0 (\chi_{21} E_x + \chi_{22} E_y + \chi_{23} E_z) \qquad (11.1)$$
$$P_z = \varepsilon_0 (\chi_{31} E_x + \chi_{32} E_y + \chi_{33} E_z)$$

or

$$\mathbf{P} = \varepsilon_0 \chi \mathbf{E} \qquad (11.2)$$

where

$$\chi = \begin{pmatrix} \chi_{11} & \chi_{12} & \chi_{13} \\ \chi_{21} & \chi_{22} & \chi_{23} \\ \chi_{31} & \chi_{32} & \chi_{33} \end{pmatrix} \qquad (11.3)$$

is the susceptibility tensor. The relation between the susceptibility and permittivity tensor elements is conventionally written as

$$\varepsilon_{ij} = \varepsilon_0 (\delta_{ij} + \chi_{ij}) \qquad (11.4)$$

so that

$$\varepsilon = \begin{pmatrix} 1 + \chi_{11} & \chi_{12} & \chi_{13} \\ \chi_{21} & 1 + \chi_{22} & \chi_{23} \\ \chi_{31} & \chi_{32} & 1 + \chi_{33} \end{pmatrix} \qquad (11.5)$$

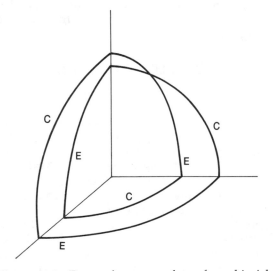

Figure 11.2 Form of wave surfaces for a biaxial crystal

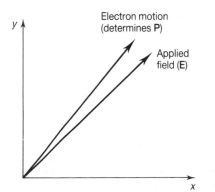

Figure 11.3 *P* and *E* for an anisotropic medium

For crystals in which the birefringence arises from only electric, and not from magnetic, effects, axes may be chosen so that the tensor of (11.3) reduces to the diagonal form

$$\chi = \begin{pmatrix} \chi_{11} & 0 & 0 \\ 0 & \chi_{22} & 0 \\ 0 & 0 & \chi_{33} \end{pmatrix} \qquad (11.6)$$

and χ_{11}, χ_{22} and χ_{33} are the *principal susceptibilities* with which are associated the corresponding principal permittivities $\varepsilon_{jj} = 1 + \chi_{jj}$, with $j = 1, 2, 3$. Provided the permittivities are those corresponding to the light frequencies (Section 8.9), then the three refractive indices referred to above are $n_j = \varepsilon_{jj}^{1/2} = (1 + \chi_{jj})^{1/2}$.

11.3   The wave surface

It should be possible to establish this with the help of Maxwell and Chapter 8. In a dielectric there will be no conduction currents ($\mathbf{J} = 0$) and eqn (8.22) becomes

$$\nabla \times H = \dot{D} = \varepsilon_0 \dot{E} + \varepsilon_0 \chi \dot{E} \qquad (11.7)$$

whence, following the procedure of Section 8.7,

$$\nabla \times (\nabla \times E) = -\mu_0 \varepsilon_0 \frac{\partial^2 E}{\partial t^2} - \mu_0 \varepsilon_0 \chi \frac{\partial^2 E}{\partial t^2} \quad (11.8)$$

This is the familiar wave equation, with plane-wave solutions of the form $\mathbf{E} = \mathbf{E}_0 \exp i(\omega t - \mathbf{k} \cdot \mathbf{r})$. Using (11.4) and noting that $n_1 = (1 + \chi_{11})^{1/2}$, etc., the three components of eqn (11.6) may be written as

$$\left(\frac{n_1^2 \omega^2}{c^2} - k_y^2 - k_z^2 \right) E_x + k_x k_y E_y + k_x k_z E_z = 0$$

$$k_y k_x E_x + \left(\frac{n_2^2 \omega^2}{c^2} - k_x^2 - k_z^2 \right) E_y + k_y k_z E_z = 0 \quad (11.9)$$

$$k_z k_x E_x + K_z K_y E_y + \left(\frac{n_3^2 \omega^2}{c^2} - k_x^2 - k_y^2 \right) E_z = 0$$

The condition that the above equations are consistent is that the determinant of the coefficients vanish, leading to an equation connecting the components of $\mathbf{k}$. Since by definition the wave surface is that for which $\mathbf{k} \cdot \mathbf{r} = $ constant, then the equation for the components of $\mathbf{k}$ depicts the form of the wave surface that is as shown in Fig. 11.2.

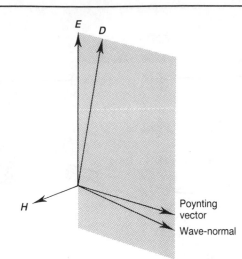

Figure 11.4   Directions of ray, wave-normal, E, H and D

The intersection between the wave surface given by eqn (11.9) and the three coordinate planes is easily shown to consist of a circle and an ellipse. When two of the three values n_1, n_2 and n_3 are equal, the surface reduces to that of a sphere and a spheroid, characteristic of a uniaxial medium. The two surfaces touch in the direction of the optic axis (Fig. 7.5).

The directions of rays and wave-normals in anisotropic media are related to the directions of the vectors E, D and H in the electromagnetic field. The ray is the direction of the Poynting vector $E \times H$ (Section 8.13). The direction of the wave-normal is that of $D \times H$ (Fig. 11.4).

11.4   Electro-optic effects

The tensor description of materials discussed above gives a good account of their behaviour under conditions where any probing radiation has negligible influence on the mean positions of the charges in the system. This applies to light beams of 'ordinary' intensity and where no electric or magnetic fields are applied. Such fields can change the charge distributions in such a way as to produce new optical characteristics. Thus an applied electric field can turn an initially isotropic material into a birefringent, uniaxial one, with the optic axis in the field direction (Kerr electro-optical effect). On the other hand, the field may change an initially uniaxial material into a biaxial one, or change the three refractive indices of an initially biaxial system. In the case of exposure of a material to very intense light beams, the linear response of the system to weak beams turns into a non-linear response, as discussed in Chapter 22.

11.4.1   Linear electro-optic response

Where the optical characteristics of a material are changed anisotropically by the application of an electric field, the changes depend linearly on the applied field: this is the Pockels effect. In this case the system responds to the combined effect of the electric vector of the light wave E_l and the applied electric field E_a. The resultant polarisation P depends on the combined effects of E_l and E_a. Since each component of P will in general depend on the nine products of the components of E_l and E_a, a third-rank tensor of 27 components would appear to be needed, of the form χ_{ijk}, with $i, j, k = 1$ to 3. In fact, $\chi_{ijk} = \chi_{ikj}$, which reduces the number to 18. This number would be needed to describe the behaviour of crystals of the lowest symmetry, viz. triclinic crystals. In other cases the crystal symmetry reduces the number of susceptibility elements needed. Thus the properties of potassium dihydrogen phosphate (KDP) may be represented by a tensor $\mathbf{r}$, related to the susceptibility tensor, where

$$\Delta\left(\frac{1}{n_r^2}\right)_i = \sum_{j=1}^{3} r_{ij}E_j \qquad (11.10)$$

where E_j are the components of the applied field, $\Delta(1/n_r^2)_i$ is the change in $1/n^2$ for the direction i and

$$(\mathbf{r}) = \begin{bmatrix} 0 & 0 & 0 \\ 0 & 0 & 0 \\ 0 & 0 & 0 \\ r_{41} & 0 & 0 \\ 0 & r_{41} & 0 \\ 0 & 0 & r_{63} \end{bmatrix} \qquad (11.11)$$

The symmetry properties of the tensor enable two, rather than three, suffixes to be used, with the following key:

$$r_{1i} = r_{i11}; \quad r_{2i} = r_{i22}; \quad r_{3i} = r_{i33};$$
$$r_{4i} = r_{i23}; \quad r_{5i} = r_{i31}; \quad r_{6i} = r_{i12}$$

KDP is uniaxial in the absence of an applied electric field, with refractive indices n_o and n_e. On applying a field in the direction of the optic axis (z direction) the principal refractive indices of the resulting biaxial crystal become

$$n_x = n_o + \tfrac{1}{2}n_o^3 r_{63}E_z$$
$$n_y = n_o - \tfrac{1}{2}n_o^3 r_{63}E_z \qquad (11.12)$$
$$n_z = n_e$$

For a suitable crystal and an appropriate applied electric field it is possible to arrange that, as a result of the induced refractive index changes, plane-polarised components with their electric vectors in the x and y directions emerge with a phase difference of $\pi/2(\pi)$ so that the system behaves as a quarter (half-)wave plate (Section 7.8). In a system such as that shown in Fig. 11.5, the voltage applied to the (KDP) crystal C to produce a half-wave plate is $\approx 15\,\mathrm{kV}$. One problem with this type of system is that the voltage is applied to the faces through which the light passes. Very thin metal films may be used, although inevitably some light is lost through absorption. Films of semiconducting oxides such as SnO_2 may be used: these are sufficiently conducting to meet the electrical need and are at the same time transparent.

KDP is one of a number of scalenohedral crystals showing useful linear electro-optic behaviour. They are transparent over a useful wavelength range (~ 0.3–$1.5\,\mu\mathrm{m}$) and have values of r_{63} in the range -10 to $-25\,\mathrm{pm\,V^{-1}}$. Another important class of crystals showing linear electro-optic effects is that of the titanates and niobates, for which the relevant r-tensor elements are r_{13} and r_{33}. These are uniaxial and, if the applied field is

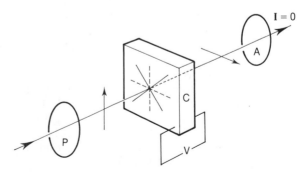

Figure 11.5 Arrangement for an electro-optic modulator

Table 11.1 Linear electro-optic materials (wavelength 600 nm)

Crystal	n_o	n_e	Extraordinary and ordinary coefficients (pm V^{-1})	
KDP	1.47	1.50	$r_{63} = -10.5$	$r_{41} = + 8.6$
KDDP[a]	1.47	1.51	$r_{63} = -24.0$	$r_{41} = + 8.8$
BaTiO$_3$	2.42	2.35	$r_{13} = + 8.0$	$r_{33} = +28.0$
LiNbO$_3$	2.29	2.20	$r_{13} = + 8.6$	$r_{33} = +31.0$

[a] Potassium dideuterium phosphate

Table 11.2 Kerr constants for a wavelength of 600 nm

Material	K at 20°C (m V^2)
Nitrobenzene	4.4×10^{-12}
Water	5.2×10^{-14}
Carbon disulphide	3.6×10^{-14}
Chloroform	-3.7×10^{-14}

along the optic axis, the birefringence changes (although the crystals remain uniaxial). The change in the birefringence is given by $\Delta n = \frac{1}{2}(n_e^3 r_{33} - n_o^3 r_{13})$. Typical values of refractive indices and electro-optic coefficients are shown in Table 11.1.

11.4.2 Quadratic electro-optic response

In an isotropic medium, e.g. a liquid, an amorphous solid or a cubic crystal, the direction of polarisation will lie in the direction of the applied electric field. The material will become uniaxial in whatever direction the field is applied and the elements of the permittivity tensor will depend on the square of the field strength, rather than linearly. The induced (uniaxial) birefringence $|n_3 - n_o|$ is thus proportional to E_a^2, where E_a is the applied field. The Kerr constant K is defined by the relation

$$n_e \text{-} n_0 = K\lambda E_a^2 \qquad (11.13)$$

and may be positive or negative. Typical values of K are given in Table 11.2.

If the electric field is applied in a direction perpendicular to that of the light beam (Fig. 11.6), the induced birefringence has its optic axis in a direction appropriate to creating waves with differing velocities in the direction of the beam. Thus a Kerr cell may be used to modulate the polarisation of an incident beam, as in the case of the Pockels effect. It has been used for a very long time as a light shutter. When placed between crossed polaroids, no light is transmitted in the absence of an applied field. If a voltage sufficient to create a half-wave plate is applied in a direction 45° to that of the electric vector of the incident beam, then the emerging beam passes the second polariser.

From the values given in Table 11.2 it is easily seen that for plates 10 mm apart and 50 mm long in a Kerr cell containing nitrobenzene, the potential difference required to produce a half-wave retardation for visible light is of the order of tens of kilovolts. Nitrobenzene has the highest known Kerr constant, so even higher voltages are required for other materials. For this reason, the Pockels cell is generally preferred as an optical modulator. There are possibilities for solid state Kerr modulators using ferroelectric crystals which, analogously with ferromagnetic materials, can display permanent electric moments. When operated close to their Curie temperature, such materials can display very large Kerr constants, such that optical modulation can be achieved with potential differences of a few tens of volts. It is, however, necessary in such cases to 'pole' the crystal with a (steady) biasing field of the order of a few kilovolts per centimetre.

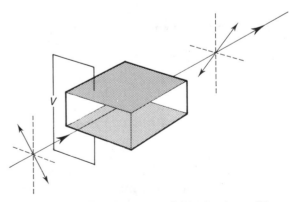

Figure 11.6 Transverse electro-optic (Kerr, Pockels) effect

11.5 Magneto-optic effects

11.5.1 Cotton–Mouton effect

This effect is the magnetic analogue of the Kerr electro-optical effect. When light traverses the medium in the presence of a transverse magnetic field B, the induced birefringence is

$$n_e - n_o = C\lambda B^2 \qquad (11.14)$$

(cf. eqn 11.13). Apart from the fact that such an effect would be expected, it has not proved to be of great importance or practical use compared, for example, with the electro-optic effects discussed above, or the effect to be described in the next section.

11.5.2 Faraday effect

If a magnetic field is established in the same direction as that of a beam of plane-polarised light, a rotation of the plane of polarisation from its original direction is sometimes observed. For a path l in a magnetic field H, the rotation θ may be written as

$$\theta = VH \cdot l \qquad (11.15)$$

where V is known as Verdet's constant. An idea of the magnitude of V for a selection of materials is given in Table 11.3. For a positive Verdet constant the direction of rotation is the same as that of a current that would produce the magnetic field.

An important feature of this phenomenon —the Faraday effect—is that the direction of rotation of the plane of polarisation is *independent of the direction of the light beam*. If the E-vector of the light beam is seen to rotate through $+45°$ as it travels away from

Table 11.3 Verdet constants (rad T^{-1} m^{-1})

Material	Wavelength	
	600 nm	10.6 μm
Soda-lime silicate glass	4.5	
Schott SFS-6 glass	29.5	
Dy^{3+} in Al silicate	−70	
Carbon tetrachloride	4.7	
Indium arsenide		865
Indium phosphide		1870
Oxygen (STP)	0.0017	

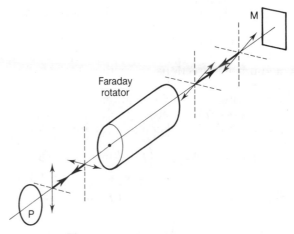

Figure 11.7 Faraday isolator

the observer, and if the light is then reflected back through the system, the light will arrive with its E-vector perpendicular to the initial direction. This is the basis of the *Faraday isolator* (Fig. 11.7)—essential in a high-power laser amplifying system such as that mentioned in section 20.19, for preventing scattered light from returning through the system and thereby causing damage.

The behaviour of a material displaying Faraday rotation is readily understood by adapting the results of dispersion theory (Section 9.6) to take account of the effects of the applied magnetic field. We use eqn (9.23), but with two reservations. Since we are generally interested in transparent materials, we put $\gamma = 0$, thus ignoring damping. We also note that for other than isotropic materials the Lorenz correction $P/3\varepsilon_0$ will not be strictly correct. Taking account of anisotropy by modifying this term greatly complicates the mathematics, but adds nothing to the essential physical ideas.

The effect of an applied magnetic field B on an electron at r will be a force given by $e(dr/dt) \times B$. Thus (9.23) is replaced by

$$m(\ddot{r} + \omega_0^2 r) = e\left(E + \frac{P}{3\varepsilon_0}\right) + e\left(\frac{dr}{dt}\right) \times B \quad (11.16)$$

where E is the electric field of the oscillating light wave. We ignore the interaction of the magnetic field component of the light wave: its magnitude is v/c times that of the electric field, where v is the electron velocity, and is generally negligible. The direction of the magnetic force is perpendicular to the direction of the electron motion and it will therefore contribute to the polarisation of the medium in directions other than that of the electric field. Noting that $P = \mathcal{N}er$, and considering sinusoidal solutions of the form $\exp i(\omega t - \mathbf{k} \cdot \mathbf{r})$, where ω is the angular frequency of the light beam, eq (11.16) becomes

$$m(\omega_0^2 - \omega^2)P = \mathcal{N}e^2\left(E + \frac{P}{3\varepsilon_0}\right) + i\omega eP \times B \tag{11.17}$$

The relation between P and E can be written as

$$P = \varepsilon_0 \chi E \tag{11.18}$$

and the components of the tensor χ may be obtained from the expressions for the three components of (11.17). The results are conveniently expressed in terms of ω_0' and ω_c, where

$$(\omega_0')^2 = \omega_0^2 - \frac{\mathcal{N}e^2}{3m\varepsilon_0} \tag{11.19a}$$

(cf. eq 9.32) and

$$\omega_c = \left(\frac{eB}{m}\right)^{1/2} \tag{11.19b}$$

the cyclotron frequency for the electron in the field B. The tensor χ then has the form

$$\chi = \begin{pmatrix} \chi_{11} & i\chi_{12} & 0 \\ -i\chi_{12} & \chi_{11} & 0 \\ 0 & 0 & \chi_{33} \end{pmatrix} \tag{11.20}$$

where

$$\chi_{11} = \frac{\mathcal{N}e^2}{m\varepsilon_0}\left\{\frac{(\omega_0')^2 - \omega^2}{[(\omega_0')^2 - \omega^2]^2 - \omega^2\omega_c^2}\right\} \tag{11.21}$$

$$\chi_{12} = \frac{\mathcal{N}e^2}{m\varepsilon_0}\left\{\frac{\omega\omega_c}{[(\omega_0')^2 - \omega^2]^2 - \omega^2\omega_c^2}\right\} \tag{11.22}$$

$$\chi_{33} = \frac{\mathcal{N}e^2}{m\varepsilon_0}\left[\frac{1}{(\omega_0')^2 -)\omega^2}\right] \tag{11.23}$$

From (11.20), the components of D for a wave with the electric vector $\perp Oz$ are

$$\begin{aligned} D_x &= \varepsilon_0(1 + \chi_{11})E_x + i\varepsilon_0\chi_{12}E_y \\ D_y &= -i\varepsilon_0\chi_{12}E_x + \varepsilon_0(1 + \chi_{11})E_y \end{aligned} \tag{11.24}$$

whence

$$\begin{aligned} D_x \pm iD_y &= \varepsilon_0(1 + \chi_{11} \mp \chi_{12})(E_x \pm iE_y) \\ &= n_\pm^2(E_x \pm iE_y) \end{aligned}$$

Thus circularly polarised waves travelling in the direction of the applied field have different refractive indices:

$$[\varepsilon_0(1 + \chi_{11} - \chi_{12})]^{1/2} \quad \text{and}$$

$$[\varepsilon_0(1 + \chi_{11} + \chi_{12})]^{1/2}$$

and so travel with different velocities. This is easily seen to correspond to a rotation of plane of polarisation, since a linearly polarised wave may be regarded as the superposition of two circularly polarised waves of opposite hands.

The above analysis involves some simplifications, without which the development of a small amount of elliptical polarisation is predicted as the light beam passes through the medium. This is in fact observed, but the effect is generally small; the description 'rotation of the plane of polarisation', implying the preservation of the state of *plane* polarisation, is sufficiently accurate to describe most cases of Faraday rotation.

11.5.3   The Kerr magneto-optic effect

This effect is confined to ferromagnetic materials and entails the change in the state of polarisation of light reflected from a surface, a change that depends on the orientation of the specimen magnetisation relative to the specimen surface and plane of incidence. The effect is conveniently discussed for three cases, namely polar, longitudinal and transverse, as illustrated in Fig. 11.8. In general, incident plane-polarised light is reflected elliptically polarised, in some cases even when the electric vector of the incident beam lies in or perpendicular to the plane of incidence. This can be readily understood in

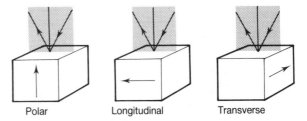

Polar   Longitudinal   Transverse

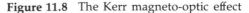

Figure 11.8 The Kerr magneto-optic effect

Figure 11.9   Magnetic domains as revealed by the Kerr magneto-optical effect

terms of the Lorenz force exerted on the electrons in the material on interaction with the internal magnetisation. The ellipticity of the reflected light and the angle between the direction of the incident electric vector and the major axis of the ellipse depend linearly on the magnetisation.

Cases are readily seen in Fig. 11.8 in which the effect is absent. Thus in the longitudinal case, the effect on the p component will vanish as the angle of incidence approaches zero, since under this condition there is no Lorenz force acting on the electron. Similarly the effect on the s component vanishes for the transverse case as normal incidence is approached.

The main application of the Kerr magneto-optical effect is for the study of magnetic domains. Where adjacent domains are magnetised in different directions, the states of polarisation of the reflected beams differ. Using the methods described in Chapter 7, these differences can be made visible. An

example of such a domain picture is shown in Fig. 11.9.

Problems

11.1 In Fig. 11.1, n_i is the refractive index of the incident medium and n_o, n_e are respectively the ordinary and extraordinary refractive indices of the uniaxial crystal. If the angle of incidence is θ_i, determine the angle of refraction of the extraordinary ray.

11.2 Verify eqns (11.9) for the plane wave $E = E_0 \exp \mathrm{i}(\omega t - \mathbf{k} \cdot \mathbf{r})$.

11.3 For the biaxial wave surface of Fig. 11.2, the velocities of the circular sections of wave-front in the $x = 0$, $y = 0$ and $z = 0$ planes are respectively c/n_x, c/n_y and c/n_z. Determine the angle between the optic axis and the z-axis.

11.4 Calculate   the   potential   difference

needed across a KDDP crystal (Table 11.1) 10 mm thick to produce a quarter-wave plate for light of wavelength 633 nm.

11.5 A miniature Kerr cell is made for an integrated circuit. The plate separation is 0.1 mm and the length is 10 mm. What potential difference must be applied to create a half-wave plate if the cell is filled with nitrobenzene (Table 11.2)?

11.6 A Faraday isolator is to be made from a glass slab 50 mm thick, the glass having a Verdet constant 520 minutes $T^{-1} cm^{-1}$. What field needs to be applied?

Interaction of Radiation and Matter

12.1 Limitation of classical ideas

In previous chapters we have generally described the way in which radiation interacts with matter by the use of terms such as refractive index, extinction coefficient, electro-optic constants, etc. In Chapter 9 we showed that an understanding of the dispersing and absorbing properties of matter could be obtained by considering the classical interaction of the electric field of the light wave with the electrons in an atom. Although this approach can account for some of the characteristics of radiation/matter interactions, there are some observations that stubbornly refuse to fit into such a picture. Thus when radiation of a suitable frequency falls on a metal surface, electrons are emitted. In principle this would appear to be explicable in classical terms. The light wave is electromagnetic in character. The associated electric field can therefore accelerate electrons. If they reach a sufficiently high velocity, they should be able to escape from the metal. (We know that a potential barrier exists at the surface—the electrons do not dribble out spontaneously—so there must be a minimum energy (the 'work function') in order to escape.)

However, the following experimental observations cannot be reconciled with this description:

1. No electrons are emitted unless the light frequency exceeds a critical value.
2. The maximum energy of the emitted electrons is independent of the intensity of the incident light. (Classically, large intensity means large electric field amplitude, larger force on electron and thus larger emitted energy.)
3. The maximum energy of the emitted electrons depends on the frequency of the incident light.

The results are summarised in Figs. 12.1 to 12.3.

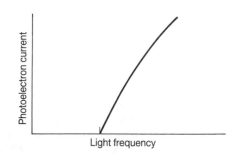

Figure 12.1 Dependence of photocurrent on light frequency

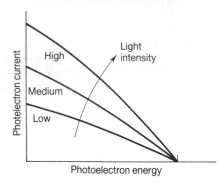

Figure 12.2 Relation between photocurrent and applied potential difference for different light frequencies

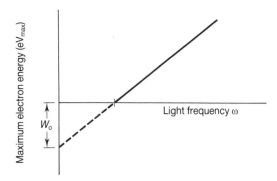

Figure 12.3 Maximum photoelectron energy as a function of light frequency

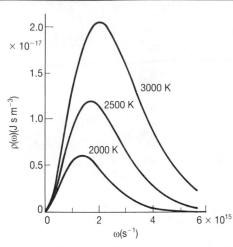

Figure 12.4 Planck radiation curves for three temperatures

12.2 The need for quanta

At the turn of the twentieth century, problems had arisen with the fact that classical ideas were unable to account for the form of the spectrum of radiation emitted by a heated body (Fig. 12.4). The classical prediction was that the radiation density $\rho(\omega)$—the energy per unit volume per unit frequency band—should increase without limit as ω increased. The difference between theory and experiment could hardly have been more dramatic. Far from increasing indefinitely the value of $\rho(\omega)$ actually *fell* with increasing ω.

Planck was able to resolve this difficulty by assuming that, when electromagnetic waves exchanged energy with matter, they did so in discrete amounts. Classically a wave can have any amplitude and so should be able to exchange energy in arbitrary, continuously variable amounts. Planck's hypothesis was

that the exchange took place in *quanta* of magnitude $h\nu$, where ν is the light frequency and $h = 6.63 \times 10^{-34}$ Js is Planck's constant. (Planck did *not* say that radiation 'consisted' of quanta: the suggestion was *not* a return to the Newtonian particle picture.)

By considering that the energy of the radiation field at a frequency ω could exist only in multiples of $h\nu = h\omega/2\pi = \hbar\omega$, and noting that in thermal equilibrium Boltzmann statistics apply, Planck deduced that the form of $\rho(\omega)$ should be

$$\rho(\omega) = \frac{\hbar\omega^3}{\pi^2 c^3} \frac{1}{\exp(\hbar\omega/kT) - 1} \quad (12.1)$$

This curve accurately follows that obtained for a 'black body' (conveniently a small hole in the wall of a furnace) if h is given the value 6.62×10^{-34} Js and $\hbar = 1.05 \times 10^{-34}$ Js.

12.3 The photoelectric effect

The idea of quanta leads to an immediate explanation of photoemission. If the incident radiation exchanges energy $\hbar\omega$ with an electron in a metal for which the work function is W_0 the energy of the emitted electron will be $\hbar\omega - W_0$. If $\hbar\omega < W_0$ no emission will occur. Thus the equation of the line in Fig. 12.3 is $eV_{max} = \hbar\omega - W_0$. Increasing the intensity of the radiation will increase the *number* of electrons emitted per unit time but *not* the energy of the electrons.

12.4 Photon momentum

If a monochromatic light beam of power P watts per unit area falls on a perfectly absorbing surface, it is found that the light exerts a pressure on the surface of magnitude P/c. This indicates that the light beam has momentum. If the beam power is P, then on absorption $n = P/\hbar\omega$ quanta per unit time are exchanged with the absorber. We may therefore associate a momentum with each photon amounting to $P/nc = \hbar\omega/c$. If a perfectly reflecting surface is used in place of the absorber, the observed pressure is $2P/c$. In this situation the momentum of the photons is reversed.

12.5 Characteristics of radiation sources

Before continuing our discussion of the interaction between radiation and matter we first consider the spectral character of the various radiation sources available. The types of spectrum obtained from most sources of radiation fall into one of two broad categories. The energy may be spread over a wide spectral range (continuous spectrum), as in the light from a heated body or a synchrotron (see Chapter 22), or it may be confined to narrow regions (line spectrum), which is the characteristic of the light from a discharge lamp. In the latter case the spectrum may be fairly simple, as in the case of hydrogen (which has a few well-spaced lines in and near the visible region), or may contain very large numbers of closely spaced lines, the characteristic of many molecules.

The spectrum of the emission from a black body is given by eqn (12.1) when $\rho(\omega)\,d\omega$ is the radiation energy per unit volume in a band of width $d\omega$ around the (angular) frequency ω.† This distribution is shown in Fig. 12.4 for three different temperatures.

In the case of line spectra, the situation is complicated. If it were possible (it is not) to hold the emitting gas atoms in fixed positions while they radiated, the observed lines

---

† Note that if $\nu = \omega/2\pi$ is used in place of ω, then since $\rho(\omega)\,d\omega = \rho(\nu)\,d\nu$ then $\rho(\nu) = 2\pi\rho(\omega)$.

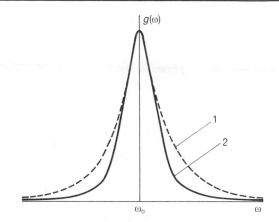

Figure 12.5 Spectrum line showing both Doppler and Lorenzian broadening

would have a Lorenzian profile of the form

$$g(\omega) = \frac{a}{(\omega - \omega_0)^2 + (\Delta\omega_N)^2} \qquad (12.2)$$

where ω_0 is the frequency at the line centre, $\Delta\omega_N$ is a measure of the line width and a is a constant. In practice the motion of the atoms imposes a further broadening of the line due to the Doppler effect (Fig. 12.5, curve 1). Away from the line centre, the shape of the line is dominated by the Lorenz term; close to the centre it is dominated by the Doppler term. The result is a profile such as is seen in Fig. 12.5, curve 2. Further complexities arise when the pressure of the gas is increased. The Lorenzian form of eqn (12.2) assumes that the atoms have sufficient time to radiate between successive collisions, a reasonable assumption at low pressures (say a few millipascals). At high pressures, the emission process is interrupted in mid-flight by collisions, resulting in a further contribution to the line width—collision broadening.

12.6 Power and line width

The power emitted by a gas discharge source increases if the current and pressure are increased, which also increases the gas temperature. From the above discussion it follows that the width of spectral lines increases with the output power of the source. With conventional line sources it is therefore difficult to obtain 'monochromatic' lines at high intensities: the action of increasing the power results in an increase of line width.

12.7 A digression into electronics

We recall one type of system in which increase in the power of a source is associated with a *narrowing* of the line width of the signal emitted. If positive feedback is applied to an amplifier, the system oscillates and produces an output in a frequency band whose width varies inversely as the output power. If we can imitate such a system in a radiation source, than a powerful, highly monochromatic output should result. The immediate requirement is for an amplifier that is capable of operating at the frequency of visible (or near-visible) light. This is much higher than the limit for any conventional amplifier based on electronics-style components. Amplification at optical frequencies is in fact possible as a result of the emission of radiation from excited atoms which is produced when they are themselves exposed to radiation of the same frequency, a mechanism first proposed by Einstein in 1917.

12.8 Radiation/matter interaction

Suppose we have a collection of atoms which possess energy levels at E_1 and E_2, in a radiation field for which the radiation density (energy per unit volume per unit frequency interval) is $\rho(\omega)$. We know that atoms in the lower energy state can absorb energy from the radiation field, thereby occupying energy level E_2. We know also that excited atoms decay spontaneously, emitting radiation of frequency $\omega_{12} = (E_2 - E_1)/\hbar$. We need to allow for the interaction between the radiation field and the atoms *while they are in the excited state*. Einstein postulated that this interaction caused the atoms to return to the lower state, emitting radiation in the process; this was termed *stimulated emission*.

With simple and plausible assumptions, we may do the book-keeping for the rates of upward and downward transitions, and for the level populations. The transition rate

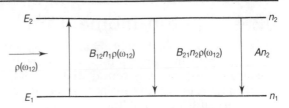

Figure 12.6 Absorption, stimulated emission and spontaneous emission

produced by the radiation for atoms in a given level will depend (1) on the radiation density and (2) on the population of the level. However, the rate of spontaneous emission from the upper level will depend *only* on n_2. We suppose simple proportionality and introduce Einstein coefficients B_{12}, B_{21} and A, defined below. The number per unit volume of atoms in state E_i is n_i ($i = 1,2$) and the equations refer to unit volume of material (see Fig 12.6):

Number of absorptions per second
$$= B_{12} n_1 \rho(\omega_{12})$$

Number of stimulated emissions
$$\text{per second} = B_{21} n_2 \rho(\omega_{12})$$

Number of spontaneous emissions
$$\text{per second} = A n_2 \quad (12.3)$$

For equilibrium, we must have

$$B_{12} n_1 \rho(\omega_{12}) = B_{21} n_2 \rho(\omega_{12}) + A n_2 \quad (12.4)$$

If the radiation and atoms are in thermal equilibrium (e.g. as in a closed cavity at a constant temperature), then n_2 and n_1 are related by Boltzmann's equation:

$$n_2 = n_1 \exp\left(\frac{-\hbar\omega_{12}}{kT}\right) \quad (12.5)$$

From (12.4) and (12.5) we see that

$$\rho(\omega_{12}) = \frac{A_2}{B_{12} \exp(\hbar\omega_{12}/kT) - B_{21}} \quad (12.6)$$

We know, however, that for thermal equilibrium, $\rho(\omega)$ is given by the Planck radiation law:

$$\rho(\omega) = \frac{\hbar\omega^3}{\pi^2 c^3} \frac{1}{\exp(\hbar\omega/kT) - 1} \quad (12.7)$$

valid for all ω and hence for $\omega = \omega_{12}$. This indicates therefore that $B_{12} = B_{21} = B$ and that

$$\frac{A}{B} = \frac{\hbar\omega^3}{\pi^2 c^3} \quad (12.8)$$

12.9 Relative rates of stimulated and spontaneous emission

Before considering how stimulated emission can be used to produce amplification, let us estimate how the numbers of $E_2 \to E_1$ transitions are divided between stimulated and spontaneous emission processes.

From (12.3) we have, noting that $B_{21} = B$ and that we are considering thermal equilibrium,

$$R =$$

$$\frac{\text{number of stimulated emissions per second}}{\text{number of spontaneous emissions per second}}$$

$$= \frac{B\rho(\omega)}{A} \qquad (12.9)$$

$$= \frac{1}{\exp(\hbar\omega/kT) - 1} \qquad (12.10)$$

Values of R for different spectral regions and temperatures are given in Table 12.1.

It is immediately clear that at any easily accessible temperature and under thermal equilibrium conditions, stimulated emissions are extremely rare for visible frequencies, although abundant at radio-frequencies. Thus it is not surprising that the first device, the ammonia maser, to use the amplification that stimulated emission makes possible operated in the microwave region (24.9 GHz). How can the stimulated emission be enhanced for the visible region? The crucial difference between stimulated and spontaneous emissions is that stimulated emissions increase with radiation density whereas spontaneous emissions do not. We shall see that, by the use of a resonant cavity, high values of $\rho(\omega)$ can be produced. Firstly,

however, we need to examine what is needed in order to ensure that amplification can occur.

Returning to eqn (12.3), we note that since $B_{12} = B_{21} = B$, the number of stimulated emissions per second will exceed the number of absorptions per second if $n_2 > n_1$. Since under thermal equilibrium conditions n_2 is necessarily *less* than n_1 (eqn 12.5), this condition is referred to as a *population inversion*. When this condition is satisfied, then the radiation density in the system increases, i.e. the radiation is amplified. The source of the energy required to ensure amplification is that which is fed into the system to create the population inversion.

We need to look at the question of the direction of the radiation emitted by the excited atoms. The spontaneously emitted radiation has no preferred direction: spontaneous radiation from a large collection of excited atoms is emitted with equal intensity in all directions. Radiation emitted by stimulation by an incident beam emerges in the direction of the stimulating beam.

12.10 Amplifying material in a resonant cavity

Suppose a collection of atoms in which a population inversion has been created is enclosed between parallel plane mirrors, one of which is slightly transparent (Fig. 12.7). Some of the excited atoms decay spontaneously so that radiation with frequency $(E_2 - E_1)/\hbar$ will be present, travelling in all directions. Since there is an inverted population, amplification will occur but (1) it will be very small (for visible light) because the

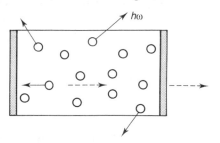

Figure 12.7 Excited atoms in a cavity. With population inversion, significant amplification may occur along the axis but not in other directions

Table 12.1

Region	$\omega(\text{s}^{-1})$	Stimulated/spontaneous emission rate	
		$T = 300$ K	$T = 3000$ K
Visible	3.8×10^{15}	1.09×10^{-42}	6.3×10^{-5}
Near infrared	3.8×10^{14}	6.3×10^{-5}	0.61
Far infrared	3.8×10^{12}	0.91	103
Microwave	3.8×10^{10}	1.03×10^3	1.03×10^4

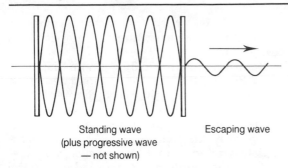

Figure 12.8   Standing waves in a cavity

majority of the atoms decay spontaneously and (2) most of the radiation will escape from the open sides of the cavity. The crucial exception is the radiation that travels along the direction normal to the cavity mirrors. For simplicity, assume that the mirrors are spaced by a distance equal to an integral number of half-wavelengths of the radiation in the cavity. Then some of the axially directed radiation will be able to form a standing-wave pattern in the cavity (Fig. 12.8). On its to-and-fro transits, the radiation will be amplified, so the amplitude of the standing-wave pattern will increase. The larger the amplitude of the standing-wave pattern, the larger is the value of $\rho(\omega)$ *for that direction*. The larger the value of $\rho(\omega)$, the higher is the rate of stimulated emission and hence the amplification, which increases the standing-wave field, which increases $\rho(\omega)$ which .... What determines the limit? The higher the rate of stimulated emission, the more rapidly are atoms removed from the upper (E_2) energy level. When this level is emptied at a rate equalling that at which it can be filled, the process saturates.

Since one mirror is slightly transparent, some flux is emitted through this. In the steady state, a steady standing-wave field is established in the cavity together with a progressive wave. Since the high value of $\rho(\omega)$ and hence the amplification occur only for the direction along the system axis, the emitted radiation is confined to a very narrow, unidirectional beam.

12.11   Establishing a population inversion

Among possible methods of producing an excess population in the upper of two states are:

1. By excitation through the absorption of radiation
2. By electron collisional excitation and
3. By the production of excited molecules through a chemical reaction

Note that it is not possible to produce an inversion simply by pumping atoms from one level to another by absorption of radiation. The net number of absorptions per second is given by $B\rho(\omega)(n_1 - n_2)$: as n_2 increases, $n_1 - n_2 \rightarrow 0$. As the populations approach equality, no further absorption occurs. The material becomes transparent.

12.12   Laser systems

12.12.1   Doped-crystal systems

Crystals containing traces of certain elements, viz. transition elements, lanthanides or actinides, show broad absorption bands in some spectral regions and sharp fluorescent emission at longer wavelengths. Thus the energy spectrum of ruby (Cr^{3+} in Al_2O_3) is as shown in Fig. 12.9. When exposed to intense blue/green radiation, excitation to the broad bands G and B occurs. From here, the chromium atoms relax into the state F before returning to the ground state GS with the emission of radiation at a wavelength of 694 nm. If the pumping intensity is sufficiently high, the population in

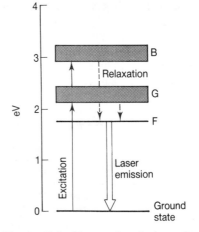

Figure 12.9   Energy levels for ruby

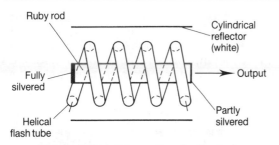

Figure 12.10 Schematic of ruby laser

level F exceeds that left behind in the ground state, and a population inversion results. Ruby formed the working material for the world's first laser, built by Maiman in 1960. A schematic of the arrangement is shown in Fig. 12.10. The parallel, plane ends of the ruby cylinder were coated with silver layers, one slightly transparent, and the crystal was exposed to intense radiation from a helical gas-discharge tube, activated by the discharge from a capacitor. A brief, intense, highly collimated flash of red ($\lambda = 694$ nm) light emerged from one end of the crystal. The key feature of the ruby is that the time for which the atoms are in the levels B and G is very short ($\sim 10^{-7}$ s) whereas that for level F is relatively long (10^{-3} s). Thus it is possible to pump atoms from the ground state to F (*via* B, G) sufficiently fast to create a population inversion. Of the isotropically emitted spontaneous radiation as the atoms return to the ground state, some travels along the cylinder axis, is amplified repeatedly on successive traverses and contributes to the emission through the transmitting end mirror, as discussed in Section 12.10.

In other types of doped crystal, laser action occurs for a transition between a level such as F and a lower level which lies above the ground state. Such systems can be more efficient and may operate continuously. The ruby laser is restricted to producing pulses: attempts at continuous operation would generally lead to a molten crystal.†

12.12.2 Gas lasers

If a current is passed through a gas, the effect

† By the use of a small-diameter ruby crystal and massive cooling, CW operation by a ruby laser has in fact been achieved—a veritable *tour de force*

of electron collisions is to produce a large concentration of excited atoms, albeit in a rather unselective fashion. In many cases, the excited atoms relax to lower states with the emission of radiation. However, states exist ('metastable states') from which relaxation by radiation is not possible. Such metastables wander around uncomfortably until an opportunity arises of getting rid of the excess energy. One possibility is for the metastables to collide with the walls of the container. An alternative is to collide with another type of atom which can readily absorb just the amount of energy that is available. In this case an excited atom of the second species is produced in one particular energy state. Thus we have a process of *selectively* exciting to a given state.

This forms the basis of the operation of the helium–neon laser. The helium atom possesses metastable levels with energies of 19.9 and 20.7 eV above the ground state. Neon possesses groups of excited states with energies close to these values (Fig. 12.11), from which radiative transitions to lower states exist. Collisions between helium metastables and ground state neon atoms result in an exchange of energy. The helium

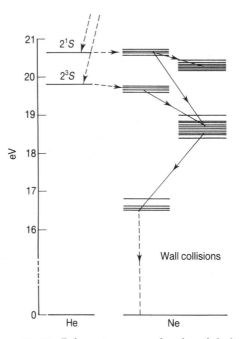

Figure 12.11 Relevant energy levels of helium and neon

atoms return to the ground state and the neon atoms are excited to the nearby levels. The coincidence in energy levels does not have to be very precise: a *small* difference (~0.01 eV) can be compensated from the kinetic energy of the colliding atoms. Note that there is no process involved which dumps a large amount of kinetic energy into the system, as is the case with the ruby laser. Thus the helium–neon laser works quite happily in a continuous mode and is not restricted to pulsed operation.

The helium–neon laser was the first of what is now a vast number of gas lasers, producing wavelengths from the far ultraviolet to the submillimetre region of the spectrum. These and other types of laser will be explored more fully in Chapter 20.

12.13   Characteristics of laser output

Two features distinguish laser light from that from any other type of source. These are (1) directionality and (2) coherence.

The directionality of the output from the laser is a consequence of the need for a resonant cavity to enhance the value of $\rho(\omega)$, the radiation density, in order to enhance the stimulated emission rate compared with the spontaneous rate. For a cavity with plane-parallel mirrors, reinforcement to produce an intense standing-wave pattern can occur only for the direction normal to the mirrors. Thus the laser output is confined to this direction. It will be seen in Chapter 20 that with a somewhat different cavity which employs spherical mirrors (the confocal cavity) not only is the output highly directional but the diameter is also confined. The narrow beam produced in the common helium–neon laser is generally from a confocal system.

The high degree of coherence of the laser output is a consequence of the nature of the stimulated emission process. Care is needed in the discussion of this point. In considering individual atoms and photons it is tempting to ascribe to an individual photon the property of phase appropriate to a classical field. The statement that 'the phase of the stimulated photon is the same as that of the incident photon' has no meaning. We may ascribe a phase ϕ to the field associated with a number N of photons. The uncertainty principle imposes the following limitation on the uncertainties $\Delta\phi$ in ϕ and ΔN in N, namely:

$$\Delta N \, \Delta\phi \approx 1 \qquad (12.11)$$

If we assert that there are *exactly two* photons, then $\Delta N = 0$ and the phase ϕ is completely uncertain.

Consider a collection of atoms, with inverted population, in a cavity in which a standing wave is sustained. Radiation emitted by stimulated emissions in the axial direction of the system will contribute to the standing-wave field only if the phase of the resultant field has the appropriate value in relation to that of the stimulating field. The standing wave will build up only if the resultant phase of the waves emitted by the collection of excited atoms is the same as that of the radiation causing the stimulated emission.

The output of the laser will lie at the frequency for which the system of atoms and cavity display an exact resonance. The width of the laser line will therefore be extremely small; the associated wave trains (Chapter 5) will therefore be very long. The phase of the output at a given point in space will therefore be highly correlated with that at a much later time. The radiation will display high *temporal coherence*.

From the geometry of a plane-parallel cavity resonator, the phase of the radiation in all points over the plane perpendicular to the axis will be the same. Thus the output from the system will display a large *spatial coherence*.

Problems

12.1   Determine from eqn (12.1) (either graphically or by the Newton–Raphson method) the frequency at which the maximum of $\rho(\omega)$ occurs for a temperature of 6000 K. Convert this into a wavelength. Think of the Sun—and worry.

12.2   The threshold frequency for the emission of electrons from beryllium is 318 nm. Calculate the work function.

12.3 What will be the maximum energy of photoelectrons emitted from caesium (work function 1.9 eV) by radiation of wavelength 256 nm?

12.4 For what wavelength will the rates of stimulated and spontaneous emission by equal for thermal equilibrium at 1000 K?

12.5 How many nodes form in the standing-wave pattern in a helium–neon laser 200 mm long ($\lambda = 633$ nm).

Holography

13.1 Background

When one looks at a three-dimensional object illuminated by monochromatic light, the eye receives a complicated wave-front which is the resultant of the waves scattered from all parts of the object within lines of sight from the eye. For white light illumination, there will be a superposition of such wave-fronts corresponding to the different wavelengths present. If one could record the information contained in the wave-front (necessitating the storage of both *amplitude* and *phase* information) in such a way that the original wave-front entering the eye could be re-created, then we should possess the means of producing an undistorted, three-dimensional image of the original object.

Can we not do this with a lens? We saw in Chapter 3 that a lens forms an image of a planar object in a conjugate plane. It should therefore be capable of forming a three-dimensional image of a three-dimensional object. We did not mention *phase* in the discussion of the behaviour of a lens: we merely noted that light rays travel in straight lines. Why bring phase into the argument?

Consider a converging lens in air. It produces a transverse magnification M_T for a given object and image distances equal to l_2/l_1. Distances *along* the system axis will also be subjected to magnification. From eqn (3.17), and noting that $f_2 = -f_1$, we see that the *longitudinal* magnification dl_2/dl_1 is given by l_2^2/l_1^2. Thus the transverse and longitudinal magnifications are different so that undistorted reconstruction of an object is not possible.

When light is recorded on a photographic plate, the blackening is a function of the total energy falling on the plate, which is proportional to the square of the amplitude of the electric field strength. If the field at a given point is represented by $E \exp(i\phi)$, the blackening depends on E^2: the plate records *no phase information*. Since the information about the depth of an object is contained in the phase differences between waves scattered from different points, then failure to record phase information means that three-dimensional reconstruction cannot be achieved.

How can differences of phase be converted into differences of irradiance, which the photographic plate *can* record? *Provided* we have a light source with a sufficiently high degree of spatial and temporal coherence (Chapter 5) then we can combine the wave scattered by the object with a second ('reference') wave. This results in sets of interference fringes that may be recorded directly on a photographic plate. We shall see that if such a fringe pattern ('hologram') is recorded, then on illuminating the plate with a wave identical with the reference wave, a wave-front is created that is identical with that originally produced by the object. If this wave enters the eye, the viewer will not know whether he or she is looking at the original object or at the reconstructed wave-front. The process—known as holography—is sometimes referred to as 'wave-front reconstruction'. It should be noted that the form of the reference wave is immaterial *provided* that the same wave is used for reconstruction as was employed in forming the hologram.

13.2   Gabor's proposal

If a monochromatic plane wave is incident normally on a photographic plate and a minute glass bead is placed in front of the plate, a spherical wave is radiated from the bead (Fig. 13.1). This wave interferes with the plane wave and produces, on the photographic plate, a series of concentric fringes. The irradiance distribution will be similar to that of Newton's rings (Section 4.10). When the plate is developed the transmission will vary radially in the way indicated in Fig. 13.2. The radii at which the half-maximum points occur correspond to those of a zone plate (Section 6.17). In place of the clear and opaque zones of the zone plate, we have gradual transitions from minimum to maximum transmission. The normal zone plate possesses a focusing property, with an 'infinite' number of focal lengths, both positive and negative. Our 'graded' zone plate possesses just *two* foci, corresponding to the

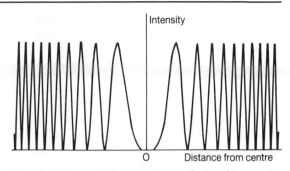

Figure 13.2 Radial variation of irradiance in zone-plate fringes

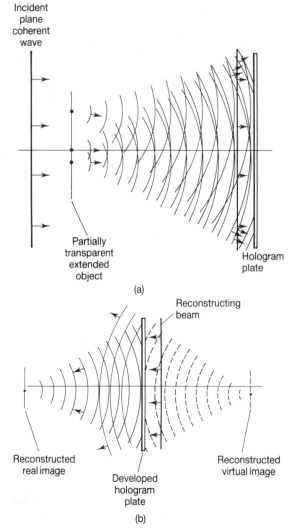

Figure 13.3 (a) Formation of a hologram of two-dimensional, transparent object. (b) Reconstruction. For clarity, only one point in the image is shown

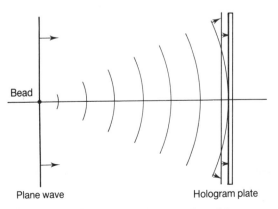

Figure 13.1 Formation of zone-plate fringes

largest positive and negative focal lengths of
the normal zone plate. Thus when the developed plate is illuminated normally with a
plane light wave of similar wavelength to
that used to make the hologram, a virtual
image of the original bead is formed at the
point where the bead was located. In
addition a real image is formed on the
opposite side of the plate from that where
the bead lies. If the bead is now replaced by
an extended object (Fig. 13.3), each point
may be thought of as a 'bead'. A complicated
superposition of 'zone-plate' fringes will be
formed. When the developed plate is
illuminated, each of the sets of rings will
reproduce a point of light and the emerging
wave will be identical with that from the
complete object. This was Gabor's original
proposal (introduced not in the context of
light waves but of electron waves, with the
aim of improving the performance of electron microscopes). The hologram so formed
is termed a Fresnel, or zone-plate, hologram.
A disadvantage of this arrangement is that
the waves from the real and virtual images
superpose and cause confusion.

13.3   Enter the laser

Gabor's proposal lay dormant for some four-
teen years. Before the arrival of the laser, the
only way of producing a light wave with a
reasonably high degree of spatial coherence
was to illuminate a minute pinhole. The
plane wave required for the arrangement of
Fig. 13.3 was produced with the help of a
lens. The irradiances at the object and on the
plate were pathetically small, so that long
exposures were necessary. Since a shift of
position of either object or plate of one half-
wavelength would convert a dark fringe into
a bright one, extreme rigidity of the whole
system was obligatory.

With the arrival of the laser, and consequent large fluxes of highly coherent light,
holography immediately became a highly
practical technique. In Gabor's experiment,
available sources had poor temporal
coherence so that imaging was restricted to
planar objects. The laser's high spatial and
temporal coherence immediately enabled
true three-dimensional reconstruction to be
achieved.

13.4   Side-band holograms

The confusion of the effects of superposition
of light waves from the real and virtual
images inherent in the arrangement of
Fig. 13.3(b) may be removed by the method
introduced by Leith and Upatnieks. The reference wave impinges on the hologram plate
at an angle α to the normal (Fig. 13.4a). From
a point on the object, a spherical wave will
combine with the plane reference wave to
create an off-centred zone plate. When the
developed plate is illuminated by a plane
wave at normal incidence (Fig. 13.4b), point
images are formed on lines making angles α

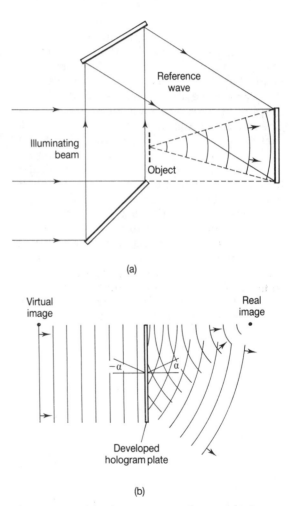

Figure 13.4 (a) Arrangement for producing a
side-band hologram. (b) Reconstruction. The
waves from the real and virtual images are separated at the position of the real image

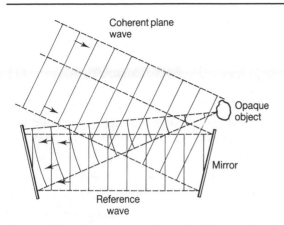

Figure 13.5 Arrangement for obtaining a hologram of an opaque object

and $-\alpha$ with the normal to the plate. Since the direction of the waves forming the real and virtual images is now different, the confusion inherent in the 'in-line' case is avoided.

Although thus far we have considered transparent objects viewed by transmission, holograms may just as easily be obtained of opaque objects. A suitable arrangement is shown in Fig. 13.5.

13.5 The reconstruction is truly three dimensional

Each small region of a hologram contains the interference pattern formed by the reference wave together with the object wave *as seen from that position*. Thus if the head is moved behind the hologram in the reconstruction process, the image seen from different points

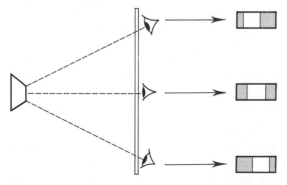

Figure 13.6 The hologram preserves perspective

is exactly what would be seen if the object itself were being viewed. Perspective is maintained (Fig. 13.6).

13.6 Analysis of the side-band hologram

In the arrangement of Fig. 13.4, the reference wave is a plane wave incident on the hologram plate at an angle of incidence α. (It should be reiterated that *any* form of wave will serve as a reference wave *provided* that the identical wave is used for the reconstruction.) The arguments of Section 13.4, whereby the imaging of a single point is described, plausibly lead to the idea of imaging an extended object. The following analysis confirms that such imaging will indeed occur.

Consider first the disturbance over a plane P in Fig. 13.7 from a wave scattered by an object O. The wave amplitude in this plane may be written†

$$E_0(x, y) = A_0(x, y)\cos\left[\omega t + \phi_0(x, y)\right] \quad (13.1)$$

where A_0 and ϕ_0 will be complicated functions of position in this plane. If a reproducing system can create a disturbance identical with that of eqn (13.1), then an eye placed to the right of the plane P will receive a wave

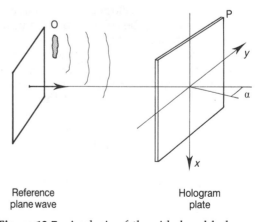

Figure 13.7 Analysis of the side-band hologram

† The analysis is approximate inasmuch as it treats the wave as a scalar quantity, whereas an electromagnetic wave entails vector fields. In general the angles between the representative vectors are sufficiently small for the approximation to be a good one.

identical with that from the original object —i.e. it will 'see' the object.

Suppose the reference wave to be a uniform plane wave whose wave-normal lies parallel to the $y = 0$ plane. The form of the field distribution on the plane P will then be

$$E_R(x, y) = A_R \cos[\omega t + (\varkappa \sin \alpha)x] \quad (13.2)$$

where α is the angle between the normal to the plane P and the wave-normal. If a photographic plate placed at P is exposed for a time T, the energy deposited per unit area on the plate at (x, y) is

$$W(x, y) = \int_0^T c\varepsilon_0 [(E_0(x, y) + E_R(x, y)]^2 \, dt$$

$$= \frac{c\varepsilon_0 T}{2} \{A_0^2 + A_R^2$$

$$+ 2A_0 A_R \cos[\phi_0(x, y) - (\varkappa \sin \alpha)x]\} \quad (13.3)$$

If we arrange that $A_R^2 \gg A_0^2$, this may be written as

$$W(x, y) =$$

$$\frac{c\varepsilon_0 T A_R^2}{2} \left\{ 1 + \frac{2A_0}{A_R} \cos[\phi_0(x, y) - (\varkappa \sin \alpha)x] \right\}$$

$$\quad (13.4)$$

$$= C\{1 + \eta(x, y)\} \quad (13.5)$$

When the plate is developed, it will have a density D which will be related to the amplitude transmittance τ by the Harter and Driffield (H&D) curve of the photographic plate (see Fig. 16.12). On the linear part of the curve, we have $D = -2 \log \tau = \text{constant} + \gamma W$. If τ_0 is the transmittance at the centre of the linear portion, then

$$\tau = \tau_0 \left(1 - \frac{\gamma C \eta}{2} \right)$$

where γ is the slope of the H&D curve. Thus

$$\tau(x, y) =$$

$$\tau_0 \left\{ 1 - \frac{\gamma C A_0(x, y)}{A_R} \cos[\phi_0(x, y) - (\varkappa \sin \alpha)x] \right\}$$

$$\quad (13.6)$$

When the hologram is illuminated by the reference beam, the amplitude transmittance

$T(x, y)$ is given by τE_R, i.e.

$$T(x, y) = \tau_0 A_R \cos[\omega t + (\varkappa \sin \alpha)x]$$

$$\times \left\{ 1 - \frac{\gamma C A_0(x, y)}{A_R} \cos\{\phi_0(x, y) - (\varkappa \sin \alpha)x\} \right\}$$

$$\quad (13.7)$$

The first term represents the transmitted reference beam. The second term may be written as

$$- \tau_0 \gamma C A_0(x, y)\{\cos[\omega t + \phi_0(x, y)]$$

$$+ \cos[\omega t + 2(\varkappa \sin \alpha)x - \phi_0(x, y)]\} \quad (13.8)$$

The first of these terms is simply a constant times $A_0(x, y)\cos[\omega t + \phi_0(x, y)]$, precisely the wave which originally arrived at the plane P from the object. The second term represents a wave travelling in a different direction from that of the reconstructed object wave.

Note that throughout the above analysis, the use of a plane reference wave in a direction α is manifest by the term $(\varkappa \sin \alpha)x$. For any other reference wave profile $\phi_R(x, y)$, the $(\varkappa \sin \alpha)x$ term is simply replaced by $\phi_R(x, y)$. The reconstructed wave-front corresponding to the object is still present, independently of the form of $\phi_R(x, y)$.

13.7 How good is the reconstruction?

As seen in Section 6.17, the successive rings in a zone plate get closer together as we go away from the centre (Fig. 13.8a). In the recording of the spherical wave from an object point to form a hologram, the number of rings recorded will depend on the resolution of the photographic plate, which is limited by the size of the silver grains produced on development. The record will therefore be a truncated zone plate, with the outermost rings missing. Such a truncated zone plate corresponds not to a *point* source but to one of finite diameter. This effect therefore limits the spatial resolution of the reconstructed image.

The above limitation can to some extent be offset by the use of a spherical, rather than a plane, reference wave, as shown in Fig. 13.8(b). For points on the object such that OH ≈ MH, fringes formed between the

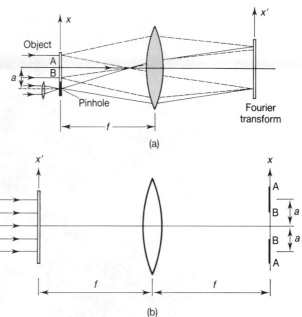

(a)

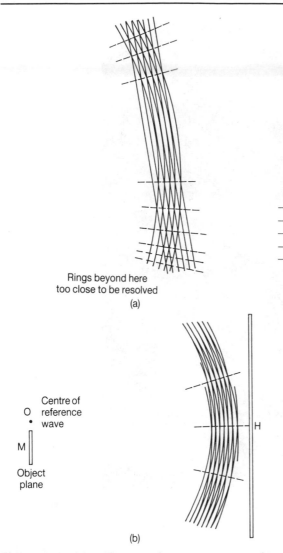

Rings beyond here
too close to be resolved
(a)

Figure 13.9 (a) Formation of a Fourier transform hologram. (b) Reconstruction

distribution of the object wave in this plane is the Fourier transform of the distribution of the wave leaving the object, yielding a Fourier transform hologram. (It should be noted that the hologram produced by the arrangement of Fig. 13.8(b) is effectively a Fourier transform hologram.)

With the arrangement of Fig. 13.9(a), the amplitude distribution in the object plane is given by $A_0(x, y)$; the amplitude distribution in the back focal plane of the lens is the Fourier transform $T(x', y')$ of the amplitude distribution in the object. The distribution on the hologram plane of the plane waves produced by the lens L of the light from the pinhole P is given by $\exp(-2\pi iax')$, where the paraxial approximation is implied and the amplitude of the object wave is normalised to that of the reference wave.

The irradiance on the hologram plane is thus

$$I(x', y') = C\,|\exp(-2\pi iax') + T(x', y')|^2$$
$$= C\{1 + |T(x', y')|^2$$
$$+ T(x', y')\exp(2\pi iax')$$
$$+ T^*(x', y')\exp(-2\pi iax')\} \quad (13.9)$$

where $C = c\varepsilon_0/2$.

Figure 13.8 (a) Plane reference wave. (b) Spherical reference wave

object and reference waves will be widely spaced and hence less subject to the limited resolving power of the photographic emulsion. Clearly this condition can be met only by a near-planar object.

13.8 Fourier transform holograms

With the arrangement of Fig. 13.9, a hologram is formed by a plane reference wave and the object wave distribution in the back focal plane of the convex lens. The amplitude

Figure 13.10 Interference hologram of a helicopter. Reproduced by permission of R. K. Erf, United Technologies Corporation

By a similar argument to that used in Section 13.6, we see that the amplitude transmission $\tau(x', y')$ of the developed hologram plate is given by

$$\tau(x', y') = \tau_0 \left\{ 1 - \frac{\gamma C}{2} [\, | \, T(x', y') \, |^2 \right.$$
$$+ T(x', y') \exp(2\pi i a x')$$
$$\left. + T^*(x', y') \exp(-2\pi i a x')] \right\} \quad (13.10)$$

If the developed hologram plate is placed adjacent to a converging lens, as shown in Fig. 13.9(b), the amplitude distribution in the back focal plane will be the inverse Fourier transform of the distribution given by (13.10). From the shift theorem† of Fourier transforms, we see that the transforms corresponding to the third and fourth terms of (13.10) will be proportional to $A_0(x - a, y)$ and $A_0\{-(x+a), -y\}$. These correspond to two reconstructions of the original wavefield. $A_0(x - a, y)$ is identical to the object field but is displaced a distance a in the x direction. The other image is inverted and displaced a distance a in the $-x$ direction.

The Fourier transform hologram suffers a difficulty that does not arise in the Fresnel ('zone-plate') hologram. The hologram plate receives both the direct, zero-diffraction

---

† If $g(x')$ is the Fourier transform of $f(x)$, then the Fourier transform of $f(x + a)$ is $e^{2\pi i x' a} g(x')$.

order, light from the object in addition to that arising from the spatial detail. This results in a very large range of irradiance, which may exceed that over which the photographic plate has a linear response.

13.9 Interference holography

With the development of high-power lasers, it has become possible to record the hologram of an object in an extremely short time—perhaps as little as a fraction of a nanosecond. It has also become possible to persuade a laser to produce two pulses separated by a short time interval. Examine what happens if a hologram is made from such a two-pulse laser of an object that is vibrating. From the two superposed holograms, two images may be reconstructed in the same region of space by two sets of *coherent* waves. If the object has changed shape in the interval between pulses, then the reconstructions will not be identical, and the waves forming them will interfere with one another. If a point on the object has moved one half-wavelength (or an odd multiple of a half-wavelength) the interference will be destructive. Thus the reconstruction will be seen to be crossed by bright and dark fringes, so giving an instantaneous picture of the way in which the object is vibrating. A striking example, showing that this technique is not an idle laboratory curiosity, is shown in Fig. 13.10 displaying the vibrations of the panels of a helicopter during a rivetting operation. The potential of this technique is enormous. Thus the vibration modes of an aero-engine turbine blade may be examined while the engine is running: the reader is invited to devise any other method for examining, with interferometric precision, the state of vibration of a 1-metre long blade rotating at 2000 revolutions per minute!

13.10 Holographic diffraction gratings

If a hologram is made of a plane wave, with the reference wave inclined at a small angle, then the intensity distribution will consist of an array of parallel fringes with a $\cos^2$ intensity distribution. The spacing between the fringes is given by λ/α, where λ is the wavelength and α the angle between the wavefronts. If α is appropriately adjusted the separation can be made corresponding to the rulings on a diffraction grating (Section 6.13). In fact the hologram will act as a diffraction grating with a $\cos^2$ transmission characteristic rather than that of clear and opaque strips. If the silver grains in the hologram are chemically bleached, the hologram becomes completely transparent, but with an alternating optical thickness across the fringes—i.e. it will be a *phase* grating.

Such a grating is of great interest inasmuch as it directs a very large fraction of incident light into first-order spectra. This method of producing gratings has many obvious advantages over that involving ruling with a diamond point—no wear problems and a process so rapid that long-term thermal stability of the ruling equipment is not needed.

Problems

13.1   If a maximum in intensity occurs at the centre of the hologram plate, what will be the radius of the first bright ring if $\lambda = 633$ nm and the bead-plate distance is 100 mm?

13.2   In an arrangement for producing holograms, why is *either* extreme mechanical rigidity *or* a very short exposure time essential?

13.3   A hologram is made using helium–neon laser light ($\lambda = 633$ nm). In making the reconstruction, an argon laser ($\lambda = 488$ nm) is used. Describe the reconstruction in relation to the original object.

14

Waveguides, Fibres and Optical Communications

14.1 Introduction

Many methods of communication entail the passage of electromagnetic waves along structures whose cross-section perpendicular to the direction of propagation remains constant. The electrical engineer has long made use of metallic waveguides. More recently considerable interest has arisen in the transmission of electromagnetic waves along guides made entirely of dielectric materials. All have in common the feature that, for propagation with very small loss the electromagnetic field distribution can have only certain well-defined forms, leading to propagation only in certain well-defined *modes*.

14.2 The metallic waveguide

We can illustrate mode propagation in a one-dimensional waveguide consisting of two perfectly conducting large (infinite) plates (Fig. 14.1). Can a wave propagate in the z direction? If so, what are its characteristics? We investigate a possible solution to the wave equation of the form $E_y = E_{y0}$ exp $i(\omega t - \varkappa z)$, noting that for an infinitely

wide guide, $\partial/\partial y = 0$. We also ask whether a transverse wave solution ($E_z = 0$) is possible.

From equation (8.29a), and with the above constraints, we obtain

$$\frac{\partial^2 E_y}{\partial x^2} = (\beta^2 - \omega^2 \mu_0 \varepsilon_0) E_y = \gamma^2 E_y \quad (14.1)$$

where we assume vacuum between the plates. (If not, then we replace $\mu_0 \varepsilon_0$ by $\mu \varepsilon$). The solution of (14.1) is of the form

$$E_y = A \cos \gamma x + B \sin \gamma x \quad (14.2)$$

The required boundary conditions are that the field should vanish at $x = 0$ and $x = a$, since perfectly conducting plates have been assumed. This dictates that $A = 0$ and that $B \sin \gamma a = 0$ or $\gamma = m\pi/a$. Thus the form of the

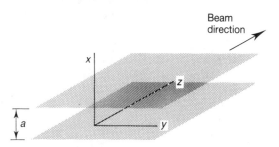

Figure 14.1 Propagation between infinite conducting planes

field distribution is

$$E_y = E_{y0} \sin\left(\frac{m\pi x}{a}\right) \qquad (14.3)$$

with m an integer. From eqn (8.24c) it is readily seen that associated with the y component of E is an x component of H, given by

$$H_x = \left(\frac{i\gamma}{\omega\mu_0}\right) E_{y0} \sin\left(\frac{m\pi x}{a}\right) \qquad (14.4)$$

Modes in such a one-dimensional guide are characterised by a mode number m. In the case of a guide bounded in two directions—e.g. a guide of rectangular cross-section—*two* mode numbers are needed. In this case, the field distributions would contain terms of the form $\sin(m_1\pi x/a)$ $\sin(m_2\pi y/b)$, where a and b are the guide dimensions in the x and y directions and m_1, m_2 are integers.

In addition to modes such as those described above, where the direction of the E-vector is perpendicular to the direction of propagation ('TE modes'), a second set of modes exists where the H-vector is perpendicular to the propagation direction ('TM—transverse magnetic—modes').

14.3 The dielectric waveguide

When the guide is bounded by dielectrics, the appropriate boundary conditions are simply those of continuity of the tangential components of the electric and magnetic field strengths, as used in Chapter 8 for calculations of reflection and transmission coefficients. In this case there will be penetration into the adjacent media. For a sinusoidal oscillation in the guide interior we may have either sinusoidal or exponentially decaying fields in the adjacent media. When the fields decay exponentially in all the regions surrounding the guide, then we have a guided mode in which waves will propagate without loss (unless the medium of the guide is itself absorptive). For this situation to arise, the refractive index of the material of the guide must be larger than those of the adjacent media.

We consider a one-dimensional dielectric waveguide consisting of a uniform layer of

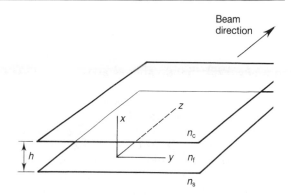

Figure 14.2 Propagation in an infinite dielectric slab

material of refractive index n_f on a substrate of index n_s. The refractive index of the medium above the layer is n_c. For a plane wave propagating in the xz plane (Fig. 14.2) at an angle θ to the normal in the film, the spatial variation will be given by $\exp\left[-ikn_f(\pm x \cos\theta + z \sin\theta)\right]$. Thus for a solution of the form $\exp i(\omega t - kz)$ we have $\beta = kn_f \sin\theta$.

The condition that multiply-reflected plane waves travel without loss along the guide in the z direction is that the phase change on one 'round-trip' should be a multiple of 2π. There are phase changes δ_c and δ_s on reflection at the guide boundaries (Section 8.20). The values of δ_c, δ_s depend on the direction of the electric vector in relation to the plane of incidence. We thus have, for a mode to

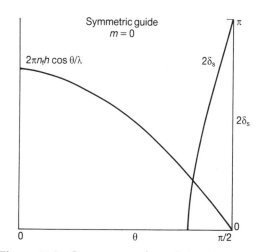

Figure 14.3 Occurrence of $m = 0$ in symmetrical guide

occur,

$$2n_f hk \cos \theta - \delta_c - \delta_s = 2m\pi \qquad (14.5)$$

where h is the guide thickness and m is an integer. Will a mode always arise, regardless of thickness? Generally not, although in the special case of $n_c = n_s$ ('symmetric' guide) the answer is 'yes'. From (14.5) the $m = 0$ mode will occur if $2n_f hk \cos \theta = 2\delta_s$. Figure 14.3 shows curves of $2n_f hk \cos \theta$ and $2\delta_s$, plotted as functions of θ. They will always intersect.

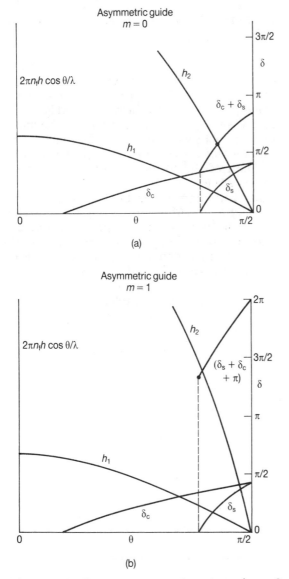

(a)

(b)

Figure 14.4 Occurrence or otherwise of $m = 0$ mode in unsymmetrical guide

In contrast, note what happens if the refractive indices of the bounding media are different. In this case, for the $m = 0$ mode, $2n_f hk \cos \theta$ must equal $\delta_c + \delta_s$, plotted in Fig. 14.4(a). Curves of $2n_f hk \cos \theta$ are plotted for two values of h. For h_1 there is no intersection and hence no solution. For the larger value h_2, an intersection does occur and an $m = 0$ mode can propagate. There is a minimum value of h/λ for a mode to be possible. A similar situation arises for the $m = 1$ mode (Fig. 14.4b).

14.4 Field strengths in the guide and the adjacent media

14.4.1 Guided TE modes

There will be three equations of the form of (14.1), with $\mu_0 \varepsilon_0$ replaced by the values of $\mu\varepsilon$ appropriate to the three media. For optical frequencies, $\mu = \mu_0$ and we may generally write $\varepsilon = n^2$. Thus $\omega^2 \mu\varepsilon = n^2 k^2$, where $k = 2\pi/\lambda_v$, with λ_v the wavelength in vacuo.

For the medium of incidence,

$$\frac{\partial^2 E_y}{\partial x^2} = (\beta^2 - n_c^2 k^2) E_y = \gamma_c^2 E_y \qquad (14.6)$$

Since E_y decays exponentially with distance into this medium, γ_c is real. In the guide,

$$\frac{\partial^2 E_y}{\partial x^2} = (\beta^2 - n_f^2 k^2) E_y = -\eta_f^2 E_y \qquad (14.7)$$

Here the solution is sinusoidal and η_f is real. In the substrate,

$$\frac{\partial^2 E_y}{\partial x^2} = (\beta^2 - n_s^2 k^2) E_y = \gamma_s^2 E_y \qquad (14.8)$$

with γ_s real for exponential decay of E into the substrate.

The solutions to Eqns (14.6) to (14.8) have the forms

Region $n_c : E_y = A \exp(-\gamma_c x)$,

$$x \leqslant h$$

Region $n_f : E_y = B \cos \eta_f x + C \sin \eta_f x$,

$$0 < x < h$$

Region $n_s : E_y = E \exp(\gamma_s x)$,

$$x \leqslant 0$$

$$(14.9)$$

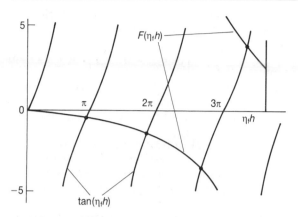

Figure 14.5 Occurrence of TE modes in asymmetrical guide

The tangential components of the electric and magnetic field vectors must be continuous at $x = 0$ and $x = h$. Since the x component of H is proportional to $\partial E_y / \partial x$, then the boundary conditions may be expressed as continuity of E_y and $\partial E_y / \partial x$. The boundary conditions are satisfied only if

$$\tan (\eta_f h) = \frac{\eta_f (\gamma_c + \gamma_s)}{\eta_f^2 - \gamma_c \gamma_s} \qquad (14.10)$$

The right-hand side of (14.10) is a function F of the variables n_f, n_c, n_s, k and h. The relation will be satisfied for certain values of $\eta_f h$, and hence of θ, since $\eta_f^2 = n_f^2 k^2 - \beta^2 = n_f^2 k^2 - 4n_f^2 h^2 k^2 \cos^2 \theta$. Figure 14.5 illustrates a plot of $\tan(\eta_f h)$ and of F against $\eta_f h$. The function F is real only up to a limiting value of $\eta_f h$, so the number of possible modes is finite. They are indicated on Fig. 14.5.

14.4.2   Guided TM modes

Essentially similar arguments apply to TM modes as for TE modes. For TM modes, eqn (14.10) is replaced by

$$\tan(\eta_f h) = \frac{n_f^2 \eta_f^2 (n_c^2 \gamma_s + n_s^2 \gamma_c)}{n_c^2 n_s^2 \eta_f^2 - n_f^4 \gamma_c \gamma_s} \qquad (14.11)$$

and the possible modes may be explored using a plot similar to that of Fig. 14.5.

14.5   Dielectric waveguides in integrated optical systems

In integrated optical devices it is necessary to direct optical signals to different regions on the surface of a crystal, to beam-splitters, modulators, demodulators and other processing devices. This is generally done by creating dielectric channels on the surface of the crystal surface. The analysis of the propagating modes in such two-dimensional channels follows the procedure discussed in Section 14.4.

14.6   Propagation along optical fibres

For long-distance transmission of optical signals, small-diameter glass or quartz fibres are used, generally in the form of a core (Fig. 14.6) which is surrounded by a cladding of slightly lower refractive index; this is termed a 'step-index' fibre. Alternatively, the fibre may have a refractive index that varies radially with distance from the axis.

In the case of the step-index fibre, certain of its characteristics can be understood in terms of a ray picture, although this approach does not bring out the mode characteristics. Consider a cone of radiation falling on the end of the core of a fibre (Fig. 14.7a). For up to a limiting angle of incidence, rays within the fibre will suffer total reflection at the core/cladding boundary and will be multiply-reflected throughout the length of the fibre, emerging finally at the opposite end. However, the total optical

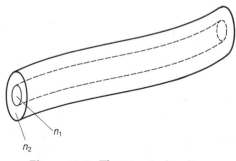

Figure 14.6 The step-index fibre

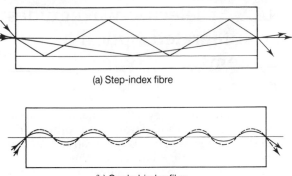

Figure 14.7 (a) Dispersion in the step-index fibre. (b) Ray paths in the graded-index fibre

path of rays making a large angle with the axis will be larger than those only slightly inclined. Thus if a short-duration pulse of light is passed into a fibre, such as are used in a digital transmission system, the pulse emerging will be broadened. Thus the repetition rate at which the fibre can handle trains of pulses will be limited. In the case of a graded-index fibre (Fig. 14.7b), although the *geometrical* path of the more steeply inclined rays is longer than that of the axial rays, the longer path is through material of lower refractive index and the velocity is therefore higher. By the use of a judicious refractive index profile, the dispersing effect characteristic of the step index can be eliminated.

The cases considered above are for rays in planes containing the fibre axis. For rays in other planes ('skew rays'), the detailed analysis is very cumbersome but the essential ideas remain valid.

14.7   Modes in cylindrical fibres

For this problem, the appropriate coordinate system is that of cylindrical polars, (r, ϕ, z). In place of eqn (14.1), we have, for a field component $\mathscr{F}$ (E or H):

$$\frac{\partial^2 \mathscr{F}}{\partial r^2} + \frac{1}{r}\frac{\partial \mathscr{F}}{\partial r} + \frac{1}{r^2}\frac{\partial^2 \mathscr{F}}{\partial \phi^2} + (n_1^2 k^2 - \beta_1^2)\mathscr{F} = 0 \quad (14.12)$$

where n_1 is the core index and the variation of field with z is as $\exp(i\beta_1 z)$. For the cladding region, n_1 is replaced by n_2. Solutions exist of the form $\mathscr{F} = F(r) \cos \nu\phi$. The variables are separable and the equation for F becomes

$$\frac{d^2 F}{dr^2} + \frac{1}{r}\frac{dF}{dr} + \left[(n_1^2 k^2 - \beta_1^2) - \frac{\nu^2}{r^2}\right]F = 0 \quad (14.13)$$

immediately recognisable as Bessel's equation, whose relevant solutions (the ones that remain finite at $r = 0$) are Bessel functions of order ν.

The general solution of this problem is algebraically cumbersome. For systems of practical interest, $n_1 \doteqdot n_2 = n$. For a guided mode, $n_2 k < \beta < n_1 k$, so we may put $\beta \doteqdot nk$ without serious error. Under these conditions the guided modes have no z component either of E or H: the waves are transverse. The forms of the field components inside the core are

$$E_x = E_{x0} \, J_\nu \, (\eta r) \begin{pmatrix} \cos \nu\phi \\ \sin \nu\phi \end{pmatrix}$$

$$H_y = nE_{x0} \frac{\beta_\nu}{|\beta_\nu|}\left(\frac{\varepsilon_0}{\mu_0}\right)^{1/2} J_\nu(\eta r) \begin{pmatrix} \cos \nu\phi \\ \sin \nu\phi \end{pmatrix} \quad (14.14)$$

where $\eta^2 = n_1^2 k^2 - \beta_\nu^2$ and β_ν is the propagation constant for the νth mode. The fields in the cladding decay exponentially away from the interface.

From (14.14) and from the corresponding expressions for the fields in the cladding region, the value of the Poynting vector can be calculated and integrated over the core and cladding regions to determine the fraction of the power carried in the core.

14.8   Optical fibres in communications

We can envisage two ways in which we might use light as a means of communication. We could follow the example of radio communication and impress the information on a *carrier wave* or we could send the signals in digital form in the way that the navy uses Morse code. In the first case the rate at which information can be sent depends on the carrier frequency: the higher the carrier frequency the greater the possible rate of information transmission. For digital transmission the requirement for fast transmission is that the pulses should be of a very short duration and be sent with a very high repetition rate.

In addition we need a medium for transmission that is transparent to the signals radiated.

Communication by radio waves relies on the fact that sources of *coherent* radio waves have long been available, with frequencies up to the 10^{12} Hz region. Although light waves (in the visible region) have frequencies some three orders of magnitude higher than this, light sources of pre-laser days were *incoherent*. The rapid random phase fluctuations of an incoherent source used as a carrier wave constitute noise, so that transmission by modulating the carrier wave, as is done in the radio region, is not possible. In addition the pre-laser light source was of extremely low radiance compared with the laser. If a pulse is to be detected, a certain minimum number of photons is necessary. If the pulse is to be of a very short duration, then a very high radiance is required and these were not available.

With the development of very intense *coherent* sources at optical frequencies, the possibility of high-rate optical communication became possible, either by modulation of the light wave in the way of radio transmission or by using the fact that pulses of very short duration and a very high repetition rate could be produced.

The question of the transmission medium is an important one. Whereas radio waves are for the most part totally unaffected by rain or fog, this is emphatically not the case with light waves. A communications system that broke down at the onset of inclement weather might be found to be somewhat inconvenient. The optical fibre, described in Sections 14.6 and 14.7, forms a satisfactory transmission medium provided the glass is of sufficiently high purity to ensure low attenuation.

For long-distance transmission to be practicable, fibre losses need to be kept below ~ 1 dB km^{-1}, for which the transmission of a 10 km length of fibre would be 1%. Since an impurity level of parts in 10^9 of certain elements (copper, nickel, iron) produces an attenuation of 1 dB km^{-1}, the need for very high purity is paramount.

In Section 14.6, the effects of pulse spreading was mentioned, arising partly from the variation of optical paths. In fact

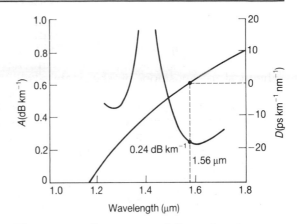

Figure 14.8 Absorption and dispersion characteristics of an optical fibre

the dispersion of the refractive index of the material of the fibre also plays a part. Although the light source is *almost* monochromatic, the line width is finite and hence dispersion in the fibre is important. Ideally one needs a fibre that produces no pulse spreading (measured in picoseconds per kilometre per nanometre) and has minimum absorption loss. Figure 14.8 indicates what can be achieved. If, however, the radiation intensity in the fibre is sufficiently high (if necessary by pumping power in along its length) then propagation of undistorted pulses ('solitons') is feasible (see Section 15.12).

The communication needs a source of radiation at the appropriate wavelength (to match the minimum absorption loss of the fibre) which can be modulated at a sufficiently high frequency to produce pulse

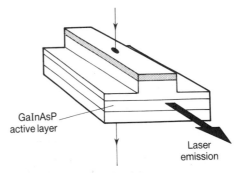

Figure 14.9 Laser source for optical communication—the 'ridge wave guide' laser

trains capable of transmitting information at $\sim 10^9$ bits per second. The semiconductor laser (Chapter 20) is such a device. The emission wavelength can be tuned by varying the chemical composition of the materials of which the laser is made. The construction of a typical laser source is shown in Fig. 14.9. At the receiving end of an optical fibre communications system, a suitable detector is needed. Silicon p–n (or PIN) diodes are satisfactory for this purpose.

14.9 Optical signal processing

In addition to transmitting signals along an optical fibre, a communication system must provide suitable means for modulating, demodulating and generally steering the signals through processing 'circuitry' at each end of the communications channel. There will generally be many sets of signals using the same fibre channel and means for separating these are needed. Modulation can be effected by using one of the electro- or magneto-optical effects discussed in Chapter 11. In situations where it is necessary to couple optical radiation from one channel to another, the properties of strip waveguides, discussed in Section 14.5, are used. With the configuration shown in Fig. 14.10, the exponentially decaying field from guide 1 has a significant field strength at the boundary of the adjacent guide 2. Continuity at the boundary means that a sinusoidal mode will be generated in the second guide. By adjustment of the dimensions, any desired transfer of power from one guide to the next may be

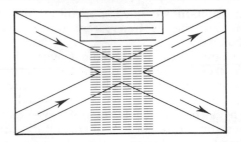

Figure 14.11 Use of diffraction at an acoustic wave to switch channels. Light from each of the left-hand channels can be switched into either right-hand channel

effected. For switching between channels, an arrangement such as that of Fig 14.11 may be used. Normally light entering the two guide channels passes directly across the central region. When an acoustic surface wave is generated it creates a sinusoidal diffraction grating in the intersecting region and this diffracts each of the incident beams into the adjacent guide.

14.10 Fibre-optic sensors

Optical fibres can be used in measuring systems in two ways. Where an optical transducer is used to effect a measurement, the fibres may simply carry the optical signals to a suitable measuring or processing location. In this case, the fibre acts simply as a connecting channel and the method is invaluable if, for example, the measuring equipment is at a high electrical potential, or in an otherwise hostile environment. The second possibility is where the optical properties of the fibre itself are affected by the physical conditions being studied. Thus the arrangement of Fig. 14.12 enables the current flowing in a high-voltage cable to be measured. Plane-polarised radiation is sent along the fibre. In the neighbourhood of the power cable, the magnetic field produced by the current causes a rotation of the direction of the plane of the electric vector (Faraday effect; see Chapter 11). Measurement of the extent of the rotation yields the value of the current flowing. Since glass fibres are excellent insulators, the actual measuring instrument operates at ground potential.

Fibre sensors used in conjunction with interferometric techniques are capable of

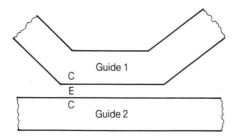

Figure 14.10 Coupling between adjacent regions of strip guides. In the regions marked C, continuous wave solutions exist. In region E the form is exponential

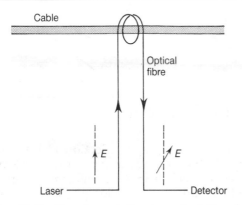

Figure 14.12 Faraday rotation in optical fibre for current measurement in high-voltage cables

achieving extremely high sensitivities in the measurement of any physical parameter that can affect fibre propagation—for example temperature, pressure and mechanical strain. In an interferometer such as the Mach–Zehnder (Section 17.6), it is possible to detect phase differences, for beams traversing the two alternative paths, of the order of 10^{-8} radian. For a typical optical fibre 1 metre long a temperature change of $1°C$ produces a phase difference of 10^{-2} radian. A pressure change of 1 bar produces a phase change of 0.1 radian. A mechanical strain of 10^{-6} yields a phase change of 10 radians. With the indicated sensitivity of phase change measurement, optical fibres thus constitute extremely powerful tools for precision measurement of the quantities indicated.

A further example of the application of optical fibres is that of the ring laser gyroscope, referred to in Section 20.20.

14.11   Integrated optics

In the foregoing sections we have laid the foundations of an optical communications system of enormous potential. We have eliminated the effects of atmospheric scattering and absorption through the use of optical fibres. We have also foreshadowed the possibility of carrying out all the processes needed for a communications system in a completely integrated manner. Light generation, beamsplitting and modulation can all be achieved in a single solid state device. At the receiving end, amplification, demodulation and detec-

tion can be similarly effected. Such a system possesses great advantages over one in which the light beam is generated, modulated and generally processed by separate, discrete components, where relative movement of the components and atmospheric distortion of the (unconfined) beams conspire to impair performance. Not only does an integrated system avoid these difficulties but it readily lends itself to the incorporation of active optical elements. Moreover, when beams of quite modest powers are confined in structures such as the two-dimensional dielectric waveguide discussed in Sections 14.3 to 14.5, the power densities may be sufficiently high for non-linear effects to be exploited.

14.12   A few of the problems

If we envisage an integrated system in which signal processing is carried out as indicated in Section 14.9, then we need an efficient method of coupling radiation from the source into the dielectric waveguides that are to channel the signals. The cross-sections of suitable guides are only a few wavelengths across. Direct focusing of the beam from a laser on to the end of the guide is impracticable. If the lens used provided a convergence such that the direction of the radiation corresponded to propagating modes in the guide, the image on the end face of the guide would be enormously larger than the section of the guide and only a very small fraction would be collected. If a short-focus lens produced a small image of the source, the large angular convergence would mean that much of the incident flux would strike the sides of the guide at angles below the critical and so would not contribute to the guided modes.

We solve this difficulty by making use of the evanescent wave which occurs in the second medium when total reflection occurs. If a laser beam is directed into a prism as shown in Fig. 14.13, the evanescent wave in the gap between the prism and the surface waveguide generates a continuous wave mode in the guide. The wave amplitude increases with distance and—for a sufficiently long path adjacent to the prism—would then

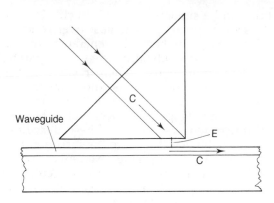

Figure 14.13 Coupling a laser beam into a guide by frustrated total reflection

begin to couple radiation back into the prism. However, once the end of the prism is passed, no such coupling is possible and the radiation is trapped in the guided mode.

An interesting alternative is to arrange for the source to be an integral part of the crystal carrying the waveguide. By implanting ions in the crystal it is possible to create a medium suitable for a semiconductor laser. However, how can one produce the necessary mirrors at the ends of the region? The effect of mirrors can be obtained by creating diffraction gratings on the crystal surface (e.g. by etching) such that the Bragg reflected waves return in the direction opposite to the incident direction. The 'channel' forming the active laser medium can then lead directly into an optical waveguide, as indicated in Fig. 14.14.

To direct part of the flux in the guide into a different direction, use can again be made of the diffraction grating, as shown in Fig. 14.15. If the guide is carrying beams of several wavelengths, then gratings of different spacings will enable the beams to be separated. An interesting possibility arises if

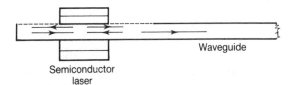

Figure 14.14 Schematic of integral semiconductor laser source and waveguide

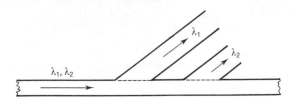

Figure 14.15 Output coupling by gratings

a holographic grating (Section 13.10) is made by combining plane and cylindrical reference waves. Such a grating will have a spacing which varies along its length, with the result that different wavelengths are diffracted out at different points of the system.

If it is required to divert the flux in a guide on command, then an ultrasonic grating may be created on the surface of the system, as illustrated in Fig. 14.11. If the amplitude of the acoustic wave that creates the grating is rapidly varied, then we have a method of modulating the beam that is being coupled out. If the material of the crystal is such that non-linear effects occur then frequency mixing and parametric amplification (Section 15.5) may occur.

The above serves to outline the way in which the processing of optical signals, e.g. for a communication system, may be effected in a single, integrated device. One general feature of the types of device described is that the components used are either completely transparent or have at best very weak absorbing properties. They therefore have the potential of highly efficient systems, with very little loss.

Problems

14.1 What is the implication of the factor 'i' in the expression for H in eqn (14.4)?

14.2 Determine the value of θ for a symmetric slab guide with an index of 1.58 surrounded by cladding of index 1.52 where the guide thickness is equal to the (free-space) wavelength.

14.3 Section 14.4.1 refers to a limiting value for $\eta_f h$ beyond which modes cannot occur. Calculate this limit for $\lambda = 633$ nm, $h = 1500$ nm and $n_f = 1.58$.

14.4 If light is to be focused into a fibre of the type shown in Fig. 14.6 and $n = 1.63$, $n = 1.59$, what is the maximum convergence angle to be usefully employed?

14.5 Apply the boundary conditions of continuity of E_y and dE_y/dx to eqn (14.9) and thence derive eqn (14.10).

14.6 For a stepped-index fibre with core index n_c and cladding index n_s, show that the maximum path difference Δl between skew and axial rays is given by $l(n_c/n_s - 1)$ where l is the axial length. What is being assumed?

14.7 What is the maximum time spread of a pulse for 1 km of fibre with $n = 1.528$ and $n = 1.520$?

14.8 A digital communications system consists of a pulsed source with an average output power of 1 mW, a length of optical fibre with an attenuation of 1 dB km^{-1} and a detector that requires pulses of 10^{-16} J for satisfactory detection. What is the maximum fibre length that could be used for a transmission rate of (a) 10^7 pulses s^{-1}, (b) 10^9 pulses s^{-1}?

14.9 If the fibre of Problem 14.8 produces a spread in pulse duration of 3 ns km^{-1} and if the spreading must not exceed $1/(2 \times$ pulse rate), what will the maximum fibre lengths be for 10^7 and 10^9 pulses s^{-1}?

15

Non-linear Optics

15.1 Polarisation at high field strengths

In Chapter 8 we referred to the fact that the linear relation between polarisation P in a medium and the applied electric field E does not hold for large values of E. This is as expected. Polarisation results from the relative displacement of the positive and negative charges in the material consequent upon the application of the field. For small values of E the displacement is small compared with the distance between atoms and is linearly proportional to E. When the displacement is large enough for adjacent atoms to have significant influence, a more complicated relationship applies.

In the case of an anisotropic material, the direction of the (vector) polarisation P will in general not be that of the applied field E, although P and E will be parallel for certain special directions (e.g. parallel and perpendicular to the optic axis of a uniaxial crystal; see Section 1.2). To understand some of the physical manifestations of non-linearity, it suffices to consider the behaviour of a material when the directions of P and E are parallel. We may then deal simply with the magnitudes of P and E. The effects of anisotropy will be considered later.

The non-linearity between P and E may be

encompassed by writing P as a power series in E:†

$$P = \varepsilon_0(\chi_1 E + \chi_2 E^2 + \chi_3 E^3 + \cdots) \quad (15.1)$$

where χ_2, χ_3 are the second- and third-order susceptibilities. Their values are such that the second and higher terms of (15.1) are completely negligible compared with the first at 'ordinary' optical field strengths—e.g. corresponding to sunlight for which E is of the order of hundreds of volts per metre, or to radiation from most non-laser light sources. For many materials of importance in the field of non-linear optics, the higher-order terms in (15.1) may begin to be of importance for values of E of the order of $10^6 \, \text{V m}^{-1}$. Such values are easily obtained by the use of laser sources (Chapter 20).

15.2 Consequences of non-linear susceptibility

15.2.1 Harmonic generation

Consider first the case where only the first two terms of (15.1) are significant. What will

---

† We neglect interaction with the magnetic field of the wave; this is generally completely negligible.

be the form of the polarisation when a sinusoidally varying field of angular frequency ω is applied? If $E = E_0 \cos \omega t$, then

$$P = \varepsilon_0 (\chi_1 E_0 \cos \omega t + \chi_2 E_0^2 \cos^2 \omega t) \quad (15.2)$$

$$= \tfrac{1}{2} \varepsilon_0 \chi_2 E_0^2 + \varepsilon_0 \chi_1 E_0 \cos \omega t + \tfrac{1}{2} \varepsilon_0 \chi_2 E_0^2 \cos 2\omega t \quad (15.3)$$

Thus in contrast to the linear response situation, in which the polarisation P simply oscillates with frequency ω, we have in addition both a steady ('d.c') polarisation term and also one at double the input frequency. The oscillating polarisation will therefore produce an emitted wave at frequency 2ω, a process known as 'second harmonic generation' (SHG). Far from being a mere optical novelty, the production of radiation of second harmonic frequency proves of immense importance in the provision of radiation sources. There are, however, things needing to be done to achieve good efficiency in the process: these are discussed in Section 15.3 below.

The idea can clearly be extended to the case of higher harmonics. Thus if the polarisation needs to be represented by all three terms of (15.1), then, on applying a sinusoidal field, we have

$$P = \varepsilon_0 (\chi_1 E_0 \cos \omega t + \chi_2 E_0^2 \cos^2 \omega t$$
$$+ \chi_3 E_0^3 \cos^3 \omega t) \quad (15.4)$$

that is

$$P = \tfrac{1}{2} \varepsilon_0 \chi_2 E_0^2 + \left(\varepsilon_0 \chi_1 E_0 + \frac{3\varepsilon_0 \chi_3 E_0^3}{4} \right) \cos \omega t$$

$$+ \frac{\varepsilon_0 \chi_2 E_0^2}{2} \cos 2\omega t + \tfrac{1}{4} \varepsilon_0 \chi_3 E_0^3 \cos 3\omega t \quad (15.5)$$

indicating the generation of second and third harmonics, in addition to the d.c. term, which is referred to as 'optical rectification'.

The existence or otherwise of particular higher-order terms in the expression for the polarisation depends on the symmetry of the material. Substances like glass, which are isotropic, will not exhibit second harmonic generation (although they may be made anisotropic by the application of a static electric or magnetic field—see Chapter 11). Similarly for those crystalline materials which possess a centre of symmetry there is no second-order (χ_2) term. For harmonic generation,

crystals must be of symmetry lower than cubic.

15.2.2 Frequency mixing

Harmonic generation can be regarded as the mixing, through the material's non-linear characteristics, of two incident waves of the same frequency. In this sense it is a special case of the mixing of incident waves of two or more different frequencies. In a material subjected simultaneously to waves of frequencies ω_1 and ω_2, the polarisation will contain a term of the form $\cos \omega_1 t \cos \omega_2 t = \tfrac{1}{2}[\cos(\omega_1 + \omega_2)t + \cos(\omega_1 - \omega_2)t]$. Thus both sum and difference frequencies will be generated. Since materials are dispersive, the susceptibilities associated with the different frequencies will be different. Associated with electric fields $E(\omega_1)$ and $E(\omega_2)$ there will be polarisation components given by

$$P^{(2)}(\omega_1 + \omega_2) = \varepsilon_0 \chi_2(\omega_1 + \omega_2) E(\omega_1) E(\omega_2)$$

and

$$P^{(2)}(\omega_1 - \omega_2) = \varepsilon_0 \chi_2(\omega_1 - \omega_2) E(\omega_1) E(\omega_2)$$

$$(15.6)$$

where, as before, we consider the special case where P is parallel to E. More generally, the second-order polarisation $P^{(2)}$ will be expressed as

$$P^{(2)}(\omega_1 \pm \omega_2)$$
$$= \varepsilon_0 \{ \chi_{xxx} E_x(\omega_1) E_x(\omega_2) + \chi_{xxy} E_x(\omega_1) E_y(\omega_2)$$
$$+ \chi_{xxz} E_x(\omega_1) E_z(\omega_2) + \chi_{xyx} E_y(\omega_1) E_x(\omega_2)$$
$$+ \chi_{xyy} E_y(\omega_1) E_y(\omega_2) + \chi_{xyz} E_y(\omega_1) E_z(\omega_2)$$
$$+ \chi_{xzx} E_z(\omega_1) E_x(\omega_2) + \chi_{xzy} E_z(\omega_1) E_x(\omega_2)$$
$$+ \chi_{xzz} E_z(\omega_1) E_z(\omega_2) \} \quad (15.7)$$

with two sets of values of χ_{ijk}, one for $\omega = \omega_1 - \omega_2$ and the other for $\omega = \omega_1 + \omega_2$. Two corresponding expressions exist for $P_y^{(2)}$ and $P_z^{(2)}$. Although 27 values of χ_{ijk} appear to be needed, there are in fact only 18 independent values: the existence of crystal symmetry elements leads to the vanishing of some of the components of χ_{ijk}, in a fashion similar to that for the linear susceptibility tensor discussed in Chapter 11.

In principle, sum and difference frequencies can be generated from a given pair of (e.g. laser) frequencies, and extensive use of this feature is made in the establishment of the standards of length and time, as discussed in Section 22.6. It is, however, essential that the material to be used for the sum- and difference-frequency generation shall be transparent at the frequencies concerned.

15.3 Phase-matching in second harmonic generation

Although second and higher harmonic generation will occur for any direction of radiation in a suitable crystal, the intensity is generally low. A reason for this is that materials are dispersive and that the velocity of the second harmonic wave will usually be lower than that of the exciting wave, and the wavelength in the crystal will be less than one half that of the exciting wavelength. The consequence is as shown in Fig. 15.1. At a given instant, sources of second harmonic radiation at A, B, C, ... emit SH waves. Whereas the waves of frequency ω from A, B, C, ... arriving at the exit face of the crystal will all be in phase, those of the SH waves will not. A vector diagram representing their resultant is shown in Fig. 15.1(b). If only there were a way of ensuring that all the SH

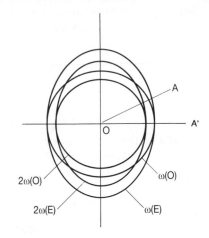

Figure 15.2 Ordinary and extraordinary wavefronts for fundamental and second harmonic; not phase matched

waves arrived in phase, a large resultant would be achieved, but this appears to require that the velocities of the ω and 2ω waves should be equal.

Consider now the behaviour of a uniaxial crystal. Wave-fronts from an imaginary point source in such a crystal consist of spheres and spheroids (Section 11.2): for waves of frequencies ω and 2ω, there will be *four* such surfaces, as shown in Fig. 15.2. Note that for the wave direction OA′, the velocity of the ordinary wave of frequency 2ω is the same as

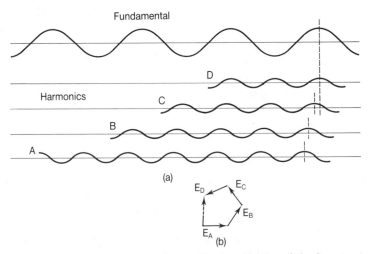

Figure 15.1   (a) Second harmonic generation. Generally the velocity of the harmonic is less than that of the fundamental. The second harmonic waves arrive at the end of the crystal out of phase. (b) Vector diagram of SH waves

Figure 15.3 Ray directions of second harmonic; not phase matched

that for the extraordinary wave at frequency ω. Thus for this direction in the crystal, the condition of equality of the ω, 2ω wave velocities is satisfied. There is a difficulty, however. Whereas the directions of ray and wave-normal for an ordinary ray are identical, this is not generally the case for the extraordinary ray wave-normal. Thus if the exciting radiation gives rise to a wave along OA (Fig. 15.2), the directions of the emitted 2ω rays will be inclined to that of the incident radiation. The SH emission will be spread over an elongated cross-section (Fig. 15.3), instead of being concentrated in a beam identical with that of the incident light.

If we were fortunate enough to find a crystal for which the direction OA happened to be perpendicular to the optic axis (Fig. 15.4), then the ray direction for the second harmonic would not exhibit the behaviour shown in Fig. 15.3, but would be everywhere parallel to that of the exciting radiation. It would seem unlikely that Nature would be so obliging. In fact, she is not so unhelpful. It is found that, for some materials, the birefringence of the crystal for the two frequencies of interest changes with

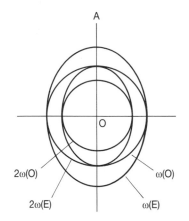

Figure 15.4 As for Fig. 15.3, but with phase-matching

temperature in such a way that the desired condition ('phase-matching') does in fact exist at a particular temperature. For a uniaxial crystal at this temperature, and for radiation travelling in a direction perpendicular to the optic axis, the phases of the fundamental and second harmonic waves are matched, giving rise to the term 'phase-matching'. As the ω wave traverses the crystal, it is converted into 2ω radiation. If the crystal is long enough, a very large proportion of the incident radiation can be converted into the second harmonic.

15.4 A practical SHG system

For efficient SHG we need not only to achieve phase-matching, as described above, but also to use as high a flux of exciting radiation as possible (subject to not blowing up the crystal). The place where one finds the largest possible flux at a given frequency is *inside* a laser cavity. This is therefore a highly suitable place to insert the crystal to be used for SHG. It must be suitably orientated and maintained at the appropriate temperature. How will the second harmonic generation get out? Will it not itself be for ever confined to the laser cavity? If perfect mirrors were used then the SH radiation *would* be so confined. However, by the use of high-reflecting dielectric stacks as described in Section 4.32, life is made easy. A stack of quarter-wave layers for wavelength λ_0 (frequency ω_0) will reflect strongly at λ_0. Their optical thicknesses will be almost a half-wavelength at the harmonic wavelength $\lambda_0/2$ and the stack will be practically transparent—precisely what is required for the SH radiation to get out. A suitable arrangement is shown schematically in Fig. 15.5.

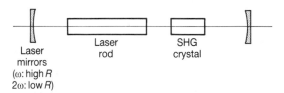

Figure 15.5 SHG system (schematic)

15.5   Optical mixing and parametric amplification

As indicated in Section 15.2, the mixing of waves of two frequencies ω_1 and ω_2 in a non-linear crystal generates both sum and difference frequencies. The efficiency of the generation process will be low unless phase-matching is achieved, as in the case of SHG. Thus although waves of frequencies ω_1, $2\omega_1$, ω_2, $2\omega_2$, $\omega_1 + \omega_2$ and $\omega_1 - \omega_2$ will be generated, it will be possible to phase-match for only one harmonic or mixed wave at a time. In practice, therefore, it suffices to deal with only one form of generation, since the lack of phase-matching for the rest means that their intensities will be very low.

In the case of sum-frequency generation, we need to know how the amplitudes A_1, A_2 and A_3 of the waves of frequency ω_1, ω_2 and $\omega_3 = \omega_1 + \omega_2$ vary as the two exciting waves progress through the crystal. We use the simplification of Section 15.1 and assume perfect phase-matching. The electric fields at the three frequencies ω_j ($j = 1$, 2, 3), for a plane wave travelling in the z direction, are given by

$$E_j(\omega_j) = \frac{1}{2}\left\{ A_j \exp i\left(\omega_j t - \frac{2\pi n z}{c}\right)\right.$$
$$\left. + A_j^* \exp\left[-i\left(\omega_j t - \frac{2\pi n z}{c}\right)\right]\right\} \quad (15.8)$$

We need to use eqn (8.27) suitably modified to take account of (1) zero conductivity and (2) the existence of higher-order terms in the susceptibility to which ε is related. If we assume plane-polarised waves with their electric vector in the x direction, then the polarisation will consist of only the first term of eqn (15.7), involving products of the form

$$E_1 E_2(\omega_3) = \frac{1}{4}\left\{ A_1 A_2 \exp i\left(\omega_3 t - \frac{2\pi n z}{c}\right)\right.$$
$$\left. + A_1^* A_2^* \exp\left[-\left(\omega_3 t - \frac{2\pi n z}{c}\right)\right]\right\} \quad (15.9)$$

with corresponding expressions for $E_2 E_3$ and $E_3 E_1$. Choosing the origin of t to make A_3 real, solution of (8.27) for the non-linear case

leads to

$$\frac{\mathrm{d}A_3}{\mathrm{d}z} = -\frac{\mathrm{i}\omega_3}{2nc}\chi(\omega_3)A_1 A_2$$

$$\frac{\mathrm{d}A_2^*}{\mathrm{d}z} = \frac{\mathrm{i}\omega_2}{2nc}\chi(\omega_2)A_3 A_1 \quad (15.10)$$

$$\frac{\mathrm{d}A_1}{\mathrm{d}z} = -\frac{\mathrm{i}\omega_1}{2nc}\chi(\omega_1)A_3 A_2^*$$

where the χ's are the non-linear susceptibilities.

A case of particular interest is that in which the crystal is exposed simultaneously to a wave of very large amplitude A_3 (the 'pump wave') and a wave of small amplitude A_1 (the 'signal wave'). Solution of eqns (15.10) reveals the growth of the signal (A_1) wave and of the wave A_2 (the 'idler wave'). The process is known as parametric amplification and provides a means of amplifying a wave of frequency ω_1 through the conversion of the energy of a photon of ω_3 radiation into two photons of frequency ω_1 and ω_2.

15.6   Phase conjugation

If a uniform plane wave (Fig. 15.6) falls normally on a perfect plane mirror, then the reflected wave is (neglecting any diffraction effects) identical in form to the incident wave, but travelling in the reverse direction. If the incident wave has the form $a\ \cos(\omega t - kz)$, the reflected wave is $a\ \cos(\omega t + kz) = a\ \cos(-\omega t - kz)$. The form is that of the incident wave, but for a reversal of the time coordinate. Similarly, the wave

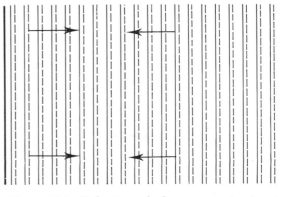

Figure 15.6 Reflection of plane wave at mirror; normal incidence

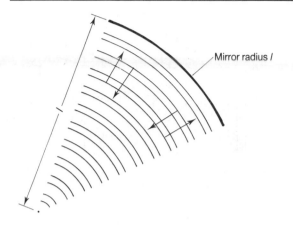

Figure 15.7 Reflection of spherical wave from point source by mirror of radius *l* and at distance *l*

from a point source P (Fig. 15.7) reflected from a mirror of radius of curvature *l* at a distance *l* is the time-reversed form of the incident wave. The reflected waves in Figs 15.6 and 15.7 are termed 'phase-conjugate' waves.

Can we make a phase-conjugate wave corresponding to the wave from an extended object? We should need a mirror whose surface contoured the wave-front of the light scattered by the object. This would appear to be difficult, but *if* we could, some surprising (and useful) effects would manifest themselves. Such a 'phase-conjugate mirror'

would reflect in the fashion shown in Fig. 15.8.

Since the form of the mirror required for phase-conjugation is determined by the form of the incident wave-front, then we need to press the wave-front into service in order to *form* the mirror. We can do this by elaborating on the idea of Brillouin scattering, discussed in Section 15.7.

15.7   Stimulated Brillouin scattering

In Section 10.9 we saw that if we sent an acoustic wave through a medium, the periodic compressions and rarefactions produced created a diffraction grating on which an incident beam could be diffracted. However, even in the absence of an *applied* acoustic wave in a specimen, such waves will exist, due to the thermal motion of the atoms present. At each point in the specimen, waves with a wide band of frequencies will be travelling in all directions. Now suppose we direct an electromagnetic wave of frequency ω_I at the material, in a particular direction. Of all the acoustic waves rushing around in the material there will be one moving in the direction of incidence whose wavelength is precisely the value required to diffract the incident light back into the opposite direction (Fig. 15.9). The acoustic wave will be weak, since it is only a small

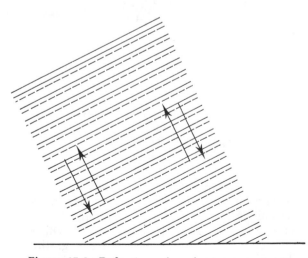

Figure 15.8 Reflection of a plane wave at non-normal incidence by a phase-conjugate mirror

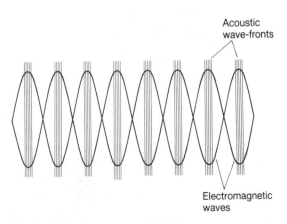

Figure 15.9 Formation of 'diffraction grating' by acoustic waves in a medium

part of the acoustic 'noise'. The back-scattered electromagnetic wave—also initially weak—will be shifted in frequency since it has been produced by diffraction at a *moving* grating. The incident and back-scattered wave will contain a standing-wave component which will, however, move through the medium, since the frequencies of the two waves are different. The 'beat' wave will move at precisely the velocity of the acoustic wave which gave rise to the scattering. Now an electric field applied to a material will, through its interaction with the charged particles of which the atoms are made, give rise to mechanical stress, a phenomenon known as 'electrostriction'. The moving 'beat' wave will therefore induce compressions and rarefactions in the material in exactly the positions of the compressions and rarefactions of the particular acoustic wave of interest. Thus the effect of the beat wave is to increase the amplitude of the acoustic wave that gave rise to the back-scattering. This increases the amount of back-scattering, which increases the acoustic wave amplitude, ... Thus as the incident wave penetrates the crystal, the back-scattered wave, *whose form is identical with that of the incident wave*, rapidly increases in amplitude. However, a wave travelling in the reverse direction with wave-front identical with the incident wave *is* a phase-conjugate wave. The process described above is that of stimulated Brillouin scattering and is one of the possible methods of producing a phase-conjugate mirror.

15.8 Degenerate four-wave mixing

This provides an alternative method for the production of a phase-conjugate wave. Consider the effects of exposing a medium simultaneously to a plane ('reference') wave A and spherical ('signal') wave S of the same frequency, as shown in Fig. 15.10(a). The resultant electric field along the lines along which the regions of constructive interference travel will, through the effects of electrostriction,† create an acoustic grating G in a fashion similar to that described in the

---

† The creation of mechanical strain by an electric field.

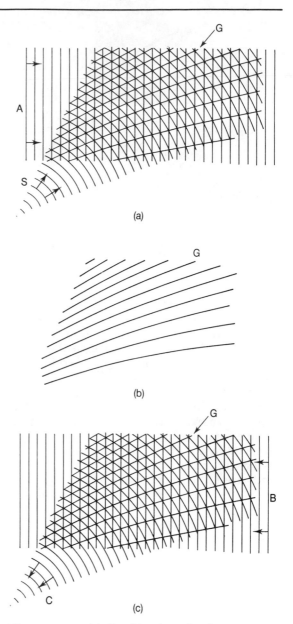

Figure 15.10 (a) Combination of reference wave A and signal wave S to form acoustic grating. (b) The acoustic grating. (c) Diffraction of oppositely directed plane wave by grating of (b)

previous paragraph, except that one is not relying on the inherent acoustic noise in the crystal. A plane wave B travelling in the direction opposite to that of the first will be diffracted by the acoustic grating (Fig. 15.10b) and will produce a wave C conjugate to that of the signal wave (Fig. 15.10c).

The combined effects of the second reference wave and the conjugate wave will increase the amplitude of the acoustic grating which will in turn strengthen the conjugate wave. In fact, a similar acoustic grating will be formed between the signal wave and plane wave B. Plane wave A will be diffracted by this grating, again producing the conjugate wave C.

This process is reminiscent of holography, as discussed in Chapter 13. Waves S and A constitute the object and reference waves which produce a hologram in the form of a pattern of varying refractive index in the active medium. The reconstructing wave B then generates the image wave, through diffraction at the refractive index pattern. Degenerative four-wave mixing is in effect a single-stage holographic process.

15.9 Application of phase conjugation

One example of the use of the phase-conjugate mirror suffices to show the importance of the device.

In Chapter 12 we discussed the laser, whose operation relied on the fact that a wave is amplified when it traverses a medium in which a population inversion is established. If there is to be no distortion in the amplified wave, then the amplifying medium needs to be perfectly homogeneous. In fact many laser materials have significant inhomogeneities so that distortion would appear to be inevitable. Consider, however, the arrangement shown in Fig. 15.11. The plane-wave signal enters the amplifying medium at A, is amplified and emerges at B with a distorted wave-front. If this is reflected at a phase-conjugate mirror, then

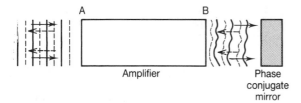

Figure 15.11 Correcting the distortions of a laser amplifier by a phase-conjugate reflector

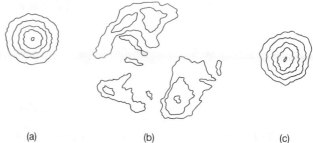

(a)   (b)   (c)

Figure 15.12 Removal of atmospheric distortion by phase-conjugation. (a) Isointensity contours of single-mode laser source beam. (b) Contours of beam received after reflection at a plane mirror 50 m away. (c) Contours of beam received after reflection at phase conjugate mirror at 50 m

the form of the wave entering the laser medium at B is precisely that which will produce a plane wave emerging at A. The wave has been amplified and although the first pass through the amplifier produces a distorted wave, the return pass of the *phase-conjugated* wave ensures the emergence of a wave identical with the input wave. Similarly, the effects of atmospheric distortion can be counteracted, as shown in Fig. 15.12.

15.10 Self-focusing

Since the susceptibility, permittivity and refractive index of a material are interrelated, then an intensity-dependent susceptibility implies an intensity-dependent refractive index. This in turn implies that the optical properties of a medium will be changed when the intensity of the traversing beam is sufficiently large. The changes induced can have dramatic effects, as shown below.

Since such effects generally arise only at the intensities produced by lasers, we consider for simplicity the effect of intensity-dependent refractive index changes on the propagation of a Gaussian laser beam. From Section 20.8 we see that the form of the radial profile of the power distribution in the beam is given by

$$I(r) = I_0 \exp\left(-\frac{2r^2}{r_0^2}\right) \qquad (15.11)$$

where r_0 is the radius at which the beam amplitude falls to $1/e$ of its axial value.

The variation of refractive index of a

medium with the intensity follows the form

$$n(I) = n_0 + n_2 I \qquad (15.12)$$

where the value of n_2 is positive. Thus the radial variation of refractive index for a material exposed to a Gaussian laser beam is

$$n(r) = n_0 + n_2 I_0 \exp\left(-\frac{2r^2}{r_0^2}\right) \qquad (15.13)$$

The refractive index is a maximum at the centre of the beam, decreasing with radial distance from the axis. Consider now a laser beam focused by a converging lens. At low intensities, for which the second term in (15.13) is negligible, the form of the wave-fronts in the neighbourhood of the beam focus is as shown in Fig. 15.13(a).† At intensities at which the second term of (15.13) becomes significant, the curvature of the wave-fronts approaching the focal region increases (Fig. 15.13b). The result is that the focal spot diameter is reduced significantly below that of the low-intensity case and the power density is increased, often to very high values. In a gaseous medium, the electric field strength in the focused wave can be sufficient to produce intense ionisation. If the focusing occurs in a crystalline material, the crystal can be damaged. The value of a £5000 crystal can be reduced to ≈ zero in a few picoseconds: a loss rate of £10^{15} per second.

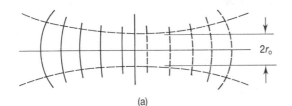

(a)

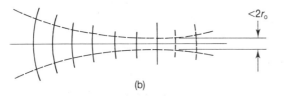

(b)

Figure 15.13 Self-focusing in laser beam: (a) low intensity, (b) high intensity

---

† In the abstraction of paraxial ray optics, rays intersect at the image point in a region of zero diameter. The more realistic wave-front picture in Fig. 15.13(a) shows that the diameter of the focal region is finite.

15.11 Optical bistability

During our discussion of lasers in Chapter 12 we noted that, for a two-level system in a radiation field, the number of absorptions per second is proportional to n_1, the lower-state population, and the number of stimulated emissions per second is proportional to n_2. The two conclusions drawn from this are:

1. If $n_2 \ll n_1$, absorption dominates.
2. If $n_2 \simeq n_1$, the *net* absorption, which is proportional to $(n_1 - n_2)$, is very small.

At low intensities of radiation, conclusion 1 applies. At high intensities, conclusion 2 applies and there is little net absorption: the medium becomes transparent to the radiation. Thus the *fraction* of the light transmitted by a system containing such a 'saturable absorber' can depend markedly on the actual intensity of the incident radiation.

If the absorbing medium is contained in a Fabry–Perot etalon, the intensity determining the optical characteristic of the contained medium is that of the standing-wave field within the etalon. If this is sufficiently low that absorption dominates, then the behaviour of the etalon may be determined by the procedure of Section 4.23, but with absorption allowed for. If the etalon spacing is l and the linear absorption coefficient α, then the amplitude of the wave in the cavity is reduced by a factor $e^{-\alpha l/2}$ after a single transit. Equation (4.31) becomes

$$E_{rc} = \sigma E_0\, e^{-\alpha l/2}(1 + \rho\, e^{-\alpha l + i\delta} + \rho^2\, e^{-2\alpha l + 2i\delta} + \cdots)$$

$$(15.14)$$

and the transmitted flux is

$$W_t = E_{rc} E_{rc}^* = \frac{\sigma^2 E_0^2\, e^{-\alpha l}}{1 - 2\rho\, e^{-\alpha l} \cos\delta + \rho^2\, e^{-2\alpha l}}$$

$$(15.15)$$

The maximum value of W_t occurs when $\delta = 2m\pi$, giving

$$W_{max} = \frac{\sigma^2 E_0^2\, e^{-\alpha l}}{(1 - \rho\, e^{-\alpha l})^2} \qquad (15.16)$$

In accordance with the discussion above, we acknowledge that the value of α is intensity-dependent, diminishing from a low-field value of α_L to a value approaching zero as saturation approaches. If the incident intensity produces a standing-wave field for which the low-field value α_L applies, and hence $e^{-\alpha_L l} \ll 1$, then the overall maximum transmission is given by

$$W_{mo} = \frac{\sigma^2 E_0^2 e^{-\alpha_L l}}{(1 - \rho\, e^{-\alpha_L l})^2} \approx \sigma^2 E_0^2\, e^{-\alpha_L l} \quad (15.17)$$

If, however, the incident intensity is sufficient to produce saturation, then $e^{-\alpha l} \approx 1$ and the saturation transmission is

$$W_{ms} = \frac{\sigma^2 E_0^2}{(1 - \rho)^2} \quad (15.18)$$

$$= E_0^2 \quad (15.19)$$

if the mirrors are non-absorbing, in which case $\rho + \sigma = 1$. Thus as the intensity increases, the *fractional* transmittance increases from a low value ($\sigma^2\, e^{-\alpha_L l}$) to $\sim 100\%$.

If we examine the way in which the transmitted intensity W_t increases with the incident flux $W_0 \equiv E_0^2$ as the value of W_0 increases, we find the behaviour shown in Fig. 15.14. The dotted line records the behaviour of an empty cavity, showing the expected linearity. The full line shows the behaviour of a cavity containing a saturable absorber. For an increasing value of W_0, the transmittance jumps discontinuously at A from a low to a high value. On decreasing W_0 from this state, the high transmittance is maintained until B is reached, when a discontinuous jump to the lower value is observed.

Such bistability has a clear role in an optical

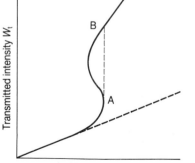

Figure 15.14 Transmission of cavity for optical bistability

computer in which binary logic is employed. The non-linear, active layer may be incorporated as the spacer layer of an interference filter, as described in Section 4.32. An important feature of the switching process in the optically bistable system is that it can be extremely fast, as is required in a high-speed computer.

15.12 Propagation of pulses along optical fibres

A pulse of finite duration and 'fixed' frequency will, due to the finiteness of the wave train, consist of a band of frequencies (Section 5.5). Since all materials are dispersive, the individual spectral components of the pulse will travel with different velocities. The duration of the pulse will therefore inevitably change as it propagates, an undesirable feature in, for example, a digital communication system in which it is essential that there is no distortion in the signals transmitted. On the face of it, it would seem that we can do nothing to avoid this effect, but this is not the case.

The above argument assumes that the behaviour of the fibre material is linear. In Sections 15.1 and 15.2 we saw that for sufficiently large values of electric field strength materials display non-linear properties. Although such effects occur only at very high values of E, the very small diameter of an optical fibre means that, for only a modest input *power*, the value of E inside the fibre can be large enough for non-linear effects to occur. The consequence is an increase in refractive index of the fibre given by

$$n = n_0 + n_2 E^2 \quad (15.20)$$

Thus for the leading edge of a pulse, where the value of E is increasing, there is an increase in refractive index. The velocity of the wave decreases and hence the wavelength increases. The spectrum shows a red-shift. At the trailing edge of the pulse, E is decreasing and so therefore is n. The velocity is increasing, the wavelength decreasing and a blue-shift results. The *spectrum* is broadened but the (temporal) shape is unaffected. This phenomenon is termed *self-phase modulation* (SFM).

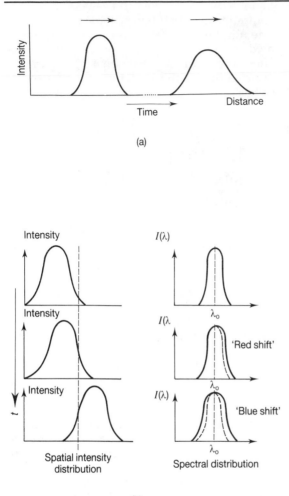

(a)

(b)

Figure 15.15 (a) Broadening of pulse during transmission: effect of dispersion. (b) Effect of self-phase modulation on temporal and spectral shape of pulse

Thus we have two effects. The 'ordinary' effects of dispersion broaden the pulse in time, due to the wavelength dependence of the group velocity (Fig. 15.15a). On this is superposed the non-linear SFM which results in the low-frequency component of the pulse arriving before the high-frequency component (Fig. 15.15b). It turns out that the frequency–time dependence is linear, producing a 'chirped' pulse.[†] Imagine now that such a chirped pulse is passed into a dispersing system which delays the blue-shifted edge by an amount corresponding to the starting pulse width. The result will be a compression of the pulse in time. A pair of diffraction gratings (Fig. 15.15) forms a suitable system. From the discussion of gratings in Chapter 4 it is readily verified that the result of the double diffraction is that all the spectral components emerge in the same direction but that the longer path of the red-shifted radiation allows it to 'catch up' the blue-shifted component. Pulse compressions of ~100 times have been achieved and pulses of duration 6 fs of light of wavelength 633 nm have been produced by this means.

The consequence of reducing the pulse duration is an increase in spectral width. With the arrangement of Fig. 15.16, input pulses of second harmonic from an Nd–YAG laser ($\lambda = 532$ nm) of duration 34 ps and spectral width 0.027 nm (Fig. 15.17) emerged from the two-grating dispersing system with duration 0.46 ps and a spectral bandwidth 1.72 nm.

[†] Strictly, 'linear chirped'. Non-linear chirping occurs in some situations.

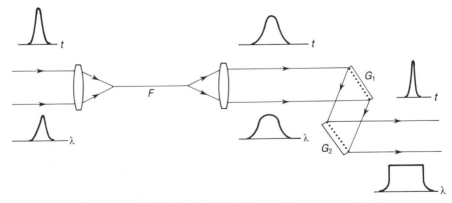

Figure 15.16 Method of obtaining pulse compression

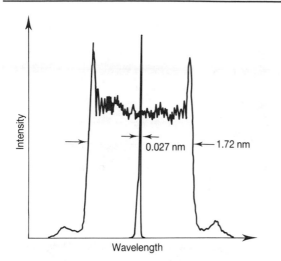

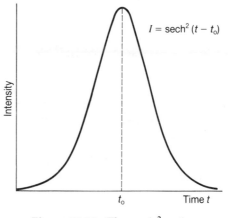

Figure 15.18   The sech^2 pulse

Figure 15.17 Example of obtaining pulse compression (spectral). Reproduced by permission of the Optical Society of America from Johnson, *Journal of the Optical Society of America*, **B2**, 619–625 (1985)

15.13   Solitons

The effects described above require that the light intensity be high enough to drive the transmitting medium into the non-linear region. Any signal propagated along a fibre will inevitably reduce in power due to the fibre absorption. Thus although we might expect fibre compression effects to occur in moderate lengths of fibre, eventually the intensity will fall so that the second term in eqn (15.20) becomes negligible. At this point self-phase modulation will cease to operate and the pulse will start to broaden.

And yet ... if we *could* somehow maintain a high enough intensity so that SFM still operates, then we should be able to maintain the small pulse width over very large distances. The non-linear equations describing pulse propagation do in fact permit solutions with exactly this character—propagation *without change of shape*. The solutions are referred to as *solitons*. It emerges that a pulse with a sech^2 distribution (Fig. 15.18) will propagate unchanged. For a very long (thousands of kilometres) fibre, this would entail pumping energy into the fibre at intervals so as to maintain the high intensity. How might this be done?

Let us consider the absorption in the fibre as being the result of transitions between two states of the system, in the manner of our discussion of Section 12.8. The amount of the absorption depends on the difference $n_2 - n_1$ between the populations in the upper and lower levels. If we could reduce the population difference we should reduce the absorption. We could achieve this by pumping atoms into the upper state. In fact, if we pumped hard enough we could produce a population inversion and so amplify any radiation of frequency ω_s where $\hbar\omega_s$ is the energy difference between the upper and lower states. We cannot produce a population inversion by pumping with radiation of frequency ω_s since the medium approaches transparency as the populations of the upper and lower state approach equality. We can, however, pump via a third level, as is done for the ruby laser. One convenient effect that can be employed is the Raman effect (Section 10.7). We pump at a higher frequency and part of the absorbed energy serves to shake up the system internally, resulting in the emission of Stokes radiation at a lower frequency. If it can be arranged that the frequency of the Stokes radiation is at the signal frequency ω_s then with the achievement of a population inversion amplification will occur at ω_s.

Consider then the arrangement shown in Fig. 15.19. Power is coupled into the fibre at intervals in opposite directions towards the centre of each fibre length so that Raman

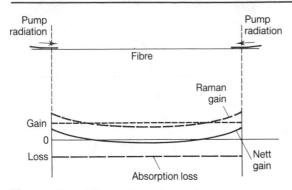

Figure 15.19 Eliminating absorption in a length of fibre by Raman gain

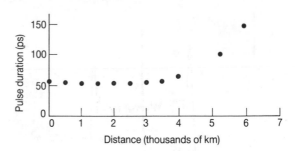

Figure 15.20 Pulse length preserved over a distance of *4000 km*!

gain occurs at the Stokes (also signal) frequency. The gain falls exponentially towards the centre of the section so that the overall gain is the sum of the two oppositely directed decaying exponentials. The ordinary loss processes in the fibre give rise to a constant loss over the fibre length. We can arrange that the total absorption loss is exactly balanced by the overall gain due to the Raman intervention.

The necessary power input could be made either by a series of semiconductor lasers either fed from a current source at the end of the system or possibly by localised solar-power supplies distributed along the route of the fibre. In this way power in the transmitted pulses can be maintained at a level such that the non-linear effects needed to ensure soliton propagation are sustained.

The effect is impressively demonstrated in the results of Mollenauer and Smith [1]. Signal pulses of duration 55 ps were launched into a 42 km fibre loop. The pulse wavelength was 1.6 μm. Pumping radiation at a wavelength of 1.497 μm was coupled into the loop. The difference in the pump and signal frequencies (433 wave numbers) corresponds to the Raman shift for the fibre material (quartz glass). The pump signal is triggered each time the train of signal pulses enters the loop. The results of measurements of the pulse width after pulses at a repetition rate of 33 GHz have travelled different distances along the fibre are shown in Fig. 15.20. The expected constancy of pulse duration over a distance of *4000 km* augers well for a high-speed, long-distance optical fibre communications system.

Reference

1. L. F. Mollenauer and K. Smith, *Optics Letters*, **13**, 675–7 (1988).

Problems

15.1 The refractive index of KDP for a wavelength of 694 nm is 1.507. The value of χ_2 (eqn 15.2) is 3.18×10^{-12} m V^{-1}. For what value of field strength will second harmonic polarisation amount to 1% of that of the fundamental?

15.2 If $n_o(\omega)$, $n_e(\omega)$ are the ordinary and extraordinary indices for the fundamental ω, and $n_o(2\omega)$, $n_e(2\omega)$ those for the second harmonic, show that the angle θ between OA′, the phase-matched direction (Fig. 15.2) and the optic axis is given by

$$\theta = \arctan \left[\frac{n_e^2(\omega) - n_o^2(\omega)}{n_o^2(2\omega) - n_o^2(\omega)} \right]^{1/2}$$

15.3 For a temperature of 290 K, the refractive indices of KDP for the wavelengths indicated are shown in the following table:

Wavelength (nm)	
514.5	257.2
1.518	1.578
1.471	1.520

15.4 The temperature coefficients of refractive index for KDP at $\lambda = 570$ nm have been determined as

$$\frac{dn_o}{dT} = 3.7 \times 10^{-5}\,\text{K}^{-1}$$

$$\frac{dn_e}{dT} = 2.4 \times 10^{-5}\,\text{K}^{-1}$$

Estimate the temperature at which phase-matching should occur. If you observed a discrepancy, what excuse would you make?

15.4 When a phase-conjugate wave is produced by stimulated Brillouin scattering, does it have *precisely* the form indicated in Section 15.6? Illustrate by considering the phase conjugation of a helium–neon laser beam ($\lambda = 633$ nm) by a material for which the velocity of sound at the relevant frequency is 1150 m s^{-1}.

15.5 For nitrobenzene the value of n_2 in eqn (15.12) is $19.2 \times 10^{-14}\,\text{m}^2\text{W}^{-1}$, where I is the mean irradiance. If significant beam distortion occurs for a refractive index change of 0.1, calculate the maximum electric field strength to which a cell containing nitrobenzene should be subjected.

Radiometry and Photometry

16.1 Symbols and definitions

Radiometry is concerned with radiation as a form of energy. Photometry evaluates radiation in respect of its usefulness in human vision. It deals only with radiation in the range 800–350 nm to which the eye is sensitive. Radiometry applies, in principle, to all wavelengths though its terminology is generally used for radiation in the range 300 μm–100 nm.

The inverse square law of the spread of radiation from a small source is the basis of calculation both in radiometry and in photometry. There are parallel quantities in the two subjects, e.g. irradiance for radiant energy per unit area and illuminance for light per unit area incident upon a surface. The defining equations and the symbols are the same in the two subjects but the names and units of measurement are different. When both radiometric and photometric units appear in the same discussion they are distinguished by the subscripts e and v (e.g. E_e for irradiance and E_v for illuminance) but when the context is such that no confusion can arise, the subscripts are omitted.

The principal quantities used in both fields are given in Table 16.1. The meanings of most of the quantities are shown by their defining equations. M is a measure of the radiation leaving a surface per unit area (called radiance) and luminance is similarly defined. Note that Φ represents total radiant energy crossing or leaving a surface per unit time and that intensity is equal to energy (or light) per unit solid angle.

16.2 Photometric units

The effect of unit energy of radiation for vision varies with wavelength. The function $V(\lambda)$ shown in Fig. 16.1 measures the visual efficiency of different wavelengths, normalised to a maximum value $V(\lambda) = 1$ at 555.02 nm for daylight vision (curve 1). This function, based on measurements for about 200 persons, has been internationally agreed. It applies to vision at light levels typical of daytime and is termed *photopic* vision. A different function $V'(\lambda)$ measures the efficiency of the eye at low levels of illuminance (scotopic vision). It is given the value 1 at 507 nm (curve 2).

In order to connect the photometric and radiometric units, the lumen (photometric unit of flux) is defined to be such that 1 watt of radiation at 550.02 nm produces a flux of 583 lumens. This is the value of K_m, the maximum luminous efficacy. For night vision $K'_m = 1700 \text{ lm W}^{-1}$ at 507 nm. Except when

Table 16.1

Symbol	Defining equation	Radiometric quantities		Photometric quantities	
		Name	Unit	Name	Unit
Q_V	$\displaystyle\int_0^\infty V_\lambda Q_\lambda d_\lambda$			Total light	lm s
Q_E	—	Total energy	Joule		
W	$W = \dfrac{dQ}{dV}$	Radiant energy density	J m^{-3}	Luminous density	lm s m^{-3}
Φ	$\Phi = \dfrac{dQ}{dt}$	Radiant power or flux	Watt	Luminous flux	Lumen
M	$M = \dfrac{d\Phi}{dA}$	Radiant excitance	W m^{-2}	Luminous excitance	lm m^{-2}
E	$E = \dfrac{d\Phi}{dA}$	Irradiance	W m^{-2}	Illuminance	lm m^{-2}
I	$I = \dfrac{d\Phi}{d\bar{\omega}}$	Radiant intensity	W sr^{-1}	Luminous intensity	$\text{cd} = \text{lm sr}^{-1}$
L	$L = \dfrac{1}{\cos\theta}\dfrac{d^2\Phi}{dA\, d\omega}$ $= \dfrac{1}{\cos\theta}\dfrac{dI}{dA}$	Radiance	$\text{W m}^{-2}\text{ sr}^{-1}$	Luminance	cd m^{-2} $= \text{lux}$
K	$\dfrac{I_v}{I_E}$			Luminous efficacy (photopic)	lm W^{-1}
$V(\lambda)$	$\dfrac{K(\lambda)}{K_m}$			Luminous efficiency (photopic)	—
K'	$\dfrac{I'_v}{I_E}$			Luminous efficiency (scotopic)	—
$V'(\lambda)$	$\dfrac{K'(\lambda)}{K'_m}$			Luminous efficiency (scotopic)	—

Abbreviations: J = joule, W = watt, lm = lumen, cd = candela, sr = steradian;
A = area, ω is a solid angle, t = time
θ = angle between line of sight and the normal to an area that is emitting light.

otherwise stated we shall be dealing with daylight vision.

These definitions are satisfactory for an engineer in lighting, who may be required to provide, for example, a certain level of illuminance on the reading desks of a library. Physical detectors of radiation whose response varies with wavelength nearly according to the $V(\lambda)$ function can be constructed (Section 16.24). They can be cali-brated using the above definitions so that they measure illuminance directly and enable the engineer to know when the specification has been met. For a scientist doing vision research the position is not so good. Even among people whose vision is classed as normal in every way, including colour vision, there are significant differences in relative visual response from the $V(\lambda)$ curve, especially in the blue.

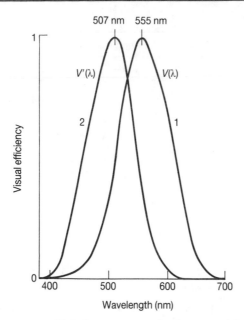

Figure 16.1   Relative response for human vision: $V(\lambda)$ day vision; $V'(\lambda)$ night vision

16.3   Luminous intensity

The procedures outlined above define the lumen and make it the fundamental photometric unit. The candela (formerly the basic unit) is now defined as the luminous intensity (in a given direction) of a source that emits monochromatic radiation of frequency† 540×10^{12} Hz and that has a radiant intensity (in that direction) of 1 lumen or 1/683 watt per steradian. The candela for any other wavelength can be obtained from the $V(\lambda)$ tables and for a mixture of wavelengths an integral (or summation) is used.

All other photometric units are similarly derived from the lumen.

16.4   The relation between excitance and luminance

In general a diffusing surface emits radiation whose intensity is proportional to cos θ, where θ is the angle between the direction of

observation and the normal to the surface. Since the area of surface whose image falls on a given small area of the retina is proportional to $1/\cos\theta$, the light per unit area of retina is independent of θ and the surface appears equally bright no matter from what angle it is viewed. The *albedo* of a diffusely reflecting surface is the fraction of incident light that is reflected. A perfectly diffusing surface has an albedo of unity. From the definitions of Table 16.1,

$$M = \frac{\mathrm{d}\Phi}{\mathrm{d}A} \qquad L(\theta) = \frac{1}{\cos\,\theta}\,\frac{\mathrm{d}^2\Phi}{\mathrm{d}A\,\mathrm{d}\bar\omega}$$

so that

$$M = \int L\,\mathrm{d}\bar\omega = \int_0^{\pi/2} L\,\cos\,\theta\,\sin\,\theta\,2\pi\,\mathrm{d}\theta = \pi L$$

$$(16.1)$$

Thus the excitance for a perfectly diffusing surface is π times the radiance (or luminance).

16.5   Noise

There are certain general considerations that affect all radiometric and photometric instruments. The study of these helps in the design of sensitive detectors and enables us to calculate the minimum amount of radiation that can be detected with a given type of instrument.

When making a measurement, a zero value is subtracted from the value obtained when the radiation falls on the detector. For some detectors the zero value changes slowly in a regular way. This is called *drift*. It need produce no error because the true zero at the time of a measurement can be obtained by interpolation from zeros taken before and after the measurement. In addition to drift both the zero and the reading obtained when the radiation is incident are subject to irregular fluctuations called *noise*. The mean value of the noise is zero and is characterised by the mean value of its square (or the RMS—root-mean-square value). The essence of noise is that we cannot predict its value at any particular time t (or even say whether it is going to be positive or negative). It therefore sets a limit to the smallest amount of radiation that can be detected and to the accuracy of measurements. The ratio of

---

† Frequency is stated to avoid doubt about wavelength in air or in vacuum. The corresponding wavelength in air is 555.02 nm.

signal to noise (SNR) is taken as an indication of the accuracy of a measurement.

16.6 Detectivity

There is a certain steady signal whose power is such that it would produce an output equal to the RMS power of the noise. This is called the *noise equivalent power* (NEP) and its reciprocal is called the *detectivity* (D). It is usual when dealing with weak sources of radiation to 'chop' the incident radiation (i.e. to interrupt it periodically). This produces a signal whose Fourier transform is centred round the chopping frequency (Fig. 16.2). The Fourier transform of the noise is usually spread more evenly over a wide range of frequencies, as shown in the figure. If the output is passed through an amplifier tuned sharply to the chopping frequency, then only the noise in its pass band is included. It is thus the RMS noise per unit pass band which is of importance; also, when comparing different detectors, the area is significant.

Therefore we define a quantity called the normalised detectivity:

$$D^* = \frac{(A \ \Delta f)^{1/2}}{\text{NEP}} = (A \ \Delta f)^{1/2} D \quad (16.2)$$

where A is the area and Δf is the frequency bandwidth.

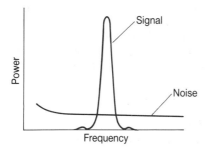

Figure 16.2 Frequency distribution for signal and noise

16.7 Sources of noise

We shall now consider some sources of noise that cannot be eliminated by good instrument design.

16.7.1 Photon shot noise

Photons are emitted, to some extent, at random. Consideration of photon statistics (Chapter 21) shows that a process that produces a pulse of average value $\langle n \rangle$ photons has a random fluctuation $\Delta n = \langle n^{1/2} \rangle$ so that

$$\text{SNR} = \frac{\langle n \rangle}{\Delta n} = \langle n^{1/2} \rangle \quad (16.3)$$

These fluctuations are significant only when the signal is very weak ($\langle n \rangle < 100$). The energy of 100 photons of wavelength 550 nm is less than 6×10^{-18} J. A millijoule pulse would contain about 1.6×10^{19} photons and the SNR would be about 4×10^8 if photon shot noise were the only source of noise.

16.7.2 Excess noise

Planck's law (eqn 12.7) gives the mean number of photons of a given frequency emitted from a unit area of a black body. For these, the statistic differs a little from that used above and we have

$$\langle \Delta n_\nu^2 \rangle = \langle n_\nu \rangle \left(1 + \frac{1}{e^{h\nu/kT} - 1} \right) \quad (16.4a)$$

where k is Boltzmann's constant and h is Planck's constant. In the optical region $e^{h\nu/kT}$ is much greater than unity and we have

$$\langle \Delta n_\nu^2 \rangle = \langle n_\nu \rangle (1 + e^{-h\nu/kT}) \quad (16.4b)$$

The first term is called photon shot noise and the second *excess fluctuations*. Although important in other connections, such excess fluctuations have no consequence for radiation detectors for the optical region and the near infrared.

16.7.3 Thermal noise or Johnson noise

In any resistance (R), whether it is carrying a current or not, there are random movements of the charge carriers. If the voltage (V) between the ends is measured, a random fluctuation of voltage is observed. Similarly, if current (I) is measured a random current fluctuation is observed. The mean square fluctuations for a frequency band Δf are

$$\langle I_r^2 \rangle = \frac{4kT \ \Delta f}{R} \quad (16.5a)$$

and

$$\langle V_r^2 \rangle = 4kRT \, \Delta f \qquad (16.5b)$$

These may be regarded as constituting a random RMS power fluctuation (P_r):

$$\langle P_r^2 \rangle^{1/2} = 4kT \, \Delta f \qquad (16.5c)$$

Note that the power is independent of R.

This noise is superposed both on the signal and on the dark current (even if this is zero). The details of the way in which this noise operates depend on the detector and associated circuitry.

16.7.4 Electron shot noise

Many sensitive detectors involve the absorption of photons causing electrons (or positive holes) to cross a potential barrier. In the photoemissive cells the electrons cross a barrier at the surface. In many semiconductor devices they cross a barrier at a *pn* junction. A probability element is involved in this process and noise—called *electron shot noise*—is associated with it. This effect is specified by giving the noise current I_n as a function of the signal current I_s:

$$\langle I_n^2 \rangle = 2eI_s \, \Delta f \qquad (16.6)$$

where e is the electronic charge $= 1.6 \times 10^{-19}$ coulombs.

The SNR is given by

$$\frac{I_s}{\langle I_r^2 \rangle^{1/2}} = \left(\frac{I_s}{2e \, \Delta f} \right)^{1/2} \qquad (16.7)$$

For a bandwidth of 10 Hz and $I_s = 3 \times 10^{-10}$ A, the SNR is about 10^4.

16.7.5 Flicker noise or contact noise

A kind of noise whose RMS value is very roughly proportional to $1/f$ (variation from $f^{-0.8}$ to f^{-2}) arises from causes that are not completely understood. Generally it arises from contacts and can be reduced by improving contacts which are already very good. It also appears in some resistors (especially carbon resistors).

16.8 Combination of sources of noise

The mean square of the total noise when several sources operate simultaneously is equal to the sum of the mean squares for the sources considered separately. From the preceding section it may be seen that photon or electron shot noise is reduced by reducing the bandwidth. The discussion of Section 5.5 shows that this involves increasing the time of observation. As we should expect, the longer observation time gives more information, i.e. a more accurate measurement.

The shot noise and the thermal noise are called 'white noise' because over the usual range of frequencies their RMS values are independent of f (as shown in Fig. 16.2).

16.9 Responsivity

Most of the detectors have a working range within which an electrical output (following amplification) is proportional to the energy of the incident radiation. Such detectors are said to be linear. The ratio of output to input is called the *responsivity*. For radiometry, detectors whose responsivity is independent of wavelength are desirable. The starting point in the design of those detectors is usually a metal strip or film coated with 'black' material which is expected to absorb nearly all the radiation that falls on it. Carbon black is satisfactory for wavelengths in the visible and near infrared but for longer wavelengths gold deposited in a suitable way is better (see p. 183 of ref. [1]). For radiometry a detector that measures energy, irrespective of wavelength, is said to be non-selective. For photometry, however, such a detector is highly selective. A different kind of detector whose response follows the $V(\lambda)$ curve is needed to measure lumens directly, i.e. to assess the radiation in terms of its value for vision.

16.10 Thermal detectors

Many detectors involve the absorption of radiation by a blackened strip and the measurement of the resulting temperature rise either by measuring the change of resistance or by a thermoelectric junction. In the

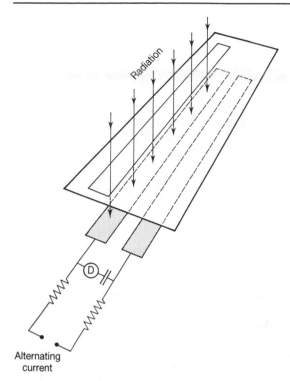

Figure 16.3 Langley bolometer. Note that only the a.c. component of the output is wanted—hence the capacitor

Langley bolometer (Fig. 16.3) a metal strip was arranged as a square array and interleaved with a second similar strip. The two were placed in the arms of a Wheatstone bridge and the resistance difference produced when radiation was allowed to pass suitable masks to fall on one strip and not on the other was measured. In the Hilger–

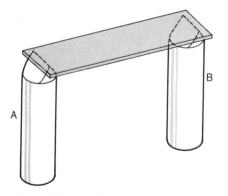

Figure 16.4 Hilger–Schwarz thermocouple. A and B are different semiconductors

Schwarz system (Fig. 16.4) the blackened strip is stretched between two different semiconductors which give a good thermoelectric voltage in response to the temperature rise. One way of calibrating these devices is to supply known quantities of heat electrically and to measure the response. A better method is to expose the device to a known amount of radiation from a black body.

16.11 Bolometers

A bolometer of very high detectivity can be made by cooling a suitable strip of metal so that it is held at the transition temperature between the normal and superconducting states. In this condition the rate of change of resistance with temperature is extremely high and a very small amount of incident radiation is sufficient to cause an easily measurable resistance change. The method detects a smaller amount of radiation than any other but only if the strip is held at the correct initial temperature within about 10^{-5} degrees. The cost and difficulties of the method are such that it is not suitable for routine use. It is appropriate for measurements where the highest possible detectivity is required.

16.12 Pyroelectric detectors

It is possible to cut from certain crystals (including $NaNO_2$ and $BaTiO_3$) a rectangular block which develops charges across two faces when radiation is incident in a perpendicular direction (Fig. 16.5). If metal plates are placed near the surface an alternating current proportional to the *rate of change* of the incident radiation is developed. Thus unless the beam of radiation is already modulated it must be chopped at a fairly high frequency. The responsivity is nearly independent of wavelength. The choice of frequency depends on the following considerations. There are two time constants: (1) a thermal time constant (τ_T), which depends on how rapidly the temperature of the crystal can follow the variation of radiation, and (2) an electrical time constant (τ_E), which depends on the input resistance to the associated electrical amplifier.

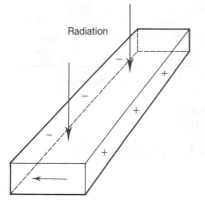

Figure 16.5 Pyroelectric detector

The responsivity $\mathscr{R}$ is proportional to

$$\frac{fAR}{[1+(2\pi f\tau_{\mathrm{r}})^2]^{1/2}[1+(2\pi f\tau_{\mathrm{E}})^2]^{1/2}} \quad (16.8)$$

Figure 16.6 shows how the responsivity varies with frequency. The responsivity falls off at low frequencies because the rate of change of temperature is small and also at high frequencies because the electrical system cuts off at approximately $f = 1/2\pi\tau_{\mathrm{E}}$. In between there is a region where output is nearly independent of frequency. The responsivity and detectivity in this range are increased by increasing the load resistance of the amplifier, but this reduces the electrical

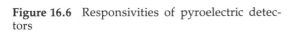

Figure 16.6 Responsivities of pyroelectric detectors

cut-off frequency so a compromise may be needed.

This device is robust, simple to construct and does not require crystals which are difficult to make so it should be fairly inexpensive. Its detectivity is not as high as that of some of the devices described above but it is sufficient for most purposes. The device may quite probably become the main device for routine uses.

16.13 Photoemissive devices

The first photoemissive devices (called *photoelectric cells*) consisted of an anode and cathode sealed in an evacuated glass envelope (Fig. 16.7a). Light falling on the cathode releases electrons and if a potential is applied a current is produced. This may be amplified to produce a suitable signal. The cell is sensitive to all wavelengths for which the photon energy exceeds the work function of the surface. With specially prepared surfaces wavelengths of up to 1.2 μm can be detected but the responsivity varies rapidly with wave-

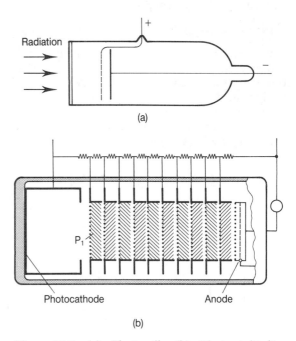

Figure 16.7 (a) Photocell. (b) Photomultiplier (schematic)

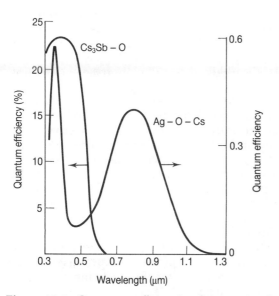

Figure 16.8 Quantum efficiencies for two kinds of cathode

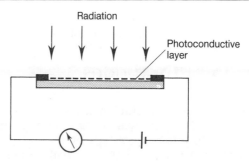

Figure 16.9 Semiconductor photoconductive detector (schematic)

16.14 Semiconductor photoconductive devices

Photoconductive devices are made by evaporating a suitable semiconductor on to a glass substrate. The layer should be thick enough to absorb most of the light that falls on it (Fig. 16.9). The absorbed photons produce charge carriers, usually by promoting electrons from the valence band to the conduction band. There is then a long-wave limit determined by the relation $hc/\lambda_m = eV_g$, where V_g is the band-gap voltage. CdS ($\lambda_m = 0.51\ \mu m$) is commonly used in light meters and PbS ($\lambda_m = 3\ \mu m$) and InSb ($\lambda_m = 7.5\ \mu m$) are used as infrared detectors.

Creation of carriers by transfer from donor impurity levels to the conduction band (or from the valence band to acceptors) also creates charge carriers and very low voltage energy gaps can be obtained leading to devices sensitive to long-wave infrared. At room temperature the amount of temperature radiation at 10 μm is appreciable so that to use photoconductive cells (or indeed any radiation-detection device) at longer wavelengths it is necessary to cool the surround.

Photoconductive cells of the kind we have described do not have a very high responsivity or detectivity but they are inexpensive to construct and suitable for a large number of practical uses including light meters for room lighting or street lighting and photographic exposure meters. A filter must be used so that the overall response is proportional to $V(\lambda)$. The infrared ones may be used (in conjunction with a source that is covered with a filter to exclude visible light)

length (Fig. 16.8). Note the much higher quantum efficiency of the visible, as opposed to the infrared, photocathode.

High detectivity cannot be obtained with this device because Johnson noise in the input resistance to the amplifier is amplified along with the signal. Accordingly an improvement called a *photomultiplier* (Fig. 16.7b) is often used. The electrons from the cathode are bent by an electric field and fall on a set of plates P_1 in the vacuum. If they are accelerated to a suitable extent more than one electron is released and these in turn are deflected on to a second plate and so on. Up to twelve plates ('dynodes') are used and at each stage there is an amplification of about 5 so that an overall amplification of the order of 10^8 can be obtained. More commonly an amplification of the order of 10^6 is used. One disadvantage is that 100 volts of potential is required at each stage (1200 V in all) and these potentials must be held constant to maintain constant amplification. The detectivity is about $10^{14}\ W^{-1}$ or about $2 \times 10^{11}\ lm^{-1}$ for a bandwidth of about 2 Hz. There is a dark current that may be reduced by cooling with liquid nitrogen and detectivities of about 100 times better are then obtained.

to provide concealed burglar alarms. The cells have a time constant of about 0.3 ms.

16.15   PIN diodes

A *p–n* diode is made by diffusing a *p*-type impurity into one side of a slice of *n*-type silicon. In the top layer the *p*-type impurity atoms outnumber the *n*-type atoms so that a *p*-type layer is formed (Fig. 16.10a). There is thus a *pn* junction within the slice. In the region of the junction the two types of impurities neutralise one another so that there is a region containing few carriers of either sign and therefore of high resistance. This is called the depletion region. If a potential is applied in one direction carriers move into the depletion region and a large current can flow in this forward direction. If the potential is in the opposite direction (reverse bias) carriers are drawn away from the edges of the depletion region, the overall current being small (Fig. 16.10b, curve 1). This depth of the depletion region is increased if a layer of intrinsic silicon separates the *p* and *n* regions. This arrangement is called a PIN diode.

Ohmic (i.e. non-rectifying) contacts to the silicon are provided as shown. Also, when the device is to be used as a photodetector the surface is covered with an antireflecting film (Section 4.30) with its maximum efficiency in the blue. This acts also as a protective layer. The *p*-type layer is so thin that light can penetrate it and ideally all the light is absorbed in the depletion layer, creating carriers of both signs. There is now a considerable current when the reverse bias is applied (Fig. 16.10b, curve 2). The insertion of the intrinsic layer reduces effective capacity and makes the response more rapid (and also more nearly linear). Rise times down to less than 0.1 ns are obtained with diodes whose active area is small ($0.2\ \text{mm}^2$ or less). Diodes with large active areas (up to $800\ \text{mm}^2$ have been made but they are slower (rise time 60 ns). A large number of PIN diodes for light detection are now on the market. Some degree of compromise may be needed in choosing the best one for a particular purpose.

16.16   Avalanche photodiodes

If a forward potential is applied to a diode with the characteristic shown in Fig. 16.10(b) a large current flows. When the forward voltage is increased to about 6 volts (in silicon devices) electrons reach voltages at which they are able to ionise neutral atoms (i.e. to produce more electrons in the conduction band). These in turn create more carriers and a breakdown 'avalanche' occurs if the current is not limited by the external circuit.

By working on the rising portion of the curve, but not so far up as to produce the 'avalanche', it is possible to make a photodetector that effectively has in-built amplification. It is, however, non-linear. If the amplification is increased beyond a certain point (typically about 50 times), extra noise increases rapidly and no advantage is gained. Owing to the non-linearity, this kind of device is not suitable for radiometry. It is very useful when a large signal is needed to work a relay and give an alarm.

16.17   Photovoltaic devices

Semiconductor diodes can also be used as

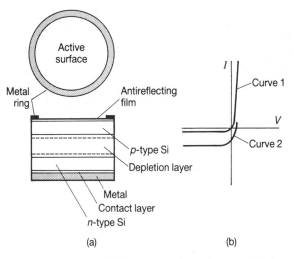

Figure 16.10   PIN diode: (a) plan view and side elevation, (b) voltage characteristic; curve 1 in darkness, curve 2 illuminated

photovoltaic devices. The new carriers created by the light produce potential differences between the ends. When used for measuring daylight this potential is sufficient to give a reading on a low-resistance microammeter. This avoids the need for a battery. They can be made more sensitive by adding an amplifier but the diode operated with a battery is usually better if a large response is required.

There are other semiconductor devices in use or under development. One promising development is in the charge-coupled devices (CCD). These are capable of storing information in a 'picture' and transferring it to a computer that can reconstruct the picture.

For further information on the devices discussed in Sections 16.10 to 16.17, refs. [2] to [5] may be consulted.

16.18 Intense sources

So much research has been done on devising detectors for very weak sources of radiation that one tends to forget that there are a few cases where the source is so powerful that it is beyond the range of an ordinary detector. Sextants for observing the height of the Sun are provided with a number of dark glasses and the observer uses them to reduce the apparent brightness of the sun to a level that can easily be tolerated. It is important that these glasses should absorb proportionately as much infrared as visible (and this applies also to sun spectacles). This method can also be used for measuring the total energy received from the Sun or from powerful sources such as high-pressure mercury arcs. A small complication arises because none of the so-called neutral filters is truly neutral. There is always some variation of transmission with wavelength. Thus it is necessary to measure the variation of transmission with wavelength for the glass using a source whose power is such that it is within the range of detector both when viewed direct and when viewed through the 'neutral' glass. Then if the distribution with wavelength of a powerful source is known approximately, the transmission of the glass for this source can be calculated.

16.19 Giant-pulse lasers

These methods cannot be used with giant-pulse lasers because the energy is so great that a piece of glass which absorbed a considerable fraction of it would be melted or shattered. Several methods have been devised of receiving the energy, or a known fraction of it, in a calorimeter and measuring the resulting rise in temperature. The calorimeter also incorporates a heating coil which can be made to deliver a known amount of energy which is varied until it causes the same temperature rise as did the radiation pulse. It should be arranged that the light pulse is not sharply focused on a small area of the calorimeter. A satisfactory way of doing this is illustrated in Fig. 16.11. The laser pulse impinges on a strongly reflecting surface many times so that only a small fraction is absorbed at each reflection. At the same time a large number of reflections occur and a negligible fraction of the incident energy is able to escape through the entry hole (rather like the black-body cavity).

Sometimes it is desired to measure the variation with time of the energy in a short laser pulse. For time resolution down to about a nanosecond this may be done by using a fast detector (e.g. a photocell or a fast diode) and displaying the response on a fast cathode ray tube. For shorter pulses the streak camera is available (Section 18.41).

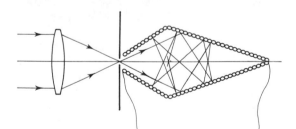

Figure 16.11 Double-cone calorimeter for measuring laser pulses. The cones are self-supporting oxidised aluminium wire which serves as the resistance thermometer

16.20 Photographic integration of light

The photographic plate is a device that responds to the integral of light received

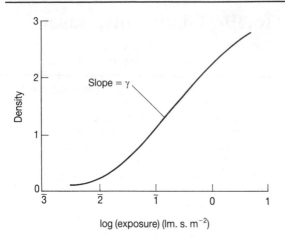

Figure 16.12   Photographic film: density/exposure characteristic

during the time of exposure. It does not continue to integrate a weak signal indefinitely because the undeveloped image slowly fades with time, but integration over several hundred hours is possible and fairly long periods (of several hours) are used by astronomers to detect weak stars. Over a considerable range the 'reciprocity law' that the photographic effect is proportional to Et (where E is the illuminance of the plate and t is the exposure time) holds good. The *density* of a photographic plate is defined as the logarithm to base 10 of the ratio of incident to transmitted flux. The variation of density with exposure, $\varepsilon = Et$, is shown in Fig. 16.12 for a typical photographic emulsion.

In the case shown the plate is very insensitive up to the point P, after which there is a linear response region of several decades. It is fairly common for manufacturers to give a preexposure so as to bring the plate to the foot of the linear region (i.e. to P). The slope γ varies a great deal between different plates. Plates can be made sensitive to about 1.2 μm in the infrared and to all wavelengths down to and including the X-ray region.

16.21   Absorption spectrophotometry

There are situations in which it is desired to compare the intensities in two beams of radi-

ation. In the most common case both beams come from the same source. One is passed through a substance whose absorption is to be measured and the other through a calibrated wedge as shown in Fig. 16.13(a). A rotating shutter (Fig. 16.13b) arranges for the beams to fall alternately on the detector.

Either by a commutator placed on its shaft or in another way, the rotating shutter notifies its position to an amplifier connected to the detector. This makes the amplifier phase-sensitive, i.e. it accepts signals only when one or other beam is completely uncovered. If the two signals are not equal the amplifier operates a relay which controls a motor so as to move the wedge until the signals become equal. The position of the wedge then determines the absorption in the other beam. This system forms the basis of a method of spectrophotometry in which light from a powerful source is passed through a monochromator whose control is varied so that different wavelengths are passed in

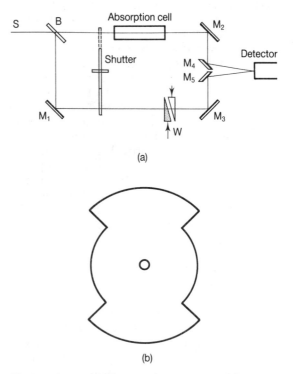

Figure 16.13   (a) Spectrophotometer, with source S; beam-splitter B: mirrors M_1 to M_5. One of the wedges W is clear glass, the other dark (neutral) glass. (b) Form of shutter

turn. The change of wavelength is slow enough for the wedge to assume the correct position, giving the absorption at every wavelength. The absorption and wavelength are both transferred to a computer which stores the information and also plots the absorption curve. This method of spectrophotometry is, however, being superseded by the interferometric Fourier method (Section 17.13).

16.22 Light amplifier

Figure 16.14 shows a device for producing an image with increased luminance. The picture to be amplified is imaged on a photoemissive film P_1. Electrons released from this film are accelerated to about 3 kV and focused to form an image on the fluorescent screen S_1. Owing to the acceleration each electron has an energy sufficient to release several photons from S_1, and more than one of these cause electrons to be released from the photoemissive film P_2. These are accelerated and focused on to a second fluorescent screen S_2. By using several stages an overall amplification of more than 10^4 is obtained and a person can clearly see a scene illuminated only by starlight. Note that noncoherent light is amplified.

If the photoemitter P_1 is an Ag–O–Cs cathode it will emit electrons when infrared radiation of up to 1.2 μm falls on it, and this infrared image can be converted to a light image by one stage of the amplifier described above. It can also be amplified by inserting further stages. This arrangement is called an *image converter*.

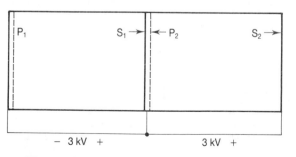

Figure 16.14 Schematic of image converter

16.23 Infrared imaging by scanning

A better method of producing an infrared image which can be used for infrared of longer wavelengths (to about 7 μm) is obtained by scanning an infrared image of the target (formed by spherical mirrors) over a small-area infrared detector. The output of the detector controls the intensity of a spot of light which is moved across a photographic plate or long-duration fluorescent screen. Thus a visible picture of temperature variations is produced. By careful design an apparatus sensitive to temperature variations of $0.006°C$ has been made. It has medical application since small local variations of skin temperature may be due to a pathological condition (e.g. a tumour) below the skin. In this form the instrument is called a *thermograph*.

16.24 Semiconductor devices in photometry

In general the semiconductor devices that are used for photometry have a maximum detectivity at about 550 nm. They are covered with a filter which gives a maximum at 555 nm and transmission falling to zero at about 800 nm. Sometimes there is a second filter which corrects the slope between 550 and 350 nm so that the overall effect is a fairly good match to the $V(\lambda)$ curve. One photometer of this type matches the $V(\lambda)$ within errors of 10% from 670 to 465 nm. Larger errors occur outside this range but the sensitivity of the eye is very low so the instrument measures illuminance quite accurately enough for most practical purposes. The vision research worker who may be interested in properties of the eye at the extreme wavelengths needs to use a correction curve (which the makers supply).

16.25 The eye as a system of radiation detectors

The eye has to provide for the brain information from which a detailed picture of the external world can be derived. To do this it has a system of two kinds of detectors, photopic (called cones) and scotopic (called rods). Each of these is a few micrometres in diameter. In the centre of the retina there is a group of 10 000 cones (and no rods). These to some extent behave like independent radiation detectors. The remainder of the retina has about 10^6 cones and 10^8 rods, but groups of these are effectively in parallel so that their sensitivity to radiation is increased.

Unlike physical receptors, the eye can adapt, i.e. its sensitivity to radiation increases if it is kept in the dark. The cones attain nearly their maximum sensitivity after ten minutes in the dark but the rods continue to adapt significantly for 30 minutes, at which stage they are about 10^4 times more sensitive than the cones. The early stages of this adaptation are due to recovery of the visual pigments so that more molecules are available to detect photons and also the background noise level falls. In the later stages the connections are changed so that signals from fairly large groups of rods are pooled. If a pulse of radiation of about 0.1 s duration and wavelength 507 nm is allowed to fall on a suitable area (neither too large or too small) it can be detected when it contains about five photons. In this condition the probability that any one rod has received more than one photon is small, so a rod must be capable of giving a signal in response to one photon. On the other hand (because of the background noise), more than one rod must be stimulated for the flash to be seen and these rods must be fairly close together so that their signals are 'pooled', i.e. recognised as associated with one event.

An average person can see, with some appreciation of detail and colour, with an ambient illuminance of 3×10^{-3} cd m^{-2}. A less detailed, colourless vision in the region outside the central $2°$ is obtained down to an illuminance of about 5×10^{-6} cd m^{-2}. Within observers still classed as having 'normal vision' the best have thresholds about three times lower and the worst about three times higher than the above figures.

16.26   Colour

In ordinary physical experiments there is an *input*, i.e. a set of conditions arranged by the experiment, and an *output* or result, both of which can be measured by instruments. In experiments on human vision we have an input that can be measured by instruments but the output is the subject's description of sensation, which cannot be measured. It is very difficult to establish meaningful relations between input and output. We can, however, go some way by means of experiments in which the subject is asked only to judge equality of sensation, e.g. to say when two patches of light appear indistinguishable.

16.27   Trichromatism

Three variables are needed to define the colour of a patch of light. Just as different coordinate systems may be used to locate a point in a solid, so there are different ways of choosing the three variables, but there are always three. One set of variables is obtained by choosing three primary colours (usually monochromatic) of which one (r) is in the red end of the spectrum, the second (g) is somewhere in the middle and the third (b) is towards the blue end. Suppose a subject is asked to view a bipartite field (Fig. 16.15) and side A is illuminated with the colour to be tested. Side B is illuminated with a mixture of r, g, b and the subject has three knobs, which each control the amount of one of these. The subject is asked to turn the knobs until side B matches side A. In general, a match can be made but a subject with normal colour vision has to use all three controls. Thus we may write:

$$C = r' + g' + b' \qquad (16.9)$$

where r' stands for an amount r of the red stimulus and so on. There is a certain range of colours for which no match is possible

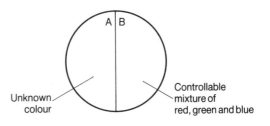

Figure 16.15   Bipartite field

(mainly purples). It is found that if the apparatus is rearranged so that one primary (usually g) can be added to the A side then a match can be obtained:

$$c + g' = r' + b' \qquad (16.10)$$

Equation (16.9) includes (16.10) if we allow g mathematically to have a negative value—even though 'negative light' does not exist.

A second way of making a match is to illuminate side B with monochromatic light plus white light. The subject again has three knobs. Two control the amount and wavelength of the monochromatic light and the third controls the amount of white. Again a normal subject has to use all three knobs and again there is a difficulty with purple-type colours. For them the monochromatic light has to be added to side A to give a match with white.

16.28 Psychometric and psychophysical terms

The second method of matching suggests a way of dividing up the concept of colour into three factors. Different names have been given for the colour perceived and the colour measured:

Colour perceived	Colour measured
Brightness	Luminance (E)
Hue	Dominant wavelength (λ_D)
Saturation	Purity

The first of these measures the total light in terms of the $V(\lambda)$ curve and the third represents the ratio of the luminances of the monochromatic light and the white light.

In many situations we are not interested in the exact value of the total luminance and then two coordinates are sufficient to define what is sometimes called the *chromaticity*.

The quantity $r = r'/(r' + g' + b')$ is plotted as ordinates and g (similarly defined) as abscissa. Since $r + g + b = 1$, the value of b corresponding to any point can be deduced.

16.29 Chromaticity

By a long series of experiments, the r and g coordinates for monochromatic light at a closely spaced series of wavelengths have

been determined for a number of subjects and a set of curves representing a standard observer has been adopted internationally. When results for one set of primaries have been measured, a fairly simple set of equations can be used to express the results in terms of another set of primaries, even though the second set may not be physically realisable (e.g. one 'primary' might consist of a certain amount of red—another amount of green). For convenience of calculation a very artificial set of primaries X, Y, Z have been adopted and the values x, y, z for each wavelength have been calculated. As before, $x + y + z = 1$ so the chromaticity is given by plotting y as ordinate and x as abscissa. The definition is such that the y coordinate represents the $V(\lambda)$, i.e. the total light is obtained by multiplying the y coordinate by a certain factor. Figure 16.16 shows the locus of the spectrum colours on an x, y diagram. The exact position of the point that represents white depends on the definition of white. The position W_1 represents the white light received from a tungsten lamp whose colour temperature is 2848 K.

Suppose colour coordinates of a paint sample under a certain illuminant are measured in London and those of another

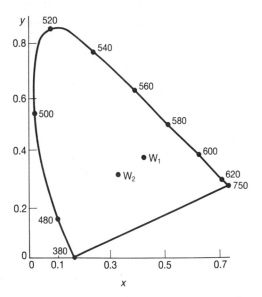

Figure 16.16 Spectrum locus on x, y, z coordinate system ($z = 1 - (x + y)$). W_1 point for white with colour temperature 2848 K, W_2 with colour temperature 6500 K

paint sample are measured in Washington under the same illuminant. Then if the measured coordinates are the same the samples should be visually indistinguishable if placed side by side. It is only recently that it has been possible to make measurements of sufficient accuracy for this to be true. The ability of the eye to distinguish small differences of chromaticity is extremely high.

16.30   Defective colour vision

About 99.5% of females and 90% of males have normal colour vision as described above. Nearly all those whose colour vision is defective fall into six groups: (1) protanopes who fail to distinguish colours in the red region and have reduced sensitivity to red light (about 1.3% of males), (2) deuteranopes who have normal $V(\lambda)$ curves (including normal sensitivity to red) but fail to distinguish red from green (about 1.3% of males), (3) tritanopes who confuse colours in the blue–green region (very rare). All of these can match certain colours with only two primaries and are called dichromats. There are also anomalous trichromats who have minor defects related to the above types, i.e. (4) protanomalous (1% of males), (5) deuteranomalous (nearly 5% of males) and (6) tritanomolus (very rare). It is desirable that more consideration should be given to defective colour vision in the design of road signs, etc. The red warning light should be given a different shape (e.g. a rectangle) from the green light.

References

1. J. Houghton and S. D. Smith, *Infra-red Physics*, Oxford University Press, 1966.
2. R. A. Smith, F. E. Jones and R. P. Chasmar, *The Detection and Measurement of Infra-red Radiation*, Oxford University Press, 1957.
3. E. Grum and C. J. Bartleson (Eds.), *Optical Radiation Measurements*, Academic Press, 1980.
4. A. Bar-Lev, *Semiconductors and Electronic Devices*, Prentice-Hall, 1984.
5. R. J. Keyes (Ed.), *Optical and Infra-red Detectors*, Springer-Verlag, 1977.

Problems

16.1   From the curve of $V(\lambda)$ in Fig. 16.1, cal-culate the flux in lumens emitted by a one-milliwatt laser (He–Ne, 633 nm).

16.2   If the beam of Problem 16.1 has a diameter at the output mirror of 0.45 mm and a divergence of 2×10^{-3} radian, calculate (a) the radiant excitance and intensity and (b) the luminous excitance and intensity.

16.3   Calculate the radiant energy density in a 4-watt laser beam of diameter 3 mm propagating in vacuum.

16.4   A 10-watt laser is mounted in a beam-expanding telescope so that the final beam divergence is 10^{-4} radians. Calculate the irradiance on the Moon's surface (Earth–Moon distance $= 3.5 \times 10^{5}$ km). If the laser wavelength is 488 nm, how many photons per second strike a detector of area 20 cm^2?

16.5   If the laser of Problem 16.4 is operated for 0.1 second, what will the signal-to-noise ratio of the detector signal? What assumption are you making?

16.6   Verify the last statement of Section 16.7.2.

16.7   If a photomultiplier photocathode has a quantum efficiency of 30% and each dynode produces on average 3.5 photoelectrons per incident electron, what is the overall gain in output electrons per incident photon for a twelve-dynode tube.

16.8   Assuming that the cones in the centre of the retina (Section 16.25) occupy 25 mm^2 and using the following simplified model of the eye, calculate, from the average separation between cones, the smallest angular separation between distant point sources that should be detectable. The eye is able to resolve point sources with an angular subtense of about 3×10^{-4} radian. Comment.

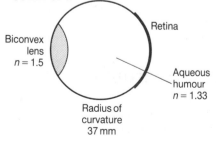

17

Interferometry

17.1 Preamble

The wavelength of visible light is of the order of 500 nm. If coherent beams from two mirrors combine to produce interference fringes, a relative shift of position of one beam in the direction of propagation of 250 nm will change the fringe intensity at a given point from a maximum to minimum. Thus the use of light waves in this manner can provide us with a method of measuring distance, displacements, angles, etc., with very high accuracy. Such devices are known as *interferometers*. We have already met the Michelson interferometer (Section 4.15). In this chapter we consider some of the interferometers introduced in the last two or three decades. We shall see that interferometers are now capable of measuring displacements of much less than $10^{-4}\lambda$. Full details of many of these instruments are to be found in refs. [1] to [5].

17.2 Sensitivity and stability

In the time of Michelson, the measurement usually involved setting a cross-wire on the centre of an interference fringe. This could be done to about one-tenth of a fringe width. Precautions to shield the interferometer from the effect of vibrations (e.g. due to a person walking in the corridor outside the room) or strain (e.g. pressure on the table supporting the instrument) were essential. Although much can be achieved by controlling the environment (e.g. by working in a cellar with good temperature control and observing from outside the room) the only completely satisfactory way of achieving the highest sensitivity is to design the interferometer so that it is insensitive to environmental changes, i.e. so that it is inherently stable.

17.3 Classification of interferometers

Three ways of classifying interferometers are commonly used.

(1) The wave-front is divided either by a beam-splitter, as in the Fizeau or Michelson instruments, or by selecting different parts of the same wave-front, as in Young's experiment and its derivatives.

(2) The distinction is made between two-beam interference (Fizeau, Michelson) and multiple-beam interference (Fabry–Perot, Tolansky).

(3) Wave-fronts are displaced in the direction of the light propagation

(Fabry–Perot) or laterally ('sheared'), as in Françon's instrument (Section 7.20 and Lebedev's interference microscope (Section 7.21).

Although all the interferometers so far mentioned involve plane wave-fronts, spherical or other wave-fronts may also be used. Thus in the testing of lenses and mirrors, interference between spherical wave-fronts is frequently used.

17.4 Lasers in relation to interferometry

With thermal sources, such as the low-pressure discharge lamp, the lengths of wave trains emitted amount to at best a few centimetres, thus limiting the maximum path difference that can be used. The high *temporal* coherence of a stabilised laser leads to coherence lengths of the order of kilometres, which in principle would allow enormous path differences to be employed. In practice the maximum measurable path difference is limited by the mechanical stability of the instrument rather than the coherence length of the laser. The high *spatial* coherence of lasers makes more light available for the division of wave-front interferometers. A line from a stabilized laser gives light confined to a very narrow range of wavelength, but interference fringes formed with such a laser are not necessarily very narrow. Their width in relation to the distance between fringes (i.e. the finesse, Section 4.23) depends on the type of interferometer, and we shall next consider methods by which fringes can be located to within a small fraction of their width.

17.5 Location of fringe position

When fringes constitute a sinusoidal fluctuation of intensity about a mean value, a cross-wire can be set on a maximum to within about $0.1d$, where d is the distance between adjacent maxima. Electronic methods can give a much higher order of accuracy.

Suppose that the fringes are scanned across a sensitive detector which is covered

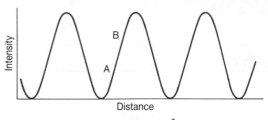

Figure 17.1 The $\cos^2$ fringes

with a narrow slit parallel to the fringes. Then a curve such as Fig. 17.1 can be displayed on an oscilloscope. The corresponding information can be stored in a computer which can use the observations over a wide range to determine the position of the maximum. This, however, is not the most accurate way to locate a fringe because the intensity varies very slowly in the neighbourhood of a maximum. A more sensitive method is to scan between two points (A and B) on either side of the region of maximum slope and to feed the signal into a detector tuned to the scanning frequency (e.g. phase-sensitive detector). A mid-point can then be determined with an accuracy of 10^{-6} of a fringe width. The scan may be made with a piezoelectric transducer attached to one mirror M of a Michelson interferometer (Fig. 17.2). This is suitable for moving the mirror in a direction accurately normal to its surface through distances up to about 2λ. A large mechanical movement may be measured by using the above method to obtain the fringe fraction and then moving

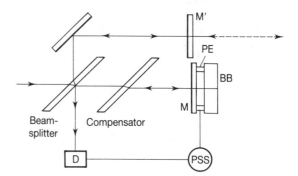

Figure 17.2 Modified Michelson interferometer. The mirror M is connected by a piezoelectric element to the firmly mounted block BB. D is the detector and PSS is a phase-sensitive electronic system

the mirror M′ through the distance to be measured. This may be so large that thousands of fringes pass. These are counted automatically and the new fraction is measured.

17.6 Laser long-path interferometers

Figure 17.3(a) shows the layout of a laser interferometer that can be used for cali-

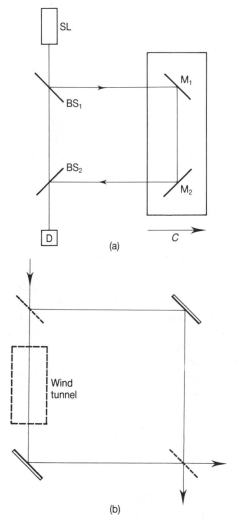

(a)

(b)

Figure 17.3 (a) Long-path interferometer with laser source. (b) Mach–Zehnder interferometer. In a wind tunnel, eddying, turbulent air produces distortion of the wave-fronts traversing the test chamber. These wave-fronts interfere with the undistorted waves traversing the alternative path

brating a high-quality metre scale. Light from a stabilized laser (SL) is divided at the beam-splitter BS_1 and recombined at BS_2 so that interference fringes are formed at the detector D. The scale under test is mounted on a carriage that also carries M_1 and M_2. The carriage is smoothly traversed in the direction shown by the arrow. A microscope objective forms an image of the scale on a slit that is in front of another detector D′ (not shown). When a division of the scale crosses the slit D′ it gives a signal to a computing system which also receives from D signals that give both the number of the fringe and the fraction. The carriage moves on continuously from one division to another traversing the whole metre scale in about 15 minutes. The computer prints out the error (to within 1 μm) of each millimetre division. The layout of an early long-path interferometer (Mach–Zehnder) used, for example in wind-tunnel studies is shown in Fig. 17.3(b).

17.7 Adjacent-path interferometers

The Michelson interferometer shown in Fig. 17.2 is more stable then the one shown in Fig. 4.10 because the paths are adjacent over a good deal of the length and are affected nearly equally by a small bending of the bed. The Rayleigh refractometer (Fig. 17.4a), despite its venerable age, would be regarded as a good stable adjacent-path interferometer if it were invented now. Four beams of light go through the tubes T_1, T_2 and T_3, T_4 (below the plane of the diagram). Two sets of fringes, one above the other, are formed and it is arranged that they touch one another. Two plates (P_1 and P_2) inclined to one another at a small angle are in the paths through T_1 and T_2 respectively. When they are rotated they insert a small path difference between the beams through T_1 and T_2 so that the upper set of fringes moves relative to the lower set. When the instrument is used as a gas refractometer, the tubes are first evacuated and the compensator is used to bring the two sets of fringes into exact registration. White light is used to identify the central fringes of zero path difference and monochromatic light to make the setting.

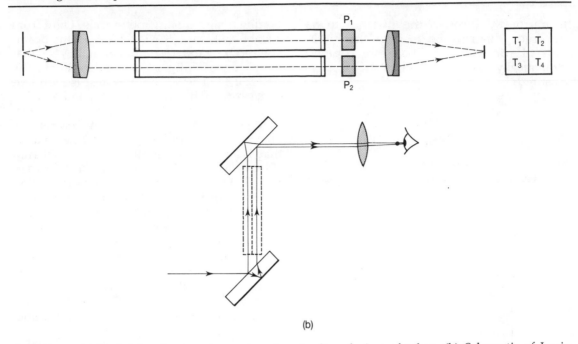

Figure 17.4 (a) Rayleigh refractometers: plan view (and) end view of tubes. (b) Schematic of Jamin interferometer

Then gas is admitted to T_1 and the compensator turned so as to restore the zero path registration. The compensator can be calibrated so that its angular movement gives the path difference $(\mu - 1)l$ due to insertion of the gas. A slightly different procedure is used to determine a small difference of index between two liquids.

The main paths are close together and are enclosed in one rigid piece of metal so that the instrument has good mechanical stability. Its chief disadvantage is that, to obtain spatial coherence between the beams through T_1 and T_2, a narrow slit source must be used and this limits the light available. In a modern version a laser would provide the coherent light and the registration of the fringes would be done electronically. An alternative way of arranging adjacent paths is shown in Fig. 17.4(b)—the *Jamin* interferometer.

17.8   Common-path interferometers

In an interference microscope (Section 7.2) the two beams follow exactly the same path except for the short length where one has to be diverted to pass through the specimen. Similarly, in the arrangement devised by Françon (Section 7.2) most of the paths are identical. Such interferometers are called common-path interferometers, even though a small portion of the paths are different. Generally, wave-shearing interferometers are common-path interferometers, but they can be used only when the structure to be examined is fairly simple. Otherwise the resulting interference pattern, which is really an overlap of two patterns, becomes too complicated to interpret.

17.9   Scatter interferometer

Figure 17.5 shows an interferometer (due to J. M. Burch) used for testing a concave mirror M. Two plastic replicas (R_1 and R_2) are made from a piece of lightly ground glass. Light from a pinhole source S forms a small image at the centre P of the mirror M. Reflected light from P and also scattered light from points on the mirror such as P_1 and P_2 are reflected by the beam splitter B and focused on R_2, which is reversed so as to give point-

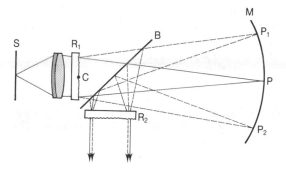

Figure 17.5 Scatter interferometer. S_1 and S_2 are identical replica scattering plates, M is the mirror under test and C its centre of curvature

to-point coincidence between R_2 and the image of R_1. C is the centre of curvature of the mirror M. An observer looking in the direction $C'P_2$ sees a large patch of light due to scattering in R_1 which comes from the whole area of M. Superimposed is a large patch due to scattering in R_2 of light from P whose wave-front is truly spherical since it has come from paraxial rays. The two beams are coherent and interference between the deformed wave-front (if M is not perfectly spherical) and the spherical wave-front is seen. The fringe pattern shows directly the departure of M from sphericity. The chief disadvantage of this interferometer is that a bright image of the central spot is also seen and the light in the fringes is comparatively small. This disadvantage is removed in a polarization interferometer of Dyson (p. 138 of ref. [1]).

17.10 Polarizing interferometers

Interferometers can be made by using a Wollaston prism or the Savart–Françon plate (Section 7.20) to form two beams. These beams are polarized at right angles to one another. They are made to interfere by imposing a relative phase difference between the components of one of them so as to rotate the plane of the light vector through a right angle. In order to obtain coherent beams it is necessary to insert a polarizer at the beginning. The beams usually follow neighbouring paths and sometimes common paths.

17.11 Compensated interferometer

Common-path interferometers are stabilized against variation of axial movement but not against rotation of mirrors (especially a mirror that is translated as in the Michelson interferometer). A very stable interferometer, due to Dyson is shown in Fig. 17.6. The principal foci of L lie in the planes through M_1 and M_2. Polarized light entering along DS is split by the Wollaston prism W into two beams, both of which go first to M_1, then to the small mirror M_2 and back to M_1. They leave along the path S'E. The quarter-wave plate P is traversed twice by one beam and so functions as a half-wave plate. It rotates the plane of polarisation of the beam which traverses it through $\pi/2$ so that both beams leave the system with the same polarisation. Interference occurs between the beam traversing the path SLOLM$_2$LOLS and SLO'LM$_2$LO"LS. Examine the effect of tilting M_1 and M_2 on this difference of path: you will see that compensation occurs

The interferometer may be used to measure very small variations in thickness of the plate O–e.g. to measure its thermal expansion—and an expansion of only $10^{-5}\lambda$ can be detected. Alternatively, if each of the three paths between L and M_1 is enclosed in a tube, the instrument may be used as a sensitive refractometer.

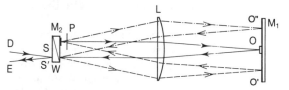

Figure 17.6 Stabilised interferometer (Dyson). M_1, M_2 and mirrors, W. Wollaston prism, P, quarter-wave plate, O, object whose dilatation is to be measured

17.12 Spherical Fabry–Perot interferometer

The original Fabry–Perot interferometer (Section 4.23) requires extremely accurate parallelism of the plane reflectors, particularly when a large separation is used so as

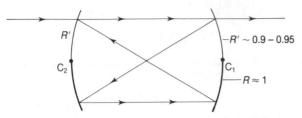

Figure 17.7 Spherical Fabry–Perot (FPS). Lower half of each mirror is totally reflecting

to obtain a high resolving power. An alternative, more easily adjusted, system was devised by Connes and is shown in Fig. 17.7; it is known as the spherical Fabry–Perot (FPS). The lower half of each mirror is covered with a completely reflecting film and the upper half with a film of reflection coefficient 0.9–0.95. The centre of curvature of each mirror lies on the surface of the opposite mirror. The diagram shows the path of rays incident in the direction parallel to $C_1 C_2$. Redraw the figure with an incident ray *not* parallel to $C_1 C_2$ and thereby convince yourself that inclined rays leave the system *together*, with the same path difference, and are not merely parallel to one another. Thus the system may be used with a converging beam of light. in contrast, the plane Fabry–Perot interferometer, when used as a scanning instrument, requires the incident light to be accurately collimated, which severely limits the amount of light that can be used.

For the plane interferometer the order of interference decreases as we go outwards from the centre, and the difference in fringe number is proportional to θ_n^2 (eqn 4.36) when θ_n, the fringe angular radius, is small. For the FPS, the difference (which arises from spherical aberration) is proportional to θ_n^4, which is an inconveniently rapid decrease for some purposes. If, however, the system is defocused by bringing the plates slightly closer together then there is a region where the difference is proportional to θ_n.

The advantage of FPS is that its performance is unaffected if one of the mirrors rotates a little. In this case the centre of curvature simply moves to another point on the opposite mirror so defining a new axis. The

resonant condition is preserved and there is no requirement for precise parallelism, as is the case for the plane mirror interferometer. In addition the FPS system is insensitive to small axial or sideways displacements. Both instruments are suitable for use in examining the spectral characteristics of the light from powerful lasers since the light flux is spread over a large area of mirror, thus reducing the danger of damage

17.13   Fourier transform spectroscopy

One method of investigating the distribution of energy with wavelength of a source of light is to form a spectrum with a large grating and to scan the spectrum with a detector covered by a slit. For high spectral resolution both the entrance slit of the spectrograph and the slit covering the detector must be very narrow. In consequence the amount of light falling on the detector is a very small fraction of that emitted from the source. For each setting of the grating position a long time is needed in order to obtain sufficient photons to get a reasonably high signal-to-noise ratio. We can estimate roughly how narrow the slits must be for a high-resolution spectrum. A grating of width D can resolve lines with a wavelength difference $\Delta\lambda = \lambda^2/D$ (eqn 19.18), so that for a large grating $(D = 0.25 \text{ m})$ and for $\lambda = 500 \text{ nm}$, $\Delta\lambda = 10^{-3}$ nm. At a distance $l = 5$ m, the slit width would be < 0.01 mm. To scan the visible spectrum from 600 to 400 nm would require 2×10^5 steps. Since the slits would be set to be less than that corresponding to the limit $\Delta\lambda$, then only $\sim 10^{-6}$ of the incident power would be measured at each point. Thus except for very powerful light sources, the time required for a scan would be very long. Although there are powerful sources available in the ultraviolet and visible regions of the spectrum (e.g. tunable lasers—Section 20.14—and the synchrotron—Section 22.25) the infrared region presents particular difficulties. We shall examine a method that overcomes the above difficulties and enables high-resolution spectroscopy to be carried out in the far infrared region.

17.14 Fourier transform system

Imagine the Michelson interferometer of Fig. 4.10 illuminated with a *collimated* beam of light (Fig. 17.8). The emerging light at E will also be collimated and so can be focused, by a concave mirror on to a detector. Suppose the mirror M_2 to be accurately parallel to the reference plane R and the beams to be adjusted to be of the same amplitude. The path difference (x) between M_2 and R is varied by moving M_1 along very accurate guides in a direction perpendicular to its surface. The signal received at the detector will vary and is recorded. (In the largest instruments of this type path differences up to ~3 m can be obtained).

How will the observed variation of the interference signal relate to the spectral emission of the source? Suppose the amplitude of the emission associated with the spatial frequency range x to $x + dx$ is $a(x)dx$. The phase difference between the beams travelling via M_1 and M_2 is xx and the resultant irradiance is $|a(x)|^2 (1 + \cos xx) \, dx = g(x) (1 + \cos xx) \, dx$. The total irradiance for all wave numbers is

$$I(x) = \int_0^\infty g(x) \, (1 + \cos xx) \, dx \qquad (17.1)$$

$$= \tfrac{1}{2} I(0) + \tfrac{1}{2} \int_0^\infty g(x) \, (e^{ixx} + e^{-ixx}) \, dx$$

$$= \tfrac{1}{2} I(0) + \tfrac{1}{2} \int_{-\infty}^\infty g(x) e^{ixx} \, dx \qquad (17.2)$$

where $I(0) = 2 \int_0^\infty g(x) \, dx$.

Defining $W(x)$ as

$$W(x) = 2I(x) - I(0) = \int_{-\infty}^\infty g(x) e^{ixx} \, dx \qquad (17.3)$$

we see that $W(x)$, obtained from the measured signals at $x = 0$ and at $x = x$, is the Fourier transform of $g(x)$. Thus on transforming the curve of $W(x)$ against x, the required spectral distribution function $g(x)$ is found.

The principle of this method was known to Rayleigh and also to Michelson but a 'hand calculation' would have taken an inordinately long time (more than a century!). The method became practicable when (1) elec-

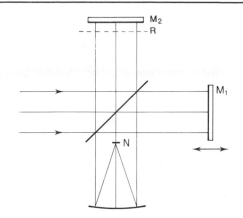

Figure 17.8 Schematic of Michelson Fourier transform interferometer

tronic computers reduced this time to a few minutes and (2) guides of sufficient accuracy were made. The instrument need not depend on the accuracy of the screw because it can be calibrated by using a helium–neon laser to produce accurately spaced oscillations in $W(x)$.

The instrument has two advantages over the grating instrument:

1. Each value of $W(x)$ depends on the sum of contributions from all wavelengths and thus the signal-to-noise ratio (for a given time of observation) is much better. This is known as the Fellgett advantage (after P. B. Fellgett, who discovered it in 1951) or the multiplexing advantage. Simple theory would suggest an improvement by a factor $N^{1/2}$ in the signal-to-noise ratio (where N is the number of observations), but when we allow for the cosine factor multiplying each contribution to the signal the gain is $\tfrac{1}{2} N^{1/2}$ which is a factor of about 500.

2. The instrument accepts a much larger fraction of the light from the source than does a grating spectroscope. It is necessary to restrict the size of the source in order to have effectively plane waves, but it can subtend an angle of a few degrees at the first lens (because the error is proportional to $1 - \cos \Phi$ which equals 0.004 for $\Phi = 5°$). This gives an angular area many times larger than that of a narrow slit and there is no restriction compared with the exit slit. This advantage is called the Jacquinot advantage.

17.15   Resolving power

The resolving power is determined by the maximum phase difference and hence by the maximum path difference so that we should expect it to be the same as that of a grating of width x_{max}. Detailed investigation confirms this so we have $\Delta\lambda = \lambda^2/36$ for an instrument with $x_{max} = 18$ cm (where $\Delta\lambda$, the minimum wavelength resolved, and λ are expressed in centimetres). Infrared spectroscopists express their results in wave numbers, $\bar{\nu}$. The wave number is the reciprocal of the wavelength in centimetres so that $\Delta\bar{\nu} = -\Delta\lambda/\lambda^2$ and hence is $1/36 = 0.028$ cm^{-1} for this instrument. There are, of course, residual effects of noise, imperfect parallelism of the entrance beam and imperfections of manufacture. The practical limit of resolution is 0.04 cm^{-1}. There is a difficulty in finding materials for the beam-splitter which are transparent over the whole of the spectral region. The parallel beam at the entrance may be obtained by using a small source and a concave aspherical mirror. The beam-splitter may be made by depositing a thin film on a very thin layer of plastic which is stretched across a plane ring.

17.16   Laser speckle

Figure 17.9(a) shows a piece of ground glass S illuminated by a laser and observed visually. Each point on the object produces an approximately Fraunhofer diffraction pattern on the retina due to diffraction at the pupil. At any point on the retina the diffraction patterns D_A, D_B from a small area AB of the object overlap. Since the light beams are coherent interference occurs. There are phase differences between light from different parts of the small area because of the irregularity of the surface. These phase differences are random and at some points on the retina there is high illuminance and at others much lower. A granular appearance known as 'laser speckle' is seen. It is obtained with laser light because the phase differences are constant over a typical observing time. A thermal source would superpose its own phase fluctuations (at a rate of 10^6 s^{-1} or more) and the observer would see a constant uniform luminance on the screen. The laser speckle can be photographed (Fig. 17.9b) but the diffraction images become smaller as the aperture of the lens increases. The speckle can equally well be observed with opaque objects and diffusely reflected light. It can also be obtained without the use of a lens under the conditions of Fresnel diffraction.

There are numerous applications of laser speckle which can be classed as interferometric. Some of these are more convenient than alternative methods of measurement. We give below two applications of speckle. Reference [3] describes many others.

17.17   The Burch–Tokarski experiment

Suppose a speckle pattern is photographed with the arrangement shown in Fig. 17.9(b) and two exposures are made, the plate S being moved sideways through a distance d between the exposures. Two superposed speckle patterns appear on the photograph. These will constitute a number of pairs of points, the separation (d) being the same for every pair. In Section 6.12 we showed that the Fraunhofer diffraction pattern for a number of geometrically similar objects with random distances between them was similar to that of a single object but N times stronger. The pairs of speckle points constitute a large number of similar objects and if

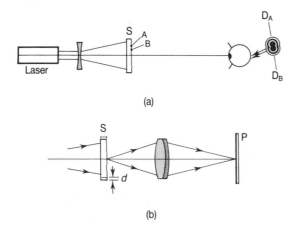

(a)

(b)

Figure 17.9   (a) Scatter plate S illuminated by expanded beam from laser L and viewed by eye. (b) S photographed on plate P

the photographic plate is put in a diffractomer a Fourier transform consisting of interference fringes similar to those observed by Young is obtained. The value of d can be deduced from measurement of the fringes.

Since there are many easy ways of measuring the displacement of S, this experiment must be regarded as a basic demonstration of the detection of movement by double speckle patterns. Modifications developed from this experiment enable displacements of different parts of a concrete structure under load to be measured. Also, detection of small vibrations of parts of an engineering structure is possible. In these applications the speckle method competes with holography. It has the advantage of not requiring a reference beam but does not always give results that are so accurate or so easily interpreted.

17.18 Astronomical applications of speckle

At most times the atmosphere contains variations of refractive index due to irregularities in local temperature and humidity. The extent of these is from less than a centimetre to tens of centimetres and they change in times of the order of 0.1 s. If a star is photographed with an exposure of a second or more the plane waves from the star are distorted in different ways during the exposure and a diffuse image is obtained. If, however, an exposure of 0.01 s is used then some of the light is received in the form of waves which are nearly plane over most of the objective but inclined at small angles to one another. A speckle pattern is obtained. Each dot is an image whose size is determined partly by the size of the objective and partly by the diameter of the star (if its angular size is large enough). If the star is double, pairs of spots with equal separations are obtained and the Fourier spectrum (obtained with a diffractometer) shows Young's fringes. The angular separation of the components can be obtained from the spatial frequency of the fringes. Also, if the components are of equal intensity, the fringes will have unit contrast (zero intensity in their minima). If the stars

are not of equal intensity the contrast will be lower and its value will yield the ratio of the intensities. By observing the double star from time to time the variation of angular separation is measured. If velocity information is also obtained by measuring Doppler shifts the complete orbit and masses of the stars can be derived.

17.19 Looking at stars

If the star is single the Fourier transform of the speckle pattern is like that shown in Fig. 17.10(a). By averaging spectra obtained with different exposures a smooth Fourier transform is obtained (Fig. 17.10b). By comparing this with the transform for a star whose angular diameter is known to be very small, the angular diameter of the star can be obtained. The computation is a little complicated because the star is brighter in the centre than at the limb. Note that by averaging the Fourier spectra all the information in the speckle is used. If we had averaged the speckle itself, the information it contained would have been thrown away. Up to 1976 Labeyrie had measured more than a hundred stars by this method with an accuracy of 0.01 seconds of arc. Stars down to the ninth magnitude were measured. It seems probable that CCD photoelectric devices (Section 16.17) will soon be developed to the point where they can record speckle images of still weaker stars. In the case of a binary star, the distribution is crossed by fringes (Fig. 17.10c): The angular separation of the two stars is obtained from the fringe spacing.

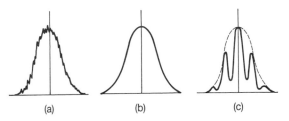

Figure 17.10 Fourier transform along a diameter of speckle pattern: (a) derived from one exposure, (b) average Fourier transform for several exposures, (c) form of trace for binary star

References

1. J. Dyson, *Interferometry as a Measuring Tool*, The Machinery Publishing Company, London, 1970.
2. M. Françon and S. Mallick, *Polarization Interferometers* (in English), Wiley–Interscience, London, New York, Sydney and Toronto, 1971.
3. M. Françon, *Laser Speckle and Applications in Optics* (in English), Academic Press, New York, San Francisco and London, 1979.
4. J. Guild, *Interference Systems of Crossed Diffraction Gratings*, Clarendon Press, Oxford, 1956.
5. K. M. Baird, in *Advanced Optical Techniques* (Ed. A. C. S. Van Heel), North Holland Publishing Company, 1967, pp. 129–41.

Problems

17.1 If the phase-sensitive detection system referred to in Section 17.5 can measure the intensity difference AB (Fig. 17.1) to within $\pm 0.01\%$, with what accuracy may the fringe position be located?

17.2 In using a long-path interferometer, such as that shown in Fig. 17.2, for calibrating one metre scales to within one micrometre, would you worry about variations in atmospheric pressure?

17.3 You wish to design a Rayleigh refractometer (Section 17.7) which will estimate the amount of helium $(n - 1 = 3.5 \times 10^{-5})$ in a mixture of helium and argon $(n - 1 = 2.8 \times 10^{-4})$. The mixture is at STP and the argon content is in the range 50–100%. The compensator range covers 200 fringes. What length of tube would you use and what accuracy would you expect $(\lambda = 500$ nm$)$?

17.4 In the Dyson interferometer (Section 17.11) in which a 10 mm thick fused quartz specimen is mounted, the number of wavelengths counted on heating from 0°C to the temperatures indicated are given in the table. If a helium–neon laser $(\lambda = 633$ nm$)$ is used determine the (quadratic) relation between the specimen length and temperature.

Temperature ($^\circ$C)	Wavelength shift from 0°C
20	180.095
50	387.046
70	557.662
85	688.783
100	714.060

17.5 A beam-splitter for the Michelson interferometer (Section 17.14) for the infrared is to be made of a thin film of organic material of refractive index 1.51. If it has *no* metallic coating on it, determine the thickness that will ensure maximum reflectance at 10 μm. Calculate the reflection coefficient for 45° incidence (note the polarisation effects) and estimate the amplitudes of the two emerging beams. Assume the two mirrors are perfectly reflecting.

18

Optical Instruments

18.1 Instruments and systems

The purpose of an optical instrument is to provide the human brain with information that cannot readily be obtained by the unaided eye. Originally all optical systems were designed to form images to be inspected by the eye and these instruments had to be matched to the capabilities of the eye. Usually the objective of a telescope or microscope produces an image containing fine detail that the eye cannot resolve. Therefore the picture is inspected through an eyepiece which magnifies it. When a picture is magnified the light per unit area is decreased and, beyond a certain point, this makes it more difficult for the eye to distinguish detail. Thus there is an optimum magnification determined by the properties of the eye. A visual instrument and the eye taken together form a system whose parts must be matched to one another. We shall see later that, within the system, the eye itself forms a subsystem whose parts are matched in a remarkable way.

18.2 Use of detectors

In recent years the relation between man and many optical instruments has changed in an

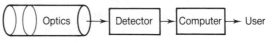

Figure 18.1 Schematic of optical system

important way. The image produced by the optics is inspected not by the eye but by a set of physical detectors. Sometimes these detectors produce an output that can be directly used by the experimenter, e.g. a graph showing the variations of the brightness of a star with time. In other cases the information produced may be so great that no human brain could handle it directly. The system must then include a computer which analyses the information and produces things like average values, mean-square deviations and correlation coefficients which the scientist may use to test theories (Fig. 18.1).

18.3 Need to consider whole system

An optical system, whether it be as simple as a hand magnifier and the eye or as complicated as a satellite telescope, must be considered as a whole. To obtain the best results the system must be optimised as a whole using *systems analysis* in which each part is considered both in regard to its own performance and in its relation with the other parts. Cost enters into this analysis because it would obviously be inefficient to spend a great sum to obtain excellence in one part so that it produced information that was lost through defects in other parts.

18.4 Use and limitation of ray theory

The analysis of optical systems involves different aspects of optical theory and each chapter in this book contributes something to this analysis. We must treat the system as a whole and then we shall find that the ultimate performance of most instruments is limited in ways that can be understood only in terms of either wave theory or photon concepts. We shall also find that there are optical systems such as interferometers and holographic devices in which ray theory plays only a small part. Nevertheless, the ray theory of many standard instruments such as telescopes and microscopes is very important in modern optics.

18.5 Aperture

For a given pair of conjugate points on-axis (A_1 and A_2 in Fig. 18.2) the cone of rays that form the image is limited either by the edges of one of the components or by a stop (P_0 in Fig. 18.2). Fig. 18.2 shows a simple system with only two lenses but the discussion which follows applies to more complicated systems if it is understood that L_1 stands for all the components on the input side and L_2 for those on the output side of P_0. It applies also, with minor alterations, to mirror systems. P_0 is called the *aperture stop*. Its conjugate in respect of L_1 is called the *entrance pupil* (P_1) and its conjugate in respect of L_2 is called the *exit pupil* (P_2). In respect of the

system as a whole P_2 is the image of P_1. All rays that pass through P_0 must also pass through P_1 and P_2 if the cone of rays is small enough for paraxial optics to be applied, but not if the image at P_2 is subject to large aberrations. The optical designer therefore wishes to correct the system in respect of P_1 and P_2 but this is often in conflict with the more important corrections in respect of the object and image points A_1 and A_2. A compromise has to be made and only moderately good correction of the image at P_2 has to be accepted.

18.6 The exit pupil

With a visual instrument, all the rays used in forming the image enter the eye if the eye can be placed in coincidence with the exit pupil and if the exit pupil is not larger than the eye pupil. On the other hand, if the exit pupil is much smaller than the eye pupil the best resolving power of the eye is not obtained. Therefore the exit pupil should be about the same size as the eye pupil, i.e. about 2.5 mm in diameter for an instrument to be used in daylight and 7.5 mm for an instrument to be used at night.

The distance of the exit pupil from the last optical component is called the *eye relief*. For an instrument such as a theodolite or a pair of binoculars about 5 mm of eye relief is sufficient, although not for wearers of spectacles. An appreciably larger eye relief is needed in a gunsight which will recoil when the gun is fired. The eye is usually guided to the right position by means of an eye-cup of plastic or (for the gunsight) of rubber. This does not permit very accurate location and it is an advantage if the rays do not make very large angles with the axis thus permitting some tolerance in eye position (compare Fig. 18.3a and b).

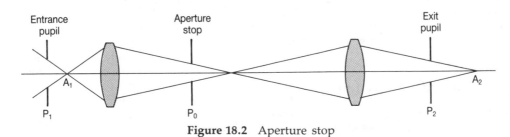

Figure 18.2 Aperture stop

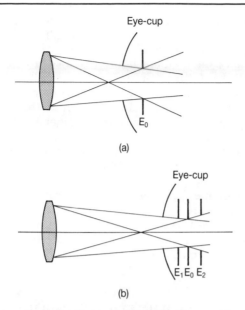

(a)

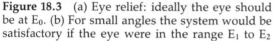

(b)

Figure 18.3 (a) Eye relief: ideally the eye should be at E_0. (b) For small angles the system would be satisfactory if the eye were in the range E_1 to E_2

18.7 Field

For an object point (B_1 in Fig. 18.4) some distance off-axis the whole of the cone of rays from the first lens may not go through the stop P or may fail to pass one of the components. This effect of *vignetting* implies that beyond a certain size of field the image becomes less strongly illuminated and beyond a certain large size no light at all gets through. Thus there is a finite size of field. The size is determined either by one of the components or by a field stop inserted in a suitable place to limit the field. In a telescope the size of the field is usually specified as the angle of view and its size is determined either by the size and position of the eyepiece or by a stop placed in the plane of the focus of the objective, as shown in Fig. 18.5(a). When the field is limited by the size of the eyepiece it may be substantially increased by a suitably placed lens (Fig. 18.5b). This lens, called a *field lens*,

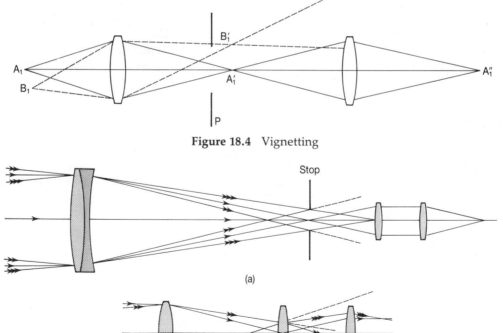

Figure 18.4 Vignetting

(a)

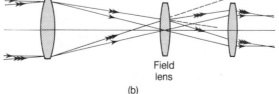

(b)

Figure 18.5 (a) Field stop. (b) Field lens

bends into the eyepiece rays which would otherwise be lost. It does not affect the focus of the instrument. If, as is usual, it is a fairly weak lens it has only a small effect on the corrections of aberrations. In specifying the field, it is necessary to remember that the field on the object side may be very different from the field on the image side. For example, a pair of binoculars with a magnification of ×10 might have a field of only 5° on the object side and 50° on the image side.

18.8 Relay systems

In certain instruments such as *endoscopes* and *cystoscopes* it is necessary to transmit the light along a tube which is long in comparison with its width. We have seen in Chapter 2 that this may be done with fibre-optic systems and these systems must be used when the tube is appreciably curved. When the tube is straight lenses may be used to give a somewhat sharper image. Figure 18.6 shows a traditional system of lenses suitable for a periscope or a cystoscope. Images of the object that, in a periscope, would be far to the left are formed at I_1, I_2 and I_3. The field lenses F_1, F_2, F_3 form images of the first relay system R_1 and on the second relay system R_2 and this keeps as much as possible of the light within the tube. The cystocope differs from the periscope in that the object is comparatively near to the objective and is in a medium of index 1.3. The arrangement of alternate relay systems and field lenses is essentially the same. The periscope may be regarded as a telescope with relays and the cystoscope as a low-power microscope with relays. The cystoscope is also limited by the necessity of making the outside diameter very small (a few milimetres). A typical cystoscope may have 20 or more lenses which implies 40 or more surfaces. About 4.3% of the light is reflected at each surface

for a glass of index 1.52 and for one of index 1.67 the loss is 6.3% (Section 8.17). Thus 40 untreated glass surfaces transmit only about 10% of the incident light. The loss of light is important, but even more serious in its effect on the performance of the instrument is the fact that a considerable fraction of the light appears in the field of view as a general haze (possibly with some bright spots) after repeated internal reflections. In the traditional cystoscope the light was supplied by a small lamp placed at the internal end of the cystoscope. The lamp had to be run at a low temperature, to avoid burning the patient. This greatly reduced the available light and made it particularly deficient in the blue end of the spectrum so that colour reproduction in the image was poor.

18.9 Modern endoscopes

Figure 18.7(a) is a schematic diagram of an endoscope designed by Hopkins. The small lenses separated by long air spaces in Fig. 18.6 have been replaced by long glass rods (with curved ends) separated by short air spaces. We may regard this as a set of very thick lenses or, alternatively, as a set of air lenses in a medium that is mainly glass. The total light transmitted (F) is proportional to the area of the entrance pupil and to the area of the region imaged so that

$$F = K\rho_0^2(n_0 \sin \alpha_0)^2 \qquad (18.1)$$

where ρ_0 is the radius of the area imaged, n_0 is the index of the medium and K is a constant that depends on the illuminance of the object and its coefficient of diffuse reflection. From eqn (3.7b),

$$\rho_0 n_0 \alpha_0 = \rho n \alpha \qquad (18.2)$$

where ρ is the radius of one of the lenses L_1, etc., and if D is the distance between successive images I_1, I_2, etc., $\alpha \doteq \rho/D$. Thus,

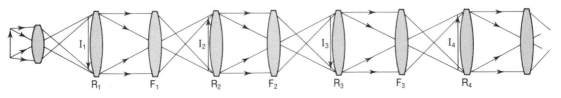

Figure 18.6 Early type of cystoscope

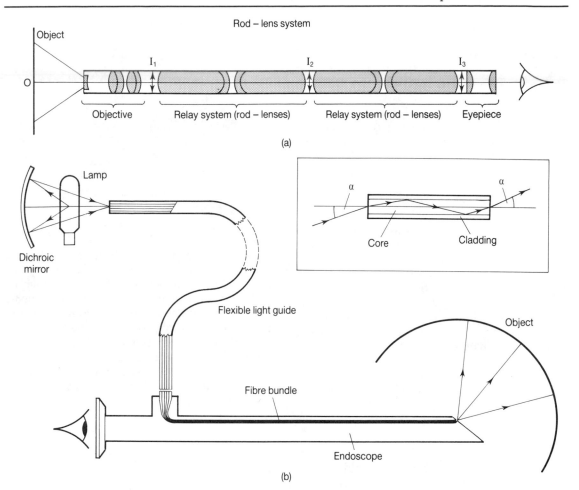

Figure 18.7 (a) Modern form of endoscope. (b) Fibre-optic endoscope. Reproduced by permission of H. H. Hopkins

substituting in eqn (18.1),

$$F = Kn^2 \frac{\rho^4}{D^2} \qquad (18.3)$$

18.10 Increasing the light throughput

Owing to the difficulty of mounting the thin lenses of the traditional endoscope it was necessary to use an inner sleeve which obstructed part of the otherwise available area. The rod lenses can be ground to fit the tube accurately and the short spacers are extremely thin. The usable radius of the lens for a given radius of the tube is increased by a factor of 1.4. Also, in the traditional endoscope $n = 1$ and in the new instrument n is between 1.5 and 1.6. Thus the light transmitted is increased by a factor $n^2\rho^4 = 1.55^2 \times 1.4^4 = 9.6$. A further gain is obtained by coating the lenses, not with the single-layer antireflecting film used in ordinary 'bloomed' lenses but by multilayer antireflecting films (Section 4.30). These reduce the reflection coefficient to 0.25% and this increases the transmission from 10 to 90%. The haze due to repeated reflections is removed and the total light is increased by a factor of about 80.

18.11 Optical fibre system

In another version of the endoscope, optical fibres are used to convey light from a quartz–iodine lamp as shown in Fig. 18.7(b). The light is taken to the fibres after reflection at a dielectric reflector (Section 4.31), which reflects less than 20% of the heat and 95% of the light. This further increases the available light.

In order to obtain a good image after transmission through 20 lenses it is necessary that the aberrations should be very well corrected. A computer aided design of the type described in Chapter 4 is used to achieve this result.

The new endoscope enables the surgeon to see a clear picture of a large area using an instrument only 4 mm in outer diameter with lenses 2.7 mm in diameter. It has been described in detail because it shows how one radically new idea (the rod lenses) supported by the logical application of modern techniques (the multilayer films, the fibre-optic light conduit and the computer correction of aberrations) can achieve an outstanding overall improvement in the usefulness of an instrument.

18.12 The visual system

The visual system can be regarded as having four parts:

1. The optic system of the eye which forms an image of the external world on the retina at the back of the eye.
2. The retina itself which contains photoreceptors. These transduce the light into electrical signals which, after some processing, are fed into the optic nerve.
3. The optic nerve which transmits information to the visual cortex (in the brain).
4. A central processor in which the information is filtered and reorganised into a form that helps to decide what action is needed.

18.13 The eye

The dioptric system of the eye, shown in Fig. 18.8, consists of the *cornea* and the lens. Between the cornea and the lens is a liquid called the *aqueous humour* and another liquid called the *vitreous humour* is between the lens and the retina. The cornea is a thin transparent window in the outer covering which encloses and protects the eye. The two

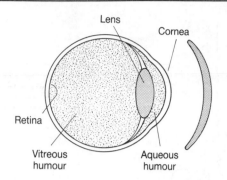

Figure 18.8 Diagrammatic view of the eye

liquids have about the same refractive index as water (1.33). The index of the lens increases from about 1.38 at the surface to 1.41 in the centre. The main refraction occurs at the cornea which has a power of about 40 D. The power of the lens can be adjusted by altering the shape of the front surface so that near or distant objects can be focused on the retina. Its mean power is about 23 D. The aperture of the eye is controlled by the iris diaphragm which responds to light so as to give an entrance pupil of just over 2.0 mm in very bright light and 7.5 mm at low illumination.

18.14 Aberrations of the eye

The eye is not well corrected either for spherical aberration or for chromatic aberration. At high luminance, the aperture is so small (about $f/12$) that spherical aberration has a negligible effect and the eye can distinguish two point sources which subtend a visual angle of one minute (1') at the eye. The chromatic aberration should still cause coloured fringes to be formed at the edges of objects but, in some way that is not understood, the central processor suppresses perception of these fringes. At low luminance the aperture is nearly $f/3$ and spherical aberation makes the image much less sharp; also the adjustment for the best focus is slightly different.

18.15 The eye's receptors

The receptors are of two kinds called *cones* and *rods* though the shapes do not always correspond with these descriptions. An area in the centre of the retina, corresponding to a diameter of about $2°$ in the visual field, contains about 10 000 cones and nearly no rods. The spacing of these cones is almost exactly that required to match the quality of the image formed by the dioptric system at high luminance. Outside this central area, which is called the *fovea*, there are both rods and cones, with the proportion of rods increasing in the outer region. Altogether there are about 10^6 cones and 10^8 rods. The optic nerve has about 10^6 nerve fibres. The central cones probably have direct connections to the cortex but in the outer region outputs from many receptors must be combined to feed one nerve fibre. This provides a system that is very sensitive to weak light but does not resolve fine detail. Owing to the aberrations the image on the retina is not as sharp when the large pupil is used so, once again, the receptor structure matches the optics.

18.16 Interpreting the visual signals

Some processing of the optical information is done in the retina and some at the relay station but most is done in the visual cortex. Experiments in which electrodes are implanted in the brains of cats and monkeys have led to a general understanding of the processing of information but what remains unknown probably exceeds what is known. One fact shows that filtering of information is very important. The optic nerve can carry about 10^7 bits of information per second; the optics of the eye and the number of receptors can supply this information. Yet when a person is asked to describe a scene that has been seen briefly, only about 40 bits (or less) are remembered. It appears that the processor is capable of extracting from the enormous flood of information just those elements that are interesting or useful in deciding action. The system—optics, receptors, nerve fibres and processor—taken as a whole is matched so that every stage is adapted to handle the output of its predecessor.

18.17 Accommodation

The power of the dioptic system of the eye can be changed by the action of certain muscles which alter the shape of the lens. The principal change is an alteration of the curvature of the front surface of the lens; the exact change in shape (since the volume does not change) and the way in which the muscles operate is still disputed (see ref. [1]). The normal observer of age 20–30 can usually focus objects at distances from infinity down to a point distant about 25 cm from the nodal point of the eye. This is called the nearest point of distinct vision, d_n. This distance varies in a complicated way in childhood when the lens is still growing and some people in the age range 15–25 can focus objects closer than 25 cm while still having good distant vision.

Three defects of the dioptic system are common; these are:

1. *Myopia*, short-sightedness, in which the distance d_n is less than normal and objects beyond a certain distance cannot be focused.
2. *Hypermetropia*, or long-sightedness, in which d_n is a good deal greater than normal but very distant objects can be focused clearly.
3. *Astigmatism*, in which lines parallel to a certain direction in an object are clearly focused but lines parallel to other directions are not.

Visual astigmatism is due to either the cornea or the lens having a non-spherical shape with different curvatures in different planes. It has nothing to do with the aberration called astigmatism which arises when non-paraxial rays are considered.

The effect is the same as when a cylindrical lens is inserted into a coaxial system of spherical surfaces.

18.18 Advancing geriatry

In addition to myopia, hypermetropia and astigmatism two other defects must be considered. Firstly, there is *presbyopia* in which the *range* of accommodation becomes less, usually by the distance d_n increasing. It is normal for d_n to increase with age from about

25 to 65. The first reaction to this change is to hold the book farther from the eye until (often about 40) the condition is reached where a patient said 'my eyes are all right; its just that my arm isn't long enough'. In advanced old age d_n ceases to increase and may actually decrease. Presbyopia is probably due partly to hardening of the lens with age and partly to weakening of the internal eye muscles.

Another defect *strabismus* is due to failure of the muscles external to the eye. In normal vision the directions of the eyes are coordinated so as to give the appropriate amount of convergence required for a near object. Surgical operations on the extra occular eye muscles offers the only hope of correcting a major failure of this system. A small failure may be corrected optically (Section 18.19).

18.19   Correction of vision defects

The use of convex spectacle lenses to correct long-sightedness began in Europe and also in China about the end of the twelfth century. The first portrait of a person with glasses is dated 1352. It is uncertain whether the discoveries in Europe and in China were independent and which was first. Concave lenses came later and the first portrait showing them is dated 1517. The power required must be calculated from eqn (3.33), i.e. allowance must be made for the separation between the spectacle lens and the first principal point of the eye.

The early lenses were biconvex or biconcave. These give clear vision when the user looks straight through the centre but there is considerable aberration when he turns his eye so as to look far off centre. Near the end of the eighteenth century it was realised that this aberration could be corrected by using meniscus lenses (Fig. 18.9). In 1804, the first meniscus lens was made in London by Dolland to the design of Wollaston (1766–1828), but most people thought the extra expense was not justified and by 1850 the meniscus lens had been abandoned. It came back from Germany, where Tscherning calculated exact formulae for the optimum radii of curvature, and is now in general use.

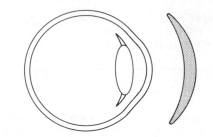

Figure 18.9   Correction by meniscus lens

18.20   Convergence

Sometimes it is desirable to correct for weakness in the extraocular muscles which have difficulty in converging the eyes for viewing near objects. This is done by adding prisms to the correction needed. Decentering the lens in an appropriate direction effectively adds a prism and the correction is made in this way. Only rather weak prisms, up to 2 or 3 prism dioptres, can be used because of the chromatic aberration introduced. This has the effect of making parts of a picture that are of different colours appear to be at different distances.

18.21   Bifocals

Bifocals, in which the upper part of the lens corrects for distant vision and the lower part for near vision, were invented by Benjamin Franklin (1706–1790), better known for his work on lightning and as one of the original authors of the United States Constitution. He used two separate lenses held together by the frame. Today it is usual to fuse a lower piece of high index glass into a piece of low index glass (Fig. 18.10). Smooth surfaces are

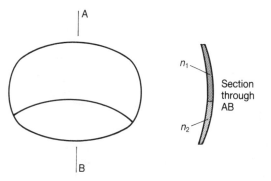

Figure 18.10   The bifocal spectacle lens

then ground and polished, the necessary increase of power in the lower section being given by the larger refractive index.

18.22 Astigmatism

G. B. Airy (1801–1892), mathematician and astronomer, studied the variation of his own visual astigmatism over a period of 60 years. He devised a corrective lens which had one cylindrical and one spherical surface. Today toroidal surfaces that have different radii of curvature in different planes (like a teaspoon) are commonly used.

18.23   Contact lenses

Originally contact lenses were quite large and covered a considerable region of the sclera. These lenses were difficult for an unskilled person to insert and remove. Also after a period of wearing them, hazy vision developed. This was due to oxygen starvation of the cornea. Today small plastic lenses which cover little more than the cornea are used. The tear fluid circulates under this small area and maintains oxygenation. A small 'lens' of liquid is imprisoned between the contact lens and the cornea. This approximately corrects corneal astigmatism. It also corrects the errors of corneas that are conical or have irregular errors of shape. This is the important medical advantage of contact lenses. For most people their use is mainly cosmetic. When there is lenticular astigmatism the lens must be inserted and remain in the correct orientation. Attempts to apply the bifocal principle to contact lenses have not been successful.

18.24   Hand magnifiers

The size of the image on the retina is proportional to the angle that it subtends at the first nodal point of the eye. This angle may be increased by bringing the object nearer the eye until the nearest point of distinct vision (conventionally taken as $d_n = 25$ cm) is reached. If a lens is held in front of the eye and the object is at the focus of this lens, the image is seen at infinity and its size on the retina is increased in the ratio d_n/f. It may be shown that if the final image is brought in to

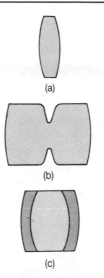

Figure 18.11 (a) Simple magnifier. (b) Coddington magnifier. (c) Hastings triplet

the near point, the magnification is increased to $1 + d_n/f$.

Simple magnifiers used by watchmakers, diamond valuers, stamp collectors, etc., are biconvex lenses. (Fig. 18.11a). These suffer from all the aberrations described in Chapter 4 including spherical aberration. They give a good image over rather a small field, with magnification of up to about ×3, though they are often used with higher magnifications.

The magnifier shown in Fig. 18.11(b) (invented by Brewster but generally known as the Coddington magnifier) gives a much clearer image over a field of diameter 30 mm at a magnification of ×7. Better chromatic correction is obtained by using a doublet or the Hastings symmetrical triplet (Fig. 18.11c) which gives a good image over a field of diameter 8 mm at a magnification of ×20. In buying a hand magnifier it is well to consider the purpose for which it is to be used and the magnification required. There is no doubt about the optical superiority of the corrected lenses as compared with the simple biconvex lens but there may be a ratio of 20 : 1 in price.

18.25   Oculars and eyepieces

In a telescope or microscope, the objective forms an intermediate image which is further magnified by an ocular or eyepiece. This divides the magnification into two stages and, when a magnification of more than about 30 is needed in a microscope, it is almost essential to do this in order to obtain a good image. In principle, the eyepiece can be a single lens and the Hasting's triplet is sometimes used in this way. More frequently the ocular consists of two separated lenses, one of which acts mainly as a field lens and the other as a magnifier. Figure 18.12(a) shows the original Ramsden eyepiece in which L_1 acts entirely as a fields lens and L_2 as a magnifier. The two lenses are of equal focal length f and are separated by a distance f which is equal to half the sum of the focal lengths so that lateral chromatic aberration, arising from the wavelength dependence of the refractive indices, is corrected. This arrangement has the disadvantage that L_1 is in focus as seen by the user. Any particle of dirt on L_1 or small bubble in the glass shows up very prominently. This defect is overcome by bringing the two lenses closer together, so that L_1 acts mainly as a field lens but contributes to the magnification. The lateral chromatism is no longer completely corrected, but this may be rectified by making L_2 (which still does most of the magnification) into a corrected doublet. This arrangement, known as the Kellner eyepiece, is shown in Fig. 18.12(b). With this eyepiece a graticule or cross-wire is usually placed in the plane of the intermediate image formed by the objective of a telescope or microscope. Focusing to suit the vision of the observer may then be done with a screw motion, as shown in the figure. If the screw fits well enough and is lightly greased, dust is excluded.

18.26   More eyepieces

The above eyepieces, in which the intermediate image is outside, are called positive eyepieces. An alternative, in which the intermediate image and the graticule are inside, was designed by Huygens (Fig. 18.13).

The Hastings triplet, used as an eyepiece, gives a good image and a flat field of $30°$. The Huygens eyepiece gives a very good field out to $25°$ and fairly good definition to $40°$. It has a very curved field and also strong longitudinal chromatism. Various improvements including a meniscus lens in front to flatten the field and a corrected doublet in place of the eye-lens have been invented. The invention of antireflecting films (Section 4.31) has altered the balance of advantage between different types of eyepieces. It is no longer so important to avoid internal reflections in the glass surfaces.

It is fairly common to design eyepieces to correct the lateral chromatic aberration of the objective of a refracting telescope or microscope. Both kinds of eyepieces may be modified to project the image on to a screen either for photography or for viewing by a group of people.

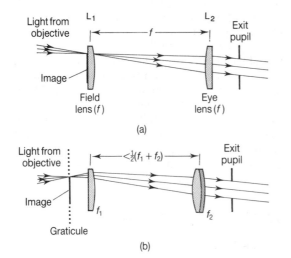

(a)

(b)

Figure 18.12   (a) Ramsden eyepiece. (b) Kellner eyepiece

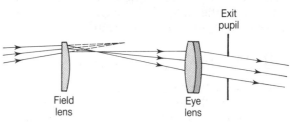

Figure 18.13   Huygens eyepiece

18.27   Plane mirrors and prisms

A great many instruments employ mirrors or mirror combinations for one or more of the purposes described below and still more use reflecting prisms instead. The prism has several advantages. It forms a mechanically solid unit which can conveniently be fixed to a metal base and hence to the main structure of the instrument. Total reflection can be obtained and though the reflecting surface needs to be scrupulously clean it deteriorates only very slowly in the atmosphere. If desired the reflecting surface can be protected by coating it with silver and protecting the silver with a fairly thick layer of copper put down by the inexpensive process of electroplating. If two or more mirrors are combined in one prism the angles between them can be made correct within a few seconds of arc and remain unchanged. There are, of course, some disadvantages. The prism requires a block of glass that is very homogeneous and free from bubbles. It is usually heavier than the equivalent mirror system though not much heavier when the weight of the mount is included. Primary chromatic effects can be compensated and usually are, but there is a secondary effect that has to be taken into account when considering correction of aberration. One of the advantages of mirror systems is that they can be used for wavelengths to which glass is not transparent.

18.28 Deflection and inversion

Mirrors, mirror combinations and reflecting prisms are used for one or more of the following purposes:

1. To deflect a beam of light through an angle that depends on the angle of incidence
2. To deflect a beam through a fixed angle independent of the angle of incidence
3. To shift a beam sideways without altering its direction
4. To invert the image in one plane
5. To invert the image in two planes, i.e. to rotate it through $180°$

In the following sections we shall illustrate prisms for each of these purposes. The ones illustrated are only specimens of the large number of reflecting prisms that have been designed.

18.29 Use of prisms

Figure 18.14(a) shows an isosceles prism used to replace a single mirror. If the angle of incidence on the lower face is less than the critical angle this face must be silvered. Since the prism is isosceles, the angle $\beta_2 = \alpha_2$ and therefore $\beta_1 = \alpha_1$ and the angle $(2A + 2\alpha_1)$

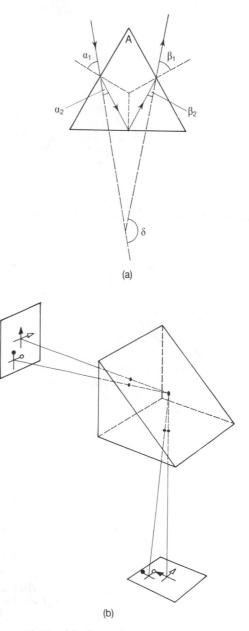

(a)

(b)

Figure 18.14 (a) Isosceles prism acting as a mirror. (b) A $45°$ prism reflector

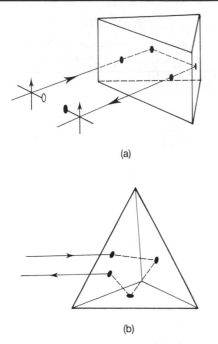

Figure 18.15 (a) Porro prism. (b) Corner-cube reflector

through which the ray is deflected is independent of the index n. It is usual to arrange that $\alpha_1 = 0$ for the central ray of a beam and the angle of deflection for the central ray is then $2A$. If a converging beam is incident the angle of convergence is not changed. Figure 18.14(b) shows the most commonly used isosceles prism (angles of $90°$, $45°$, $45°$) which turns a beam through a right angle. The image is inverted in the plane of the paper but not in the perpendicular plane.

Figure 18.15(a) shows a Porro-prism which turns the beam through $180°$ irrespective of the angle of incidence and also inverts the image in the plane of the paper provided the rays are all in the plane of the paper. Figure 18.15(b) shows a 'corner cube', which turns the beam direction through $180°$ irrespective of the angle or plane of incidence.† The 'cats-eyes' sometimes used as road markers are imperfect corner cubes formed by moulding. They return the light in directions near to the reversal of the original direction so that the driver of a car sees them as bright spots of

† One hundred of these were placed on the Moon on the Apollo mission in 1969.

light. If they were perfect corner cubes they would be visible only from the headlights themselves.

18.30 And more prisms

Figure 18.16(a) shows a prism that displaces the beam sideways without altering its direction and without inversion. Figure 18.16(b) shows a Dove prism which inverts the image without changing either the direction or line of the rays. If the Dove prism is rotated, the image rotates twice as fast.

Figure 18.17(a) shows a prism (sometimes called an Amici prism). This inverts the image in both planes and may be used in a telescope of the types shown in Fig. 3.16 to make the image erect. The same result is achieved by using two Porro prisms arranged as shown in Fig. 18.17(b) and this arrangement is generally used in binoculars.

Figure 18.18 shows the pentagonal prism which turns the beam through a right angle irrespective of the angle of incidence. Although similar properties are shown by two plane mirrors at AB and CD, it is difficult

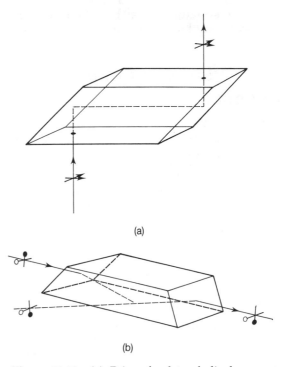

Figure 18.16 (a) Prism for lateral displacement. (b) Dove prism

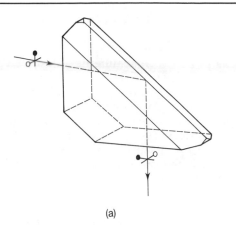

(a)

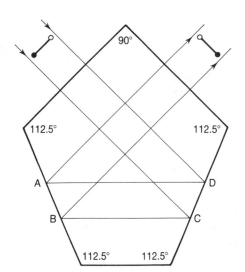

(b)

Figure 18.17 (a) Roof (Amici) prism. (b) Prism arrangement in binoculars

to mount the two mirrors so that the correct angle is obtained and maintained.

18.31 Dispersing prisms

Originally prisms were more often used (in spectrometers and spectrographs) to disperse light rather than to reflect it (Fig. 18.19). Nowadays this is usually done by means of groatings. However, to select a narrow region of the spectrum the monochromatar shown in, Fig. 18.20 is often used. The Pellin–Broca prism may be regarded as built of three prisms (as shown by the dotted lines), two acting to disperse

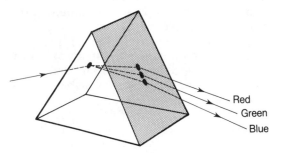

Figure 18.19 Dispersion by a prism

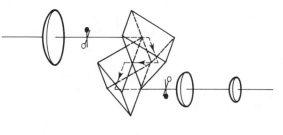

Figure 18.18 90° deflection occurs regardless of angle of incidence

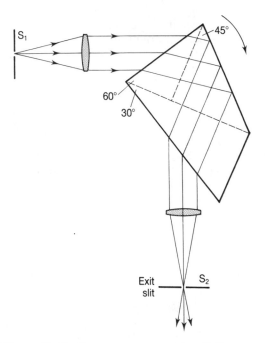

Figure 18.20 Spectrometer with Pellin–Broca prism

the light and the other one as a reflector. If the prism is rotated the spectrum passes over the exit slit S_2 so that different colours are obtained in turn while the deflection remains constant.

18.32   Small telescopes and binoculars

In this section we consider instruments that have refracting objectives not more than a few centimetres in diameter. They include monocular telescopes for amateur astronomers, coast-guards and bird-watchers, and binoculars for many purposes—naval, military and civilian. The basic instrument is the astronomical telescope shown in Fig. 3.16. The objective is usually a doublet or triplet corrected for chromatic aberration, spherical aberration and coma, and a similarly corrected eyepiece. Often the available field is fairly small so that correction for the other primary aberrations is not necessary. Except for the astronomer it is necessary to make the image erect and this is done either by an erecting eyepiece or by erecting prisms. For monocular telescopes mounted on a firm support a length of 50–100 cm can be tolerated so an erecting eyepiece or a 'straight-through' erecting prism may be used. For binoculars usually intended to be hand-held, the double Porro prism has the advantage that it shortens the tube and places the axes of the telescopes farther apart, thus increasing the stereo-effect. Binoculars are usually described by a pair of numbers such as 7×50 which means that the angular magnification is 7 and the diameter of the objective is 50 mm. The size of the exit pupil would then be $50/7 = 7.1$ mm and such a binocular would be suitable for night use, since the diameter of the pupil of the dark-adapted eye is 7–8 mm. A somewhat higher magnification and a smaller objective would be suitable for daytime use but it does not pay to increase the magnification much above 10 unless the instrument is to be firmly supported. Otherwise the tremor of the hand spoils the image.

Galilean telescopes (Fig. 3.17) have the advantage of light weight and low cost and are sometimes used as opera glasses. However, the field is liable to be very small.

18.33   Goniometers

There are a large number of uses for small telescopes in instruments designed to measure angles. These include small telescopes with objectives 25 mm diameter and length about 25 cm capable of measuring angles to $0.1°$. They also include instruments for accurate surveying over large distances and instruments used for missile guidance. These last measure angles relative to lines directed towards certain stars. The precision of reading of the scales by eye is often about $2''$ arc and by using photoelectric sensors a precision of $0.1''$ is attained. A double measurement (made on both sides simultaneously) eliminates an error due to any slight eccentricity of the mount.

The effective axis of the telescope (which is set to pass through the object point) is defined by the first nodal point of the objective and the centre-point of the graticule in the plane of the first image. In surveying, angles may have to be measured between objects at different distances. The designer must therefore ensure that changes of focus do not alter the sighting-line.

18.34   Cameras

These are used by a considerable number of different people including amateurs (still probably the most numerous), newspaper photographers, portraitists, scientists and military reconaissance photographers. To meet their different needs, an enormous variety of mechanical or electromechanical devices for automatic adjustment of aperture to take account of light conditions and of the chosen exposure, high-speed shutters, view finding, assisted focusing, etc., have been devised. These are outside the scope of this book.

Among the many photographic lenses that have been developed in response to the needs of various users, we shall mention three specialised lens systems: (1) the telephoto lens, (2) the zoom lens and (3) the soft focus lens.

18.35   Telephoto lenses

The standard camera lens usually gives a fairly wide-angle view of an extended scene.

Figure 18.21 Telephoto arrangement

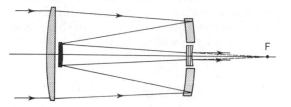

Figure 18.22 Higher-magnification telephoto system

The tangent of the angle is roughly proportional to the linear size of the plate or film divided by the focal length, and the magnification is proportional to the focal length. Sometimes the photographer wishes to increase the magnification, accepting a reduction of the area of the distant scene, and therefore wants a lens of much longer focal length, perhaps 10 times longer. For this purpose a combination of a negative and a positive lens is used (Fig. 18.21). The distance of the picture from the back of the lens system, called the *back focal length* (f_b), can be made a small fraction of the focal length of the combination (f_c). The ratio f_c/f_b, called the *telephoto magnification*, may be as high as 8. To eliminate aberrations, the front element in Fig. 18.21 is replaced by a complicated lens system and the back element may be a doublet or triplet. Much higher telephoto magnification is now obtained by adding reflecting elements as shown in Fig. 18.22. In situations where we need a very wide field, but are content with a low magnification, a *reversed telephoto system* (in which the negative element is in front of the positive element) is used.

18.36 Zoom lenses

To cover a range of needs a photographer may require three or more telephoto lens of different telephoto magnifications. Also the higher magnifications require a tripod or other support. The total cost is high and the weight to be carried is significant. It also takes an appreciable time to change one telephoto lens for another and this may cause a picture (e.g. of a distant bird) to be lost. The cinema industry also requires to be able to change the focal length quickly to produce a 'close-up'. For this purpose lens combinations whose focal length can be changed by moving one or more components relative to the others while maintaining focus on the object have been designed. These are called *zoom lenses*. The design of these lenses constitutes one of the most difficult problems in ray optics. The ideal zoom lens would be perfectly corrected over its whole range of magnification. This ideal is never achieved and the designer usually has to be content with good correction in the centre of the range and some imperfection at the ends. Thus those who wish to obtain pictures of the highest definition and different magnifications still prefer to use telephoto lenses.

18.37 Soft focus lenses

A portrait photographer sometimes does not wish his picture to be too sharp, but wants what is called a *soft focus*. The right effect is not produced by putting a well-corrected lens out of focus. A certain amount of spherical aberration is required. However, it is not sufficient just to design a lens with a little spherical aberration because the amount of aberration would vary with the size of stop used; also only one degree of 'softness' could be produced with any one aperture. Special lenses have therefore been designed in which moving one component within the lens alters the amount of aberration. The photographer is then able to obtain the degree of 'softness' that is preferred.

18.38 High-speed photography

This may be regarded as the magnification in time of the progress of transient events but the range of magnification is very much greater than the spatial magnification of optical microscopy. Ordinary cameras give exposures of down to about 1 ms and high-speed photography may be said to begin at this point. To obtain a single high-speed photograph showing one stage of a transient event is not difficult. Sources of light giving integrated light outputs sufficient to take a photograph and with durations of 10^{-6} s (flash-tubes) to 10^{-8} s (spark sources) are commercially available. It is necessary to synchronise the flash with the event to be photographed, but this is not usually difficult. Pictures may be obtained using electronic devices to amplify the light in times down to 10^{-10} s. Lasers have been used to produce light pulses of duration down to 30×10^{-15} s and these have been photographed.

18.39 Repetitive high-speed photography

In most investigations it is desired to obtain a series of pictures of different stages of a transient event. Then a second time comes into consideration—the time between successive flashes. This is usually expressed in terms of its reciprocal—the number of pictures (called frames) per second (f.p.s.). It is necessary to produce a succession of short bright flashes and also to produce relative movement of film and picture between the flashes but not during the flashes. Flash tubes are activated by discharging the energy stored in a condenser of capacity C charged

to a voltage V. This energy is $\frac{1}{2}CV^2$. The time within which the condenser can be recharged is determined by $(LC)^{1/2}$ (where L is the inductance of the charging circuit). Since C cannot be made very small (because the power is proportional to C) it is necessary to make L extremely small. Charging rates sufficient to make the intervals between pulses not much greater than the duration of the pulses have been achieved and these lead to repetition rates of up to 6×10^8 f.p.s.

18.40 Freezing the image

In ordinary cinematography the relative movement between film and picture is obtained by moving the film with a hole-and-sprocket motion so as to give 24 f.p.s. This 'pin register' method has been developed to give speeds of up to 500 f.p.s. The limit is set by the strength of the film which tears when too high an acceleration is attempted.

In the next stage the film moves continuously and a rotating prism is used to move the picture at the same rate (Fig. 18.23). During part of the rotation the prism blocks the light and so creates intervals between pictures. This gives speeds of up to 10^4 f.p.s. Still higher speeds (up to 25×10^6 f.p.s.) can be obtained with the rotating mirror system shown in Fig. 18.24. L₁ and L₂ form a relay

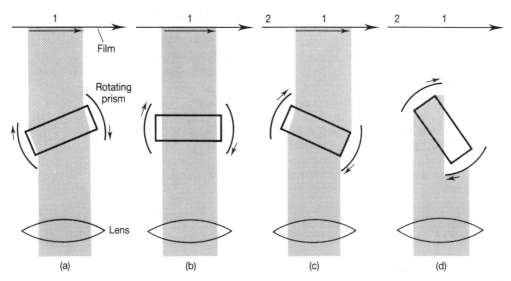

Figure 18.23 Rotating prism high-speed photographic system. Reproduced by permission of Taylor & Francis Ltd from Field, *Contemporary Physics*, **24**, 439–459 (1983)

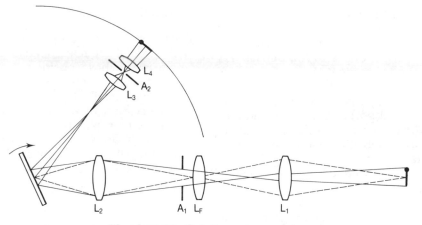

Figure 18.24 Rotating mirror system

system which focuses an image on to the film and a second image is formed by L_3 and L_4 [2].

Beyond this point it is no longer possible to rely on mechanical movement. The accelerations (and therefore the forces to produce them) are too high. Using the image tube (which for present purposes may be regarded as the television picture tube) it is possible to move the image electronically from one part of the screen to another and the persistence of the fluorescence is sufficient to enable the whole array of pictures to be photographed. Effective speeds of up to 6×10^8 f.p.s. have been achieved.

18.41 The streak camera

This is designed to give a continuous record of what is happening along a chosen line in an event rather than a series of photographs. If the point of interest moves along the axis of the slit and the film moves at right angles to this a streak is obtained on the film and the angle of the streak gives the velocity with which the event is moving. An electronic form of streak camera has photographed light pulses of 10^{-12} s, produced by compressing pulses from a laser.

18.42 Astronomical telescopes

The general scheme of a refracting astronomical telescope is shown in Fig. 3.16. Two

different arrangements for reflectors are shown in Fig. 18.25. The Newtonian arrangement shown in Fig. 18.25(a) is still very useful for certain purposes. Figure 18.25(b) shows the original Cassegrain system. The Ritchey–Chrétien system is an aplatanised form of Cassegranian usually with a hyperbolic primary. The Hubble space telescope, launched in 1990, had both primary and secondary mirrors of hyperboloidal form, but unfortunately not of the correct shape! A fourth main type, called the Coudé focal system, differs from the others in that the final image and final direction of the light is

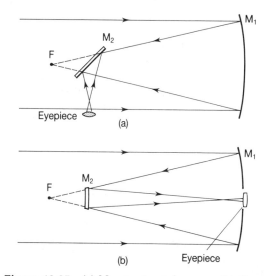

Figure 18.25 (a) Newtonian telescope. (b) Cassegrain system

fixed relative to the floor of the dome, even when the telescope is rotated. Coudé systems involve several plane mirrors, at least one of which rotates to compensate for rotation of the telescope.

All large telescopes (with primary elements more than 1 m (40 inches) are reflectors. It is much easier to support a mirror from behind, so that it does not change its shape (under its own weight) when the telescope rotates, than it is to obtain corresponding stability with a lens supported only at the edges. There is also the difficulty of obtaining the necessary homogeneity in a large block of glass. In addition, as Newton stated, the mirror system is free from chromatic aberration, at the primary stage, though generally mirror systems are used with eyepieces, cameras, corrector plates, etc., all of which have to be corrected for chromatic aberration.

18.43   Where to put them?

Nowhere in optics is the necessity of systems design based on consideration of the kind of information desired, more important than in optical astronomy. The flux collected by a telescope is proportional to d^2 (where d is the diameter of the primary mirror). Under ideal conditions the area of the image of a star would be *inversely* proportional to d^2, but ideal conditions are obtained only by a telescope mounted in a satellite or on the Moon. The atmosphere above any terrestrial site has irregular variations of density so that plane waves from a star acquire irregular corrugations or, in terms of ray theory, the parallel rays from the star suffer a multitude of small irregular deviations. This effect (known as 'seeing') increases the area of the stellar image and reduces the advantage of a large telescope. Seeing is very much better in some places than in others and, as a first stage in systems design, a detailed (and possibly costly) investigation of alternative sites for a new telescope is justified.

18.44   How are telescopes used?

There are three main ways in which astro-

nomical telescopes are used. These are:

1. The 'one-star-at-a-time' method in which the axis of the telescope is made to follow one star whose position, brightness and spectrum are measured in the greatest possible detail.
2. Observation of a star field in which the astronomer usually wishes to photograph as large an area of sky as possible to obtain information about the brightness of the stars, the spectral types and the positions. There are alternative methods of studying star fields using electrical detectors.
3. Flux-collecting, in which the object is to collect from faint stars enough light to enable their spectra to be investigated. We shall defer consideration of this function until we can discuss the appropriate methods of spectroscopic analysis.

18.45   Problems of the paraboloid

Historically astronomers have given most attention to the first of these methods. The Newtonian arrangement with a parabolic primary gives an aberration-free image for a star exactly on axis and for this reason nearly all large reflecting telescopes built before 1960 had parabolic primaries. Usually, however, when an expensive instrument with a large primary had been built it was desired to use it also for the study of star fields. Unfortunately the comatic aberration of a parabolic mirror is very bad so that coma is observable at only 1" off-axis. Various methods of 'rescuing' the older telescopes by introducing aspheric corrector plates near the prime focus or using eyepieces or cameras which nearly eliminate coma over a larger field have been used, greatly increasing the available field. (This has been done for the Mount Palomar telescope.)

18.46   Enter the hyperbola and the Schmidt plate

In recent years, the Ritchey–Chrétie design which uses a hyperbolic primary has been used, usually in the Cassegrain mode shown

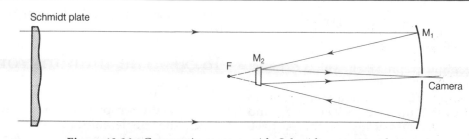

Figure 18.26 Cassegrain system with Schmidt corrector plate

in Fig. 18.25(b). Corrector plates or lenses are still needed but a much larger field up to 30', with images free from coma, can be obtained. To obtain still larger star fields a radically different design is needed. Figure 18.26 shows a design used by Schmidt in which an aspheric plate is placed at the centre of curvature of a spherical primary. The plate is shaped so that, for a beam on-axis, the combined effect of the spherical mirror and the plate on a wave-front is the same as the effect of a parabolic mirror. Thus a perfect image is obtained on-axis. However, if the beam is a little off-axis the effect of the plate is the same, apart from a cosine factor and $\cos 5° = 0.996$, and a nearly perfect image is obtained over a field of $5°$ diameter or 100 times the area of the best field that can be achieved with a Ritchey–Chrétien system. Unfortunately the tube has to be nearly twice as long to accommodate the plate and also some chromatic aberration is introduced. The cost may be 20 times that of a Ritchey–Chrétien of equal aperture. If two aspheric plates are used (Fig. 18.27) the chromatic aberration may be fairly well corrected and it has been found possible to bring the corrector plates nearer to the primary. There is of course the problem of supporting the corrector plates and this at one time restricted Schmidt

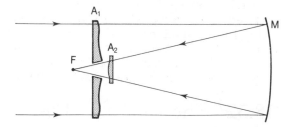

Figure 18.27 System employing two aspheric corrector plates

systems to primaries of diameter 1.5 metres (60 in), but systems up to 2.5 metres have now been made.

18.47 Making mirrors on a lathe

One recent technical advance which affects the design of many optical instruments may be mentioned here. Up to about 1975, it was very much cheaper to make a spherical surface of optical quality than an aspheric surface and the precision of the finished surface was better. Then extremely good lathes (usually with air bearings) and diamond tools with extremely sharp durable points were developed. It was found possible to make, on a lathe, mirrors so good that without any polishing they could be used in infrared instruments where scratch marks up to $0.5\ \mu\mathrm{m}$ deep could be tolerated. Since 1980, these processes have been further improved and the manufacture of mirrors for the visible region is now being achieved. Given the formula that defines the asphericity a microprocessor can program an automatic lathe to cut an aspheric surface with the same precision as a spherical surface. It does not matter if the aspheric surface departs very much from the nearest spherical surface. Hard-metal 'masters' against which lenses can be ground can be made in the same way. Thus the main technical difficulties associated with the manufacture of aspheric components has been removed. The computer also can handle easily the calculation of paths of rays through aspheric surfaces. It seems probable that these developments will lead to increasing use of aspheric surfaces in the future.

18.48   Plate measuring machines

A single photograph taken with a Schmidt telescope registers between 10 000 and 100 000 stars. The man-hours required to measure even one plate is very high and automatic plate-measuring machines have been made. One of these, called GALAXY, makes a preliminary exploration of the plate to distinguish the stars to be measured from occasional random clusters of darkened photographic grains. It then explores each star image with a fine spot of light so that it determines the exact position of the centre relative to the centres of certain neighbouring stars whose astronomical coordinates are assumed to be known with high accuracy. It also measures the amount of light received from the star. The instrument includes a computer which corrects certain errors and finally stores in its memory or prints out a star catalogue. The star catalogue produced by even ten plates is too vast for direct human analysis and further computation must be done to provide the statistical information. By comparing plates taken with different colour filters, or in other ways, a computer can also produce a list of stars that have unusual properties so that further study is justified.

18.49   Next generation telescope (NGT)

The Mount Palomar telescope, completed in 1948, with a primary 200 in (5 m) in diameter was the largest telescope in the world till about 1978 when a Russian telescope with a primary of 6 m diameter came into operation. The interval of about 30 years and the fact that the later telescope had only a 20% increase in size suggest that the end of this line of development has been reached. It is doubtful whether, with techniques now available, it would be possible to make a mirror much larger than 6 m in diameter and to move such a mirror into position. Assuming that all technical problems could be overcome, a conventional telescope with a primary mirror 25 m in diameter would cost 1000 million dollars and take over 50 years to build.

18.50   The multimirror telescope

It is generally accepted that a new design in which the single mirror is replaced by a number of smaller mirrors is needed. The individual mirrors must be controlled by accurate servo mechanisms so that, at least, they can all collect the flux from a weak star or nebula. Numerous designs have been proposed. The number of mirrors in some designs is as low as 6, in others it exceeds 1000. Many designs have got no further than a sketch of the general layout, but the Kitt Peak National Observatory, Arizona, has made detailed studies of several possibilities. Some of these will now be briefly described [3].

18.51   The large steerable dish

In this design about 600 mirrors of hexagonal shape are mounted so that, when exactly in position, they form a single mirror about 25 m in diameter. The assembly is moved as a whole so as to point approximately in a suitable direction and the servo mechanisms then point each small mirror at the object to be studied. To obtain the best results all paths from the star to the common focus must be equal within about 0.1 wavelength $(0.05 \ \mu m)$, but a somewhat less precise control will direct all the flux from a faint object on to one detector. Various arrangements (Newtonian focus, Cassegrainian focus, Coudé systems, etc.) are possible. Corrector plates and lens systems, are in principle, possible but have to be very large and very accurately positioned. As compared with the conventional design this method removes the problem of making a single large mirror, but it seems doubtful whether cost or time of construction would be less.

18.52   The rotating shoe

One design proposed is to have a very large hemispherical   bowl   which   is   fixed

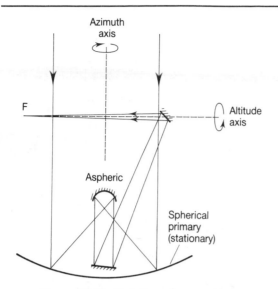

Figure 18.28 Rotating shoe system

(Fig. 18.28). The sky is scanned by rotating the apparatus placed at the primary focus as shown in the figure. In this design only a small fraction of the area of the bowl is used at any one setting. Thus, if the area used is to be 500 m² (corresponding to a conventional primary mirror of 25 m diameter) the area of the bowl must be many times larger than this and the cost would be prohibitive. It is therefore proposed to construct only a small slice of the bowl and to rotate this about a vertical axis.

The rotation scans the sky in one direction and movement of the apparatus at the focus scans in the other direction. About a quarter of the area of the mirror is used in one setting so that the total area is about 2000 m². This design has the advantage that when the 'shoe' is rotated about the vertical axis the strains due to weight do not change. However, the engineering difficulties of rotating the shoes about a fixed axis are severe.

18.53 The multitelescope designs

There are several designs in which the outputs from a number of telescopes are combined. The Kitt Peak teams have examined two extreme cases (1) where the number of mirrors is as small as possible (6) and (2) where it is large (108). Design (1) has

the advantage that the six telescopes can be on one mount which can be made very strong. The combination can be moved as a whole so that once the directions of the individual telescopes have been aligned they should remain aligned. However, this design requires the manufacture of six mirrors each 10.2 m in diameter if the total collecting area is to be equal to that of a 25 m telescope. If these mirrors are similar to the Mount Palomar mirror, the area is increased in the ratio of 4 : 1 and the weight in the ratio 10 : 1. They would almost certainly have to be manufactured on site. If they were made of glass a cooling period of several years would be needed to avoid severe strains. A system of this type with smaller mirrors equivalent to a single mirror of diameter 4.5 m has been constructed at Mt Hopkins in the United States. A similar system, with seven mirrors, each of 0.375 m in diameter, is at Preston in the United Kingdom.

18.54 The 108-mirror system

In design (2) where there are 108 telescopes, each one has a mirror 2.4 m in diameter. Many telescopes of this size (or larger) have been made and there would be no technical difficulty in making the individual telescopes. The main technical problem would be to combine the output from the units. There would probably be little difficulty in combining the flux of light but great difficulty in obtaining equal optical paths from star to detector. There would be serious difficulty in obtaining and maintaining alignment. Possibly this may be done by aligning the system on a fairly bright star and then turning it slightly to search the neighbouring sky for objects that are so faint that they could not be detected by one unit alone.

One great advantage of this design is that it could be built in stages. As soon as one unit was made, it could make useful observations. When about four units were available the method of alignment could be tested. Different nations might pay for one or more units which would be used part of the time for their own programmes and part of the time as part of the combination.

This design appears to be the only one whose cost could be reliably estimated in advance and the least likely to encounter unforeseen technical problems, at any rate in regard to combining the flux of light. It would probably not operate at long wavelengths as well as assemblies with larger mirrors.

18.55   What next?

These proposals are interesting in that they show the ingenuity of modern optical designers and their ability to produce basically new ideas. It may well be that cost will prevent a 25 m telescope being built this century and that the next step will be a 10 m telescope. To obtain its diffraction-limited performance, the giant telescope ought to be outside the Earth's atmosphere, i.e. on a satellite or on the moon! In fact an attempt has already been made to put a diffraction-limited telescope 2.4 m in diameter in orbit around the Earth. Unfortunately it did not achieve the intended performance (see Section 18.42).

18.56   Microscopes

The purpose of a microscope is to reveal detail that cannot easily be observed by the naked eye or with a hand magnifier. In Fig. 18.29, the following four parts of a microscope are shown:

1. The illuminating system which concentrates light on the object
2. The stage upon which the specimen is supported
3. The objective lens system which produces a magnified image (I) of the object at a distance of about 13 cm from it
4. The eyepiece which further magnifies the image formed by the objective.

18.57   Resolution

Despite the fact that very wide cones of rays are sometimes used, most microscopes are so well corrected that the smallest detail which can be resolved is determined by diffraction. From the analysis of Section 6.9 we see that the finest detail in an image formed by a lens

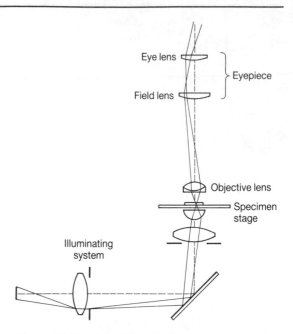

Figure 18.29   Principle of microscope operation

subtending a half-angle α at the focal plane, in a medium of refractive index n, is given by $0.61\lambda/n \sin \alpha$. Thus the number of lines per unit length that can be resolved is

$$N = \frac{n \sin \alpha}{0.61\lambda} \qquad (18.4)$$

For visible light $N \sim 3000n \sin \alpha$ lines mm^{-1}; $n \sin \alpha$ is the *numerical aperture* (NA). It is possible to make $\sin \alpha = 0.95$ (corresponding to $\alpha = 71°$) so that, in air, the maximum NA is 0.95 and when oil of index 1.5 is used the maximum NA is about 1.45. The maximum number of lines per millimetre that can be resolved is about 4400. The eye can resolve nearly 15 lines mm^{-1} at 250 mm but in order that the eye may see the detail easily magnification of up to 5 lines mm^{-1}, i.e. a magnification of $4400/5 = 880$ is useful. For visual observation any higher magnification is called *empty magnification* because it reveals no more detail and may reveal less because the illuminance is inversely proportional to the square of the magnification. To allow for grain size in a photographic plate and for losses in reproduction, photographs are often reproduced at higher magnification, up to about ×2000.

18.58 Common types of microscope

Three classes of microscope may be usefully considered even though in fact there is a continuous range. We call these (1) the beginners' microscope, (2) the standard microscope and (3) the research microscope. The costs of these are roughly in the ratio of 1 : 10 : 300.

The beginners' microscope has, for the objective, a single lens of focal length about 1 inch, giving a magnification of about ×5 or occasionally ×10. Thus the overall magnification is from 10 to 50. The NA is about 0.1 so that 300 lines mm^{-1} would be resolved if all aberrations were completely corrected. Usually to save expense, only moderately good correction of chromatic aberration is provided. 100 lines mm^{-1} can be resolved and a magnification of ×30 is justified. A great deal of the structure of both plants and small animals (such as are found in pond water) can be seen with this kind of microscope.

18.59 The standard microscope

The standard microscope has three objectives mounted on a rotating head so that when the object has been focused at low power the next objective can be turned into position and is in fairly good focus. The lowest power objective has one or two lenses; the second and third have more components and are well corrected. The third objective is often intended for use with oil-immersion. The instrument has a set of oculars and the combination of the highest power objective with the highest power ocular gives a magnification of about ×1000. Provision is made for rough adjustment of focus followed by a fine adjustment. Rack-and-pinion devices are available to move the stage so as to scan different parts of the object.

18.60 The research microscope

The research microscope differs from the standard microscope in three ways. Firstly, it is more heavily engineered and provided with better mechanical adjustments both for focusing and for moving the stage. Secondly, the optics is corrected as accurately as possible under all conditions of use. Alternative illumination systems of high quality are provided. Thirdly, it is more versatile, being quickly changed for phase-contrast microscopy (Section 19.17), for dark-field illumination and in other ways. A camera of very high quality, which is automatically in focus when the field has been focused by eye, is provided. Sometimes an arrangement similar to a zoom lens enables the magnification to be altered continuously.

18.61 Reflecting microscopes

Whereas the designer of an instrument for visible light has a wide variety of glasses at his or her disposal, the designer for the ultraviolet (350–200 nm) has only two or three media (quartz, fluorite, lithium fluoride). For some regions of the infrared and for ultraviolet below about 120 nm there are no suitable transparent substances. There are, however, metals that reflect well over a large range of infrared wavelengths and also in the visible and far ultraviolet (down to a few nanometres). This makes the design of reflecting microscopes of considerable importance. About 1947 C. R. Burch examined the problems involved and designed a series of reflecting objectives, one of which is shown in Fig. 18.30. To obtain a numerical aperture greater than 0.5 it is necessary to use at least one aspherical surface.

Even so numerical apertures as high as those obtained with oil-immersion refracting objectives are not available.

There are two very important advantages of the reflecting microscope. Firstly, it is

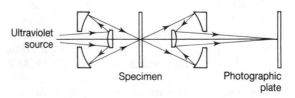

Figure 18.30 The reflecting microscope

entirely free from chromatism so that when focused for visible light it is also focused for all wavelengths. Secondly, the distance from the nearest mirror to the object is very large. This enables heated objects on the stage to be examined, even up to temperatures of over 1000°C. Until recently X-rays could be reflected only at grazing incidence and it was extremely difficult to get adequate definition even at low aperture. The development of surfaces that reflect X-rays at near normal incidence makes possible the design of good reflecting X-ray microscopes.

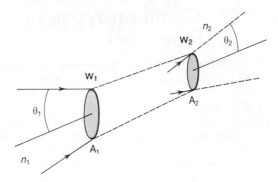

Figure 18.31 Solar collector

18.62 Flux collectors—solar energy

We now consider optical devices designed to concentrate the radiant energy received at an entrance window (W_1 in Fig. 18.31) of area A_1 into an exit window W_2 of smaller area A_2. These devices may form an image of W_1 at W_2 but do not necessarily do so. Interest in these devices has increased greatly in connection with the direct use of solar energy as a form of power and in the following paragraphs we shall discuss this application.

In an ideal concentrator all the rays that enter A_1 within a certain angle θ_1 emerge through A_2 within an angle θ_2. When the energy is derived from a source such as the Sun the incident energy is proportional to $A_1 n_1^2 \sin^2\theta_1$, where n_1 is the index of the medium immediately to the left of A_1. Equation. (3.7a) then implies, for an image-forming system,

$$A_1 n_1^2 \sin^2\theta_1 = A_2 n_2^2 \sin^2\theta_2 \qquad (18.5)$$

where n_2 and θ_2 apply to the exit rays.

Clausius (1822–1888) has shown that this relation can be derived from thermodynamic principles so that it applies also to non-image-forming systems. $A_1 n_1^2 \sin^2\theta_1$ is called the *étendue*; A_1/A_2 is the *concentration ratio R*.

Since the Sun subtends only 0.25° at the Earth it might seem sufficient to have a device that transmitted only rays falling within a small angle, even allowing for some spread by small-angle scattering. This is correct if the device is continuously turned so that it always points at the Sun but the cost of following the Sun in this way is prohib-

itive. Actual devices are fixed (or possibly adjusted about once a month) and it therefore becomes necessary to collect from as wide an angle as possible. Since $\sin\theta_2$ cannot exceed unity, it follows that $\sin^2\theta_1$ cannot exceed

$$\frac{n_2^2}{n_1^2}\frac{1}{R}$$

If $R = 10$, $n_2 = 1.7$ and $n_1 = 1$ this implies $\sin^2\theta_1 < 0.28$ or $\theta_1 < 17°$ so that the Sun can be followed over a range of not more than 34°. Other considerations generally make the angle smaller. If $R = 6$ a range of 57° is obtained.

For a device designed to use the heat from the Sun in a direct way the required value of R may be calculated in the following way. First suppose that the sunlight falls normally on a black body containing water to be used for domestic heating. Then the temperature will rise until $\sigma T^4 = S$, where σ is the Stefan constant 5.7×10^{-8} W m^{-2} K^{-4} and S is the solar energy in watts per square metre. Since S is about 1 kW m^{-2}, $T = 364$ K or about 90°C. This is sufficient for domestic heating, and devices of this type are useful in regions where the annual hours of sunshine are large enough, i.e. in a large part of the United States, in Israel, etc. In Britain and most of Western Europe the amount of heat produced is scarcely enough to pay for the cost of installing the device. The temperature is too low to run a heat engine to provide electrical power. When R is 6 the temperature is nearly 300°C and when R is 10 it is 374°C; temperatures in this range are sufficient to provide high-pressure steam for driving a turbo generator.

18.63 Imaging systems

An image-forming system has to deal with aberrations due to rays making large angles with the axis and, if no correction is made, there will be a large loss of rays due to spherical aberration and coma. It is necessary to aplanatise the system for the outer parts of the field in order to get all the rays through the exit window. Aberrations for rays in the centre of the field do not matter provided the aberrant rays remain within the window. This requirement is opposed to that of most optical systems which are designed to give the best imaging in the centre of the field. To obtain the required corrections for rays of large angle it is necessary to use several components and the cost becomes prohibitive.

18.64 Non-imaging systems

The theory of non-imaging systems is rather complicated because skew rays are important. Welford and Winston [4] give an extensive treatment and references to about 50 scientific papers. Here we shall consider briefly only two systems. The first is the reflecting cone (Fig. 18.32). The diagram shows that unless the cone is of very small angle some rays are turned back after a few reflections. On the other hand, if the cone angle is small the cone becomes very long in comparison with the size of the exit window, thus increasing costs.

The compound parabolic concentrator (Fig. 18.33) can be made to have very high collection efficiency. The paraboloid is

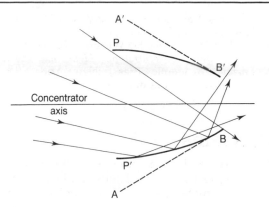

Figure 18.33 Compound parabolic concentrator. AB is the axis of the parabolic section P. The surface is formed by rotating P about the concentrator axis

formed by rotation of the parabola about the axis of the collector (not the axis of the parabola).

So far we have assumed that the receiver is planar and that the collector is concentrating the rays in two planes (three dimensions). Systems have been designed in which the receiver is a long tube and the collector concentrates in only one plane (two dimensions). The analogue of the cone is then a V-shaped trough and there is a corresponding analogue for the paraboloid. These collectors are usually cheaper and for some purposes it is convenient to have the receiver in tubular form. Altogether the prospects for producing large amounts of electrical power from solar energy are not good. The radiation incident on a square kilometre is only 1000 MW and with 60% efficiency of collection and 50% efficiency of the heat engine only 300 MW of useful power are obtained for part of the day and part of the year. A modern power station of medium size produces 1000 MW of electrical power. However, once it is constructed, the solar power station has no fuel cost!

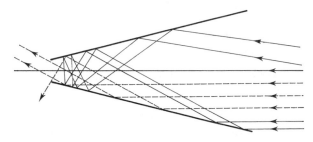

Figure 18.32 Section of a conical collector: the dotted rays arrive at the absorber whereas the rays shown by the full lines are reflected back into space

References

1. R. Kingslake (Ed.), *Applied Optics and Optical Engineering*, Several volumes, Academic Press, New York, San Francisco and London, 1965 onwards.
2. J. E. Field, 'High speed photography' *Contemporary Physics*, **24**, 439–459 (1983).

3. Kitt Peak National Observatory, Reports on the Next Generation Telescope.
4. W. T. Welford and H. Winston, *The Optics of Non-imaging Concentrators, Light and Solar Energy*, Academic Press, New York, San Francisco and London, 1978

Problems

18.1   A small surveying telescope has an objective of focal length 200 mm and diameter 25 mm. The eyepiece has a focal length of 20 mm. Determine the position of the exit pupil. Comment.

18.2   If the diameter of the eyepiece is 8 mm, calculate the angular field of the telescope of Problem 18.1. Explore what happens when field lenses of various focal lengths are placed in the position of the intermediate image.

18.3   Estimate the approximate loss of light transmitted through the system of Fig. 18.6 if the lenses are made of glass of refractive index 1.58.

18.4   Make plausible guesses as to the size of the eye of Fig. 18.8 and of the radius of the cornea. Then estimate the thicknesses of the eye lens for (a) distant and (b) close (20 cm) vision.

18.5   Why the odd shape of the Coddington magnifier (Fig. 18.11b)?

18.6   Prove that a corner cube deflects a beam on its hypotenuse face through $180°$.

18.7   Collect the following lenses from the laboratory: two convex lenses of focal lengths 20 and 5 cm; one concave lens of focal length -10 cm. Use them to construct (a) an astronomical telescope, (b) a telephoto system (Fig. 18.21) and (c) a reverse telephoto system. Compare the lengths, magnifications and fields of view of the three instruments.

18.8   In the system shown in Fig. 18.23, the image width and lens diameter are both 30 mm. The refractive index of the block is 1.52. The light is to be cut off when the prism has rotated by $45°$ from the position shown in Fig. 18.23(b). What thickness of block is required?

18.9   You could make a multimirror telescope from a close-packed array of either seven or nineteen circular mirrors. How would the amounts of light collected compare with that of a single mirror whose diameter is the largest dimension of the array?

18.10 Consider the behaviour of a Galilean telescope used in reverse, and ponder on the operation of the peephole in a door.

Assessment of Optical Images

19.1 Introduction

The usefulness of an optical instrument depends on other things beside the quality of the optical image that is formed. For example, mechanical stability and fineness of adjustment are very important in a high-power microscope. Many modern instruments have electrical outputs and depend both on the sensitivity of the electrical detectors and on a computing unit which follows. Yet the quality of the optical image is fundamental. In this chapter we consider how the wave theory deals with optical images and how it provides criteria for their assessment.

19.2 Images formed with coherent light

Suppose that a partially transparent object, such as a microscope slide, is illuminated by a parallel beam of coherent light of wavelength λ (Fig. 19.1). The light incident on the object may be represented by

$$A_0 \exp i(\omega t - \varkappa z) \qquad (19.1)$$

where A_0 is a constant whose square is proportional to the energy of light. Since we are not interested in the total amount of light we may put $A_0 = 1$. The light that has just passed through the object may be represented by

$$U_0(y) \exp i(\omega t - \varkappa z) = [1 + A(y) \exp i\delta(y)]$$
$$\times \exp i(\omega t - \varkappa z) \qquad (19.2a)$$

or

$$U_0(y) = 1 + A(y) \exp i\,\delta(y)$$
$$(19.2b)$$

In eqns (19.2) the first term represents the light which would pass if the object were not present and the second is the change due to absorption and phase changes caused by the structure of the absorption. This term carries all the information about the object.

A similar situation is obtained when an opaque object (such as a metal surface which has been polished and etched) is illuminated normally with coherent light. The above equations then represent the reflected light.

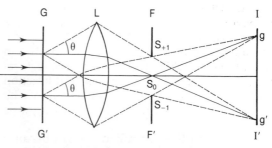

Figure 19.1 Coherent image formation

By writing $U_0(y)$ we have restricted the treatment to objects for which U_0 is a function of y but not of x (i.e. to striped patterns). The extension to the more general case $(U_0(x, y))$ is easy and involves no new physical principles although the equations are more complicated.

19.3 Amplitude objects and phase objects

In general both $A(y)$ and $\delta(y)$ vary with y but we may distinguish two special cases:

1. *Amplitude objects* for which $A(y)$ varies and $\delta(y)$ does not
2. *Phase objects* for which $\delta(y)$ varies but $A(y)$ does not

Phase objects are often obtained with a microscope slide containing bacteria which are nearly transparent but have a refractive index (μ_1) differing from the index μ_m of the medium in which they are immersed. If the thickness at the point y is $e(y)$ there is a phase difference $\varkappa(\mu_1 - \mu_m)e(y)$ which varies from point to point if either e or μ_1 vary. Since neither the eye nor any physical detector is sensitive to phase differences, such objects are difficult to discern in an ordinary microscope. We shall later (Sections 19.17 and 19.23) describe methods of making their structure clearly visible.

19.4 Sinusoidal gratings

Now consider an amplitude object for which $A(y) = A_p \cos py$ where A_p is a constant. Since $\delta(y)$ is also constant we may put $\exp i\,\delta(y) = 1$ so that

$$U_0 = 1 + A_p \cos py \qquad (19.3)$$

This is called an *amplitude sinusoidal grating* of contrast A_p. We assume $A_p^2 \leqslant 1$ (so that U_0 is never negative) and if $A_p = 1$ we have a grating of maximum contrast. A sinusoidal grating gives three spectra S_0, S_{+1} and S_{-1} in the plane FF' (Fig. 19.1) which is the focal plane of the lens L and contains a diffraction pattern that is a Fourier transform of the object when quasi-monochromatic light is used.

S_0, S_{+1} and S_{-1} are called spectra because they would be spectra (formed as explained in Chapter 6) if the light were not quasi-monochromatic. With monochromatic light they are small areas that may be regarded as points (or slits perpendicular to the plane of the paper). Interference from these sources forms the image on the plane I'I.

For a grating of low spatial frequency (wide lines and spaces) S_{+1} and S_{-1} are close to S_0 but for higher spatial frequencies the side beams diverge more and more until they no longer pass through the lens L. This means that there is effectively a circular stop in the plane FF' and that S_{+1} and S_{-1} must be within the area of this stop if their light is to pass on and contribute to the image. When S_{+1} and S_{-1} are lost, S_0 produces only a uniform field on II'. Thus in order to reproduce the structure of the grating we must have p less than a maximum value p_M, such that the first-order diffracted beam, in a direction α, is just collected by the lens. Putting $m = 1$ in eqn (4.26) we have for the angle of first-order diffraction $\sin\theta = d/\lambda$. Now $p = 2\pi/d$ for the grating and $\varkappa = 2\pi/\lambda$ so that $\sin\theta = p/\varkappa$. Thus

$$p_M = \varkappa \sin \alpha = \frac{2\pi \sin\theta}{\lambda} \qquad (19.4a)$$

and

$$d_M = \frac{2\pi}{p_M} = \frac{\lambda}{\sin\alpha} \qquad (19.4b)$$

where d_M is the minimum line space in a grating that is just resolved.

19.5 Oblique illumination

Now suppose that the object is illuminated by a parallel beam of light at an oblique angle α' so that S_0 just passes the stop on FF' and the spatial frequency of the grating is p'_M, where

$$p'_M = \varkappa \sin \alpha' \doteqdot \varkappa \sin 2\alpha \qquad (19.4c)$$

Then S_+ just passes the other side of the stop and an image of a grating of spatial frequency p'_M (nearly twice p_M) can be formed. In microscopy this advantage may be gained by illuminating the object with a hollow cone of light. We shall now show that there is a price to be paid for this advantage.

19.6   Contrast

If the object is represented by

$$U_0 = 1 + \cos py = 1 + \tfrac{1}{2}(e^{ipy} + e^{-ipy}) \quad (19.5)$$

then S_0 is represented by 1 and S_+ and S_- by the two exponentials. The distribution of light in the focal plane FF' is a Fourier transform of light in the object plane (Section 5.2). When S_0, S_{+1} and S_{-1} are all transmitted the distribution of light energy in image and object is the same and is given by

$$W = U_0 U_0^* = (1 + \cos py)^2$$
$$= \tfrac{3}{2} + 2 \cos py + \tfrac{1}{2}\cos 2 \, py \quad (19.6)$$

or

$$W_N = 1 + \tfrac{4}{3}\cos py + \tfrac{1}{3}\cos 2py \quad (19.7)$$

W_N shows the y-dependent part of the intensity variation in relation to unity background intensity, enabling the *contrast* to be estimated.

If only S_0 and S_{+1} are transmitted

$$W = (1 + e^{ipy})(1 + \tfrac{1}{2}e^{-ipy}) = \tfrac{5}{4} + \cos py \quad (19.8)$$

and

$$W_N = 1 + \tfrac{4}{5}\cos py \quad (19.9)$$

The multiplier of $\cos py$ in eqn (19.9) is only 0.6 of that in eqn (19.7). Thus, with oblique illumination, a wider range of spatial frequencies is reproduced but for all frequencies the contrast is lower. When the object is of high contrast (e.g. a black and white lantern slide) oblique illumination is better because the higher spatial frequencies are needed to reproduce sharp boundaries between black and white for other fine detail. If the initial contrast is nearly unity, reduction to 0.6 does not make the picture much less clear. If the initial contrast is low in part of the object, as often happens with biological microscope slides, the loss of contrast may make the object considerably harder to see. In an extreme case it may bring some features of the object below the level at which the eye can respond and then they are not seen at all in the image. Figure 19.2, curves (1) and (2), shows the contrast reproduction for normal and oblique illumination.

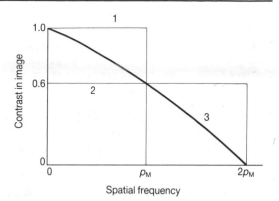

Figure 19.2   (a) Illumination at normal incidence; (b) at oblique incidence; (c) for non-coherent illumination

19.7   Image formed with non-coherent light

Figure 19.3(a) shows a system in which a sharp image of an extended source is formed on the object. A full cone of light from a point on the object fills the lens L. In place of

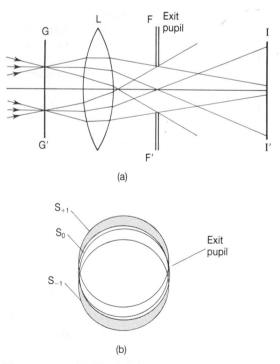

Figure 19.3   (a) Illumination by convergent, incoherent light. (b) S_+, S_- and S_0 in relation to exit pupil

S_0 we have a circle of light in the plane FF'. The exit pupil (Section 18.6) of the system is in this plane and S_0 nearly covers its area. S_+ and S_- are represented by circles whose centres are displaced from that of S_0 (Fig. 19.3b). Thus only a portion of their light passes through the exit pupil and goes on to the plane OY'. For low spatial frequencies the displacement of the centres is small but it increases as the spatial frequency increases and the amount of light corresponding to S_+ and S_- which reaches the plane OX' falls. The contrast of the image-grating formed on OY' thus gradually falls as the frequency increases, reaching zero when eqn (19.4b) is satisfied (Fig. 19.2).

19.8   Optical transfer function (OTF)

When considering the formation of images with coherent light we have to use the light vector E because light from different parts of the object can interfere. Yet both for coherent and non-coherent illumination it is the light *energy* distribution in the image that we observe. Let us now therefore consider the Fourier analysis of the energy distribution in object and image. We use a new symbol (q) for spatial frequencies in the energy distribution because, as we have seen above (eqn 19.7) and in Problems 19.2 and 19.3, the spatial frequencies in the energy distribution differ from those in the amplitude distribution.

Using a Fourier integral we may represent the object by

$$W_0(y) = \int_{-\infty}^{\infty} a(q)e^{iqy} \, dq \qquad (19.10a)$$

where

$$a(q) = \frac{1}{2\pi} \int_{-\infty}^{\infty} W_0(y)e^{-iqy} \, dy \qquad (19.10b)$$

For each spatial frequency there is a transmission factor $D(q)$ and the image is represented by

$$W_i(y) = \int_{-\infty}^{\infty} D(q)a(q)e^{iqy} \, dq \qquad (19.11)$$

$D(q)$ is called the *optical transfer function* (OTF). Sometimes $D(q)$ is complex which means that the phase of the 'grating' in the

image is not the same as that of the object, i.e. the 'grating' is shifted sideways. More usually $D(q)$ is real. In either case the ratio of the amplitude of the q-term in the image to the amplitude of the corresponding term in the object is denoted by

$$M(q) = |D(q)| \qquad (19.12)$$

$M(q)$ is called the *modulation transfer function* (*MTF*) and the value for a given frequency is called the modulation transfer factor for that frequency. Since the object is of finite size Fourier integrals might have been used in eqns (19.9) to (19.11).

19.9   Line-spread function

Suppose that the object is a very narrow line between $y = 0$ and $y = dy$. Then W_0 is an approach to a delta function (Fig 5.1d and Table 5.1) whose Fourier transform is a constant, i.e. $a(q) = A$ for all values of q. Then eqn (19.11) becomes

$$W_s(y) = A \int D(q)e^{iqy} \, dq \qquad (19.13a)$$

W_s is the distribution of light in the image of a narrow line (Fig. 19.4) and is called the *line-spread function*. If the line object is between y_0 and $y_0 + dy$ eqn (19.13a) is replaced by

$$W_s(y') =$$
$$A \int_{-\infty}^{\infty} D(q)\exp iq(y' - y'_0) \, dq \qquad (19.13b)$$

where y'_0 is the image of y_0 and y' is another line in the image plane. Thus, in either case, the line-spread function is the Fourier transform of the OTF and if either is known the other can be calculated. We shall see below that W_s is more directly useful in the assess-

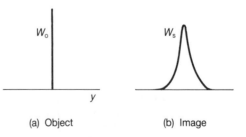

(a) Object   (b) Image

Figure 19.4   (a) A δ-function object; (b) distribution of illumination in image—the line-spread function

ment of telescopes and of spectroscopic instruments but the OTF is required for microscopes, cameras, etc.

When two-dimensional objects are involved we need the spread of light in the image of a point source (P). There is a point-spread function that represents the distribution of light round P' (the geometrical image of P). For aberration-free systems the spread is nearly symmetrical and may be described by a function $W(r_0' - r')$, where r_0' is the geometrical image of a point on the object and r' is a point on a circle of radius $r' - r_0'$ centred on r_0'. When the spread of light is not symmetrical a two-dimensional Fourier transform is required.

19.10 Measurement of OTF or of the spread function

Many methods for the measurement of either $D(q)$ or W_s have been developed. The most obvious method is to use a narrow slit as object and to sweep the image plane with a detector covered by a narrow slit parallel to the object slit. This gives W_s directly and $D(q)$ can be computed using eqn (19.13a). This is a possible, though difficult, method for an optical system with large aberrations giving a wide spread in the image plane. For a good optical system the spread is small and then both slits would have to be very narrow. Very little light would reach the detector and the signal-to-noise ratio would be very poor. Most existing methods measure $D(q)$. The instrument has three parts (Fig. 19.5). The first generates a grating of high contrast whose spatial frequency can be varied. The

second is a detector covered by a narrow slit in the image plane and parallel to the lines of the grating and the third is an electronic system, essentially a small computer, which analyses the signal. In some instruments the detector system is moved to scan the image, in others the grating object is made to drift across the field. In either case the signal gives $D(q)$. Sometimes the object is generated by using interference to give sinusoidal fringes. It is, however, easier to generate square wave gratings by mechanical methods. The computer then extracts the fundamental frequency from the signal and $D(q)$ is calculated for this frequency.

In the earlier instruments $D(q)$ was measured for different values of q in turn and the curve of $D(q)$ versus q was plotted. In more advanced instruments the spatial frequency q is altered at a rate that is slow enough to allow a measurement at one frequency to be made yet so fast that the whole curve of $D(q)$ versus q can be displayed on CRT and also plotted as a graph. The accuracy of the instruments depends on the constancy of the light source and for the highest accuracy it is necessary to have a second beam of light which goes to a second detector without passing through the lens under test. Its signals are used by the computer to compensate for small fluctuations of intensity of the source of light.

It will be apparent from the above description that the measurement of $D(q)$ or W_s requires a fairly elaborate instrument. Unfortunately the performance of a lens is not described by just one curve of $D(q)$ versus q. Even for a point on or near the axis of the lens it is necessary to measure $D(q)$ for a series of image planes to determine the plane of best focus. For regions off-axis it is necessary to measure $D(q)$ for at least two orientations of the grating (a radial direction in which the lines of the grating are parallel to the radius from the centre to the region under study and a tangential direction at right angles to this radius). Also, since there is usually some curvature of field, the plane of best focus must be determined for both orientations of the grating. For lenses to be used for aerial photography very small amounts of distortion are important and this error is usually measured separately.

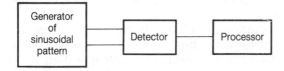

Figure 19.5 Schematic of instrument for measuring the optical transfer function (OTF) or modulation transfer function (MTF)

19.11 Rayleigh limit of resolution for a telescope

For telescopes the spread function is more directly useful than $D(q)$ because most of the objects are effectively point sources. The spread function is simply the Franhofer diffraction pattern of the objective regarded as a circular aperture (Section 6.9). Rayleigh proposed the criterion that two stars could be said to be resolved when the maximum in the pattern for one star coincided with the first minimum for the other (Fig. 19.6a). Since the two sources are non-coherent, the energy distribution in the overall pattern (solid curve) is the sum of the energies in the separate patterns (dotted curves). When the

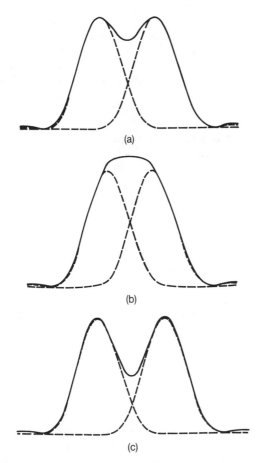

(a)

(b)

(c)

Figure 19.6   (a) Resolution: the Rayleigh criterion (RC). (b) Distribution for separation 20% less than RC unresolved. (c) Distribution for separation 20% greater than RC (easily resolved)

Rayleigh criterion is exactly fulfilled the intensity at the mid-point is 20% less than either of the maxima (Fig. 19.6a). The Rayleigh criterion is an arbitrary one but very useful. When the angle between the directions of two stars is 20% less than the Rayleigh limit, the stars are obviously not resolved (Fig. 19.6b). When it is 20% more they are very easily seen as separate objects (Fig. 19.6c).

From Section 6.9 the Rayleigh criterion is satisfied when

$$\frac{2\pi}{\lambda} R \frac{r'}{f} = 1.22\pi \qquad (19.14)$$

where R is the radius of the objective and r' is the linear separation of the two images in the focal plane (f is the focal length). Then the angular separation (α) of the two stars is

$$\alpha = \frac{r'}{f} = 1.22 \frac{\lambda}{d} \qquad (19.15)$$

where $d = 2R$ is the diameter of the objective.

19.12 Limit of resolution of the eye

Figure 18.8 shows a schematic cross-section of the human eye. The curved surface of the *cornea* (c) and the *lens* (L) constitute a dioptric system that forms an image on the *retina* (R) at the back of the eye. The retina consists of a large number of light detectors that generate digital electrical signals when light falls on them. These signals are processed to some extent in the retina, which contains a network of nerve connections immediately behind the receptors, and the resulting signals are transmitted by the million fibres of the optic nerve to a part of the brain called the visual cortex, where further processing takes place. The diameter of the *iris* diaphragm D is about 2.2 mm at daylight brightness and expands to about 7.5 mm in the dark.

Putting $d = 2.2$ mm in eqn (19.15) gives, for the Rayleigh limit of resolution of the dioptric system, a value of a little less than 1'. In the centre of the retina there is a region (corresponding to about 2° angular diameter in the visual field) where the cone receptors are small and close packed. Also in this region the number of nerve fibres is equal to the number of receptors. This enables the visual system to take full advantage of the resolving power of the dioptric system and young observers with good sight are bound to be able to resolve star images whose angular separation is 58″. This corresponds to about 0.1 mm seen at the nearest point of distinct vision (25 cm).

At lower illumination when the pupil expands aberrations of the lens system become important and the resolution of the eye is less than that given by the Rayleigh criterion. Also the cone receptors in the centre of the eye are of such low sensitivity that this portion of the eye is night blind. Vision is mediated by another kind of receptor called rods and the nervous connections are such that summed signals from groups of rods are sent along one nerve fibre to the brain. This interconnection is increased at low illumination. Thus the system is adapted to obtain maximum sensitivity, and poorer resolution at low illumination levels is inevitable.

19.13 Useful and empty magnification

Rewriting eqn (19.18), we have

$$r' = \frac{1.22}{d}\,\lambda f \qquad (19.16)$$

where r' is the separation in the focal plane of the objective of two points that are just resolved. It often happens that this distance is less than the 0.1 mm which the eye can resolve. A magnifier is therefore used. Magnification of $0.1/r'$ (r' in millimetres) just brings the smallest detail resolved by the objective within the limit of resolution of the eye. Generally a little more magnification than this is used to give 'comfortable' resolution by the eye. This magnification is called *useful* magnification. Greater magnification is called *empty* magnification. Empty magnification reduces the illumination of the image (by spreading the available light over a larger part of the retina) and usually involves aberrations and loss of contrast so that the object becomes less easy to see.

19.14 Resolving power of a spectroscope

From eqns (4.25b) and Fig. 4.11 the central maximum for light of wavelength $\lambda + \Delta\lambda$ coincides with the first minimum for light of wavelength λ (in a grating spectrum) when, for a spectrum of order m,

$$Nm(\lambda + \Delta\lambda) = (Nm + 1)\lambda \qquad (19.17)$$

i.e. when

$$R = \frac{\lambda}{\Delta\lambda} = Nm = \frac{D \sin \theta}{\lambda} \qquad (19.18)$$

where $D = Nd$ is the width of the grating. Following Rayleigh, R is called the *resolving power*. It may be shown that, for a prism spectroscope,

$$R = t\,\frac{\mathrm{d}\mu}{\mathrm{d}\lambda} \qquad (19.19)$$

were t is the thickness of the base of the prism and μ the refractive index of the prism material.

In the Littrow mounting light is reflected from the backface of the prism so that it traverses the prism twice and then $2t$ has to be substituted for t in eqn (19.19).

19.15 Resolving power of a Fabry–Perot etalon

The distribution of energy in the interference pattern produced by a Fabry–Perot etalon is very different from that produced by a grating or prism spectroscope (Fig. 19.7). The Rayleigh criterion (Section 19.11) of coinciding maxima and minima cannot be applied directly. It is, however, possible (using eqn. 4.25b) to calculate the wavelength separation required to give a '20%

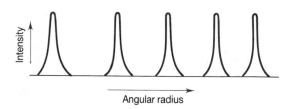

Figure 19.7 Distribution of illumination in ring pattern from Fabry–Perot etalon

dip' between the maxima due to two different wavelengths. From this equation, we see that $\lambda/\Delta\lambda$ is proportional to the separation of the plates. It also increases with the reflecting power of the surfaces but, with metal surfaces, increase of reflection beyond a certain point causes a rapid fall in the transmission and therefore in the total light in the interference patterns. There is thus an advantage in using high-efficiency dielectric reflecting films (Section 4.31), instead of metal films. The calculations of $\lambda/\Delta\lambda$ for a given transmission is done with a computer. The results are usually expressed in terms of the equivalent number (N) of beams of equal amplitude which would give the same resolving power. Values of N of 50 or more are easily obtained. With an etalon of length 10 cm (20 cm for the path difference between successive beams) the order of interference for a wavelength of 500 nm is 400 000 so with $N = 50$, eqn (19.18) gives a resolving power of 20 million. The resolving power available with an etalon is thus much higher than that normally obtained with a grating.

19.16 Reproduction of detail by a microscope

The function of a microscope is to provide a magnified image in which details of an object too fine to be seen by the naked eye become visible in the magnified image. We are therefore interested in the ability of a microscope to reproduce detail rather than in its ability to resolve point objects or line objects. In a Fourier analysis of the object the finest detail is represented by terms of high spatial frequency and we therefore first calculate the highest frequency that is reproduced. The discussion of Section 19.4 shows that when the object is illuminated by a parallel beam of coherent light in the direction of the axis, this maximum frequency is given by eqn (19.4a). If d_m is the maximum distance between the lines in a grating whose structure is reproduced then (repeating eqn 19.4b)

$$d_m = \frac{2\pi}{p_m} = \frac{\lambda_m}{\sin \alpha} \qquad (19.20)$$

In this equation λ_M stands for the wavelength of light in the medium between the

object and the objective and when this medium is air this is effectively equal to the vacuum wavelength λ. To increase p_M (and reduce d_m) it is usual, in microscopes of high magnification, to use oil immersion, i.e. the space between the object and objective is filled with an oil of index μ and eqn (19.20) is replaced by

$$d_m = \frac{\lambda}{\mu \sin \alpha} = \frac{\lambda}{N} \qquad (19.21)$$

where $N = \mu \sin \alpha$ is called the *numerical aperture*. Equation (19.4a) applies to the microscope illuminated with coherent parallel light. We have seen that, with a hollow cone of illumination, higher spatial frequencies are reproduced but with some loss of contrast.

In Section 5.14 we saw that the degree of coherence between two points in a beam becomes higher as the points become nearer. It is therefore usual that there is partial coherence between light from two points which are just resolved by a microscope of high power. However, by using a good condenser system to focus a sharp image of the source on the object it is possible to obtain non-coherent illumination.

For non-coherent illumination it is also possible to calculate the limit of resolution for point objects and the formula

$$r' = \frac{0.61\,\lambda}{N} \qquad (19.22)$$

is obtained for the shortest distance between point objects that are just resolved (Problem 19.7) The difference of the factor 0.61 between eqns (19.21) and (19.22) is due to the fact that in one case we are dealing with line objects and in the other with point objects.

The concept of useful and empty magnification (Section 19.13) applies to microscopes. For high-power microscopes used visually the limit of useful magnification is about 1000 (Problems 19.8 and 19.9). When the image is photographed the magnification must be such that the finest detail is considerably larger than the grain separation of the photographic plate. It is usual to reproduce, for publication, photographs from high-power microscopes at a magnification of $\times 2500$–5000.

19.17 Phase-contrast microscopy

A number of devices has been produced to operate on the light in the Fourier transform plane (FF′ in Fig. 19.1) so as to obtain better images of certain kinds of objects. The most important of these is *phase-contrast microscopy* developed by Zernike (1888–1966). Returning to Section 19.2 we consider the simplest amplitude object for which

$$U_a(y) = 1 + a \cos py = 1 + \frac{a}{2}(e^{ipy} + e^{-ipy})$$

(19.23)

Now consider a corresponding phase object for which $\delta(y)$ is always small compared with unity. Then

$$U_p(y) = e^{i\delta(y)} = 1 + i\delta(y) \qquad (19.24)$$

and if $\delta(y) = a \cos py$ we have

$$U_p(y) = 1 + i\frac{a}{2}(e^{ipy} + e^{-ipy}) \quad (19.25)$$

In the plane FF′ the amplitude grating produces three 'spectra' of amplitudes, l, a, a, situated at points corresponding to spatial frequencies O, p and $-p$ (Fig. 19.8a). The phase grating produces three spectra of the same amplitudes and at the same points (Fig. 19.8b). The difference is that the three 'sources' for the amplitude grating have the same phase while for the phase grating the two sources at p and $-p$ are $\pi/4$ out of phase with the source at O. Interference of the three sources that are in phase on II′ reproduces the amplitude grating. Interference of those that are out of phase gives the phase grating which cannot be seen.

Now suppose that a glass plate, plane and parallel except for a small spot in the centre, is inserted in the FF′ plane (Fig. 19.8b). The spot in the centre is coated with a thin film so as to produce a phase delay of $\lambda/4$. Then the three spectra of the phase grating are brought into phase and the *amplitude* grating is seen on II′. Thus the contrast of phase which cannot be seen is transformed into a contrast of amplitude (and therefore of energy) which can be seen. In this way objects such as bacteria which are nearly transparent but have a different refractive index from that of the medium in which they

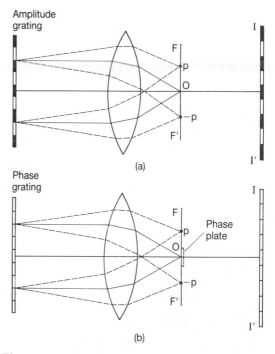

Figure 19.8 (a) Spectra from an amplitude grating. (b) Spectra from a phase grating: position of phase plate for normal incidence

are immersed are rendered clearly visible. So long as the phase difference that they produce is small, the amplitude difference on II′ is proportional to the phase difference. Thus internal structure (associated with local differences of index) or differences of thickness are represented by proportional differences of amplitude on II′.

The system we have described is known as *positive phase contrast*. If a thin film of thickness $3\lambda/4$ had been inserted, the picture seen on II′ would have been like a photographic negative of that described above and we should have *negative phase contrast*. If a hollow cone of light is used to illuminate the object, the region corresponding to S_o is an annular ring in the plane FF′ and to obtain phase contrast this annular region must be covered with the film of optical thickness $\lambda/4$ or $3\lambda/4$ (Fig. 19.9).

Ideally the ring is indefinitely narrow but in practice the ring must be of finite width and it cuts into the region for very low frequencies so that these are not completely reproduced. The result is that an object

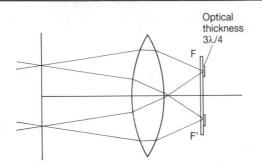

Figure 19.9 As for Fig. 19.8 (b) but with hollow annular cone illumination: position of phase ring

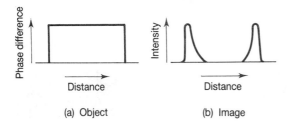

Figure 19.10 Edge distortion produced by phase plate

whose phase difference is represented by Fig. 19.10(a) is reproduced as an amplitude object shown in Fig. 19.10(b) This sharpening of the outline renders the phase object easier to see and is not a disadvantage provided that the microscopist recognises it as

an artefact and not part of the structure of the object.

19.18 Dark-ground illumination

It is not unusual to make the phase-contrast film of metal so that it absorbs some of the light of the zero-order spectrum, thereby increasing the contrast of the object. This increased visibility of faint objects is gained at the expense of some loss of fidelity of reproduction. The logical end of this process is to block out the zero-order spectrum altogether. The picture on II' then becomes dark except where interference of light from the side spectrum shows up structures in the object. This system, known as *dark-ground illumination,* is often used in metallography. A reflecting microscope views the surface of a metal which has been polished so as to be optically flat and then lightly etched. Grain boundaries and other structures are clearly seen though the picture is 'unnatural' in that the main surface of the strongly reflecting metal appears dark.

19.19 Spatial filtering

In some situations, images contain 'structure' which distracts attention from the

Figure 19.11 Picture showing raster lines. Reproduced by permission of K. Thomas

picture, and if this structure is regular it is represented by certain points or lines in the Fourier plane. By covering these points or lines in the Fourier plane the undesirable structure is removed. The resulting picture is more pleasing or more useful although it contains no more information than the original. Two cases where this method has been frequently used are (1) in the reproduction of a picture that has been printed through a screen or by rastering as in a television picture, and (2) in the reproduction of a picture that has been made by joining a number of strips that have been photographed successively, e.g. from a satellite. Figure 19.11 shows a direct photograph of a rastered picture. The lines give rise to spots in the Fourier plane (Fig. 19.12). When a filter is inserted to remove the spots, the image appears free of lines, as shown in Fig. 19.13.

Figure 19.12 Fourier transform of line structure. Reproduced by permission of K. Thomas

Figure 19.13 Picture with line structure removed by spatial filter. Reproduced by permission of K. Thomas

19.20   Schlieren method

Suppose that, as in Section 19.17, we have a phase object represented by

$$U_0 = 1 + ia \cos py \qquad (19.26)$$

where, as before, a is small so that a^2 may be neglected in comparison with 1. Then the energy distribution is

$$W = 1 + a^2 \cos^2 py \qquad (19.27)$$

and to the approximation we are using W is independent of y. Now put $\cos py$ in the exponential form so that eqn (19.25) replaces eqn (19.26) and suppose that a diaphragm with a sharp straight edge is introduced into the plane FF′ so as to cut out S_+ (corresponding to $i(a/2\, e^{ipy})$) and also about half of S_0. Then what remains is represented by

$$U_{0w} = \tfrac{1}{2}(1 + ia \cos py) \qquad (19.28)$$

and

$$W' = U_{0s}U_{0s}^* = \tfrac{1}{4}(1 + a^2 + 2a \sin py)$$
$$= \tfrac{1}{4} + \tfrac{1}{2} a \sin py \qquad (19.29)$$

(omitting a^2) so that phase differences in the object are represented by energy differences proportional to $1/p \; dU_{0s}/dy$.

The Schlieren method has been used in microscopy but generally the phase-contrast method is preferred.

In wind tunnel experiments there are differences of density in the air surrounding the test object. These differences are small, because the index of air (at atmospheric pressure) is only 1.0003, but the rate of change near the object may be fairly high so the Schlieren method is often used. Figure 19.14 shows an arrangement in which a beam of light reflected from one off-axis† paraboloidal mirror M_1 illuminates the object and is refocused by a similar mirror (M_2) at A which is the focal plane of M_2. S_0, S_+ and S_- are in this plane and the occluding diaphragm is placed here. The image on the plane II′ is photographed. The Mach–Zehnder interferometer (Section 17.6) may be used also for wind tunnel experiments.

19.21   Aberrations

We now consider the formation of the image of an on-axis point source (Fig. 19.15). A spherical wave is emitted and if L were aberration free a spherical wave C would follow L and its centre would be the image point. Another way of regarding this case is to say that the Fourier transform of the point source contains all frequencies and that S_0, S_{-1} and S_{+1} etc., for all spatial frequencies have the same phase on the sphere C and that all the

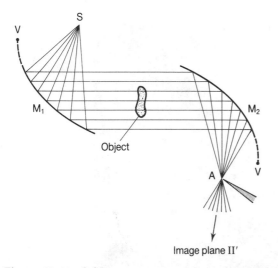

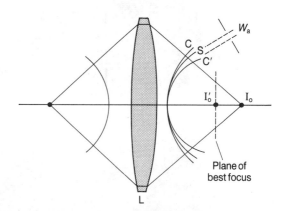

Figure 19.14   Schlieren system: M_1 and M_2 are off-axis paraboloidal mirrors (i.e. the vertices are not in the centres of the mirrors). After passing the knife-edge at A, the light enters a camera focused on the object

Figure 19.15   Aberrated wave-front for on-axis point of system

---

† The surface of an off-axis paraboloidal mirror is paraboloidal but does not include the vertex.

Huygen wavelets from the circle have the same phase at the focal point I_0.

When there is aberration the wave surface S emerging from L is no longer spherical though it touches C on-axis. The aberration W_a could be measured as the distance between S and C for any zone of the lens but in practice we are interested not in the aberration near the paraxial focus but in the aberrations in the plane of best focus.

Thus we need to measure W_a not from C but from the circle C′ corresponding to the plane of best focus. On first consideration it may seem that the plane of best focus is the one that gives a minimum value of ΣW_a^2, but this is not quite right because the area of a zone of the lens between r and $r + dr$ is $2\pi r\, dr$ and thus the outer zones have more area and have to be given more weight. However, suppose that the plane of best focus has been found and that W_a is measured from it. Then it is found that, for a point on-axis,

$$W_a = a_4 r^4 + a_6 r^6 + \cdots = W_4 + W_6 + \cdots \quad (19.30)$$

where a_4 and a_6 are constants and W_4, W_6 etc., are called the *primary, secondary, ...* aberrations. The coefficients a_4, a_6 etc., form a rapidly decreasing series and we consider only the first two. In general an optical designer aims to reduce the primary aberration to a low value and then to arrange that the secondary (and higher) abberations oppose the primary aberration over as much of the lens as possible so as to produce a very small total aberration.

For an off-axis point the emergent wavefront may be represented by a nearly spherical surface centred at the image point P′ (Fig. 19.16), and W is given by

$$W = {_0}C_{40} r^4 + {_1}C_{31} \sigma r^3 \cos\phi + {_2}C_{22} \sigma^2 r^2 \cos\phi$$
$$+ {_2}C_{20} \sigma^2 r^2 + {_3}C_{11} \sigma^3 r \cos\phi + {_0}C_{60} r^6 + \cdots$$

$$(19.31)$$

(see Ch. IV of ref [1]).

In eqn (19.31) all the terms, except the last one, represent primary aberrations. Note that the sum of the powers of σ and r is always 4. The suffixes of the constants ${_0}C_{40}$, etc., correspond with the powers of σ, r and ϕ.

The deviations of the wave surface from

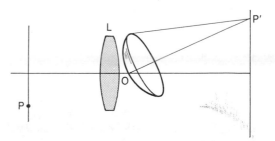

Figure 19.16 Abberrated wave-front for off-axis point P

the spherical form correspond with deviations of rays from an object point, so it is possible to begin the discussion of the primary aberrations either in terms of waves or of rays. In principle a designer should use wave theory and optimise the system taking account of both aberrations and diffraction and of their interaction. There are, however, two situations where ray-theory optimisation is appropriate. The first is where the ray aberrations can be made so small that the performance is determined almost entirely by diffraction. The second is where the ray aberrations, even when minimised, leave image defects which are still large compared with the effects of diffraction. It is only in the situation where ray aberration and diffraction produce effects of the same order that their interaction has to be considered. This situation is not very common.

19.22 Effect of aberrations on the OTF

Let us now return to the formation of an image by coherent light. In Section 19.21 we regarded S_0, S_+ and S_- as secondary sources located on the plane FF′ (Fig. 19.1). In fact they do not have exactly the same phase on that plane. If there are no aberrations, the light from these 'sources' is in phase at the point O′ in the plane II′ (Fig. 19.1). Then the waves all arrive at O′ in phase and we get maximum energy density at that point. This maximum exactly corresponds with the maximum energy in the object. The contrast of the grating image is the same as that of the object (provided S_+ and S_- lie within the entrance pupil) and the OTF = 1.

When there is aberration, the phases of S_+ and S_- differ from that of S_0 by $\varkappa W$ (where W is the aberration function) and the amplitude at O' is

$$U_0 = 1 + \tfrac{1}{2}\exp\, i(py' + \varkappa W) + \tfrac{1}{2}\exp\, i(-py' + \varkappa') \tag{19.32}$$

and

$$W = U_0 U_0^* = \tfrac{3}{2} + 2\,\cos\,\varkappa W\,\cos\,py' + \tfrac{1}{2}\cos\,2py' \tag{19.33}$$

If $W = 0$ (no aberrations) eqn (19.33) reduces to eqn (19.6) and comparing these two equations we see that the aberration has reduced the contrast in the image by a factor $\cos\,\varkappa W$. The MTF is reduced by the same factor. Since $\cos\,\varkappa W$ is always less than unity, the OTF is reduced at every frequency. It is not possible for a designer to obtain an MTF greater then that of an aberration free lens at any frequency. Since W is a function of frequency, designs may produce MTFs shown by curves as curves 2 or 3 on Fig. 19.17; i.e. a designer may aim at optimum performance at low frequencies (curve 2) or (as is more usual) at high spatial frequencies (curve 3).

From the preceding discussion it follows that the limiting frequency is determined by the aperture of the system and not by the aberration. It is possible that the aberration may reduce $D(q)$ at moderately high frequencies so much that for practical purposes the useful range of frequency is reduced.

If we considered only a sinusoidal grating object it would always be possible to draw a

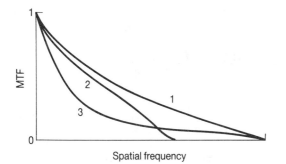

Figure 19.17 MTF for aberration-free lens (curve 1), lens designed to produce good contrast at low spatial frequencies (curve 2) and lens designed for sharp edges (good reproduction at high spatial frequencies) (curve 3)

circle through the three points where light from S_0, S_+, S_- was in focus and the full contrast (as with no aberration) would be obtained on a plane $O''y''$, where O'' was the centre of this circle. However, when there is aberration different frequencies would be focused on different planes and no one plane would give complete contrast for all spatial frequencies; i.e. we should not get true reproduction of an object containing many frequencies on any plane.

19.23   Diffraction and image formation

Zernike [1] has pointed out that there are always two stages of diffraction in the formation of an optical image. One stage is at the object itself and is well illustrated by our discussion of the formation of an image by a microscope. The other stage is effectively at the entrance pupil of the instrument that forms the image, e.g. at the objective in the case of a telescope. The Abbe theory lays stress on the diffraction at the object, though diffraction by the instrument enters in the limitation of the spatial frequencies that can be reproduced. In Rayleigh's theory of the limit of resolution by a telescope, diffraction at the objective is all important; with stellar objects the diffraction of the object is not important except when the star is so large that it is no longer a point object (Problem 19.5). Similarly, in a spectroscope the diffraction caused by the limited size of the prism or grating determines the resolving power.

When there is diffraction, the light that is deviated by diffraction carries information about the object that diffracts it (S_{+1}, S_{+2}, S_{+3}, ... and S_{-1}, S_{-2}, S_{-3}, etc., but *not* S_0). In the telescope the diffraction pattern of a point object carries information about the size of the objective. We do not want this information because we already know the diameter of the objective but we have to accept the fact that it confuses information about two neighbouring stars.

References

1. F. Zernike, *Z. Tech. Phys.*, **16**, 454 (1935); *Phys. Zeit.*, **36**, 848 (1935); *Physica*, **9**, 686, 974 (1942).

Problems

19.1 Plot a curve for W_N (eqn 19.6) and thus show that W_N is never negative.

19.2 Calculate W_N for $U_0 = A_1 \cos p_1 y + A_2 \cos p_2 y$ and show that it contains terms with frequencies equal to the sum and difference of every possible pair of frequencies (including $p_1 + p_2$ and $p_1 - p_2$, etc.).

19.3 Calculate W_N when $U_0 = \sum_{n=1}^{s} \cos p_n y$ and show that it contains terms with frequencies equal to the sums and differences of all possible pairs of frequencies. *Hint*: use the exponential expression for $\cos y$.

19.4 Find the Rayleigh limit of resolution for

(a) a theodolite with an objective 50 mm in diameter,

(b) an amateur's astronomical telescope with a primary mirror 200 mm in diameter,

(c) a 100 inch reflecting telescope and

(d) the largest telescope in existence (primary 6 m in diameter).

Give the answers in seconds of arc.

19.5 The largest angular diameter for a star is about 0.5″. For what size of telescope is the diameter of the Airy disc equal to the diameter of this star?

19.6 Calculate the linear separation in the focal plane of the primary mirror of the images of stars which are just resolved (a) for a primary of diameter 200 mm and focal length 2 m and (b) for a primary of diameter 3 m and focal length 10 mm.

19.7 A grating is 10 cm wide and the first order for a wavelength of 500 nm is diffracted in a direction $\theta = 45°$. Find the number of lines per millimetre and the resolving power.

19.8 If in the preceding problem the tenth order is diffracted through $45°$ what are the number of lines and the resolving power?

19.9 Find the number of lines in a grating whose first order has the same resolving power at 200 nm as a prism of quartz of thickness 5 cm used in Littrow mounting ($d\mu/d\lambda$ for quartz at 200 nm is 1.40×10^{-3} nm^{-1}).

19.10 Find the magnification required to make the separation of lines in a grating that is just resolved (according to eqn 19.21) equal to 0.1 mm when $N = 1$.

19.11 What value of N would correspond to a useful magnification limit of 200?

20

Lasers

20.1 Conditions for laser action

In Chapter 12, we saw that if an inverted population of atoms could be established in an optical cavity consisting of a pair of plane, parallel mirrors, then amplification and oscillation may occur and an intense, collimated, monochromatic beam result. We now put this idea on a more quantitative basis. How long must the cavity be? What reflectance of the mirrors is needed? How much bigger must n_2 be compared with n_1? On what other factors does the achievement of laser action depend? Although an accurate answer to these questions is difficult to obtain, we can get a very good idea from a rather simple model.

20.2 Parallel mirror system

We consider two plane mirrors with (energy) reflectance R and area A, separated by a distance l (Fig. 20.1). The concentrations of lower and upper state populations are n_1 and n_2 respectively and the Einstein coefficient for absorption/stimulated emission is B. The radiation density is $\rho(\omega)$ and the energy distribution in the spectral line emitted in the transition is represented by a shape factor $S(\omega)$. (This may have a Gaussian form if Doppler effects predominate, a Lorenzian form if natural lifetime effects are the most important, or a combination of these and other effects.) If the total radiation energy per unit volume is W, then we have

$$WS(\omega) = \rho(\omega) \qquad (20.1)$$

We assume (1) that we may neglect the effects of spontaneous emission, justified because we are interested only in the radiation travelling along the axis of the resonator, and (2) that the fractional gain in the amplitude of the radiation on a single pass through the system is small compared with unity.

From eqn (12.3), the net number of quanta created per second in unit volume of the material in the cavity is $B(n_2 - n_1)WS(\omega)$. In

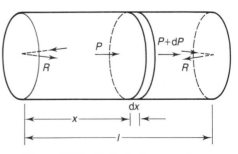

Figure 20.1 Gain in laser system

244

the volume $A\,dx$, the power generated, dP, is therefore given by

$$dP = B(n_2 - n_1)WS(\omega)\hbar\omega A\,dx \quad (20.2)$$

so that

$$P = B(n_2 - n_1)WS(\omega)\hbar\omega Al \quad (20.3)$$

How is radiant power in the cavity dissipated? Since only a fraction R of power incident on each mirror is reflected, then a radiation density W will, in the absence of amplification, decay with time with a time constant τ_c, where

$$\frac{dW}{dt} = -\frac{W}{\tau_c} \quad (20.4)$$

It is easily shown (Problem 20.1) that τ_c may be expressed in terms of l, R and the velocity v of light in the medium, by

$$\tau_c = \frac{l}{v(1-R)} \quad (20.5)$$

The total energy loss per second from the radiation field in the cavity is therefore $WV/\tau_c = WAl/\tau_c$. If laser action is to occur then the power increase due to stimulated emission must exceed this cavity loss, so that

$$B(n_2 - n_1)WS(\omega)\hbar\omega Al > \frac{WAl}{\tau_c} \quad (20.6)$$

or, using (20.5),

$$n_2 - n_1 > \frac{v(1-R)}{BS(\omega)\hbar\omega l} \quad (20.7)$$

For a gas laser, we may put $v = c$; for a solid state laser, $v = c/n$, where n is the refractive index.

20.3 A closer look at cavities

As indicated in Chapter 12, the role of the resonant cavity is to provide the feedback needed to establish a high value of radiation density, in the standing-wave field, in order to enhance the effects of stimulated emission. Amplification by stimulated emission produces an increase in the standing-wave field until saturation is reached, where the power emitted as laser radiation is balanced by the power input to sustain the population inversion (Fig. 20.2). For a total power input P,

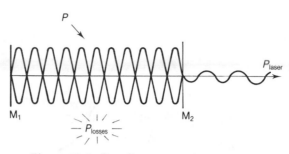

Figure 20.2 Standing waves in resonator

the laser power P_{laser} is given by

$$P_{laser} = P - P_{loss} \quad (20.8)$$

where the term P_{loss} includes:

1. Spontaneous emission at the lasing and other wavelengths
2. Scattering losses in the lasing medium
3. Losses at the resonator mirrors and
4. Lattice vibrational losses (heating) in the case of crystal lasers

The list is not exhaustive, but covers the main losses encountered in most lasers.

20.4 The cavity length

It may appear that it would be necessary to adjust the mirror separation to an exact integral multiple of the half-wavelength of the radiation to be used. In fact, for lasers in the optical region, this is not the case. The cavity length l is invariably very large compared with the wavelength. The condition for resonance is that $l = N\lambda/2$, where N is an integer. In terms of wave number $\bar{\nu} = 1/\lambda$, this condition is

$$\bar{\nu} = \frac{N}{2l} \quad (20.9)$$

The wave numbers $\bar{\nu}$ which satisfy eqn (20.9) are separated by an interval $\Delta\bar{\nu} = 1/2l$. Now the source of the radiation to be amplified is the spontaneous emission. Through natural, Doppler, collision broadening, etc., spontaneous emission lines have a finite width. How does this line width compare with the spacing $\Delta\nu_c = c\Delta\bar{\nu}$ between the resonant cavity modes? For the transition in neon which produces the familiar red helium–neon laser line, the line width is of the order

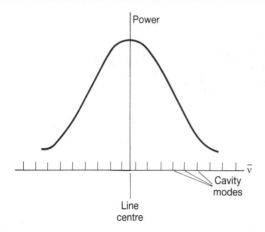

Figure 20.3 Cavity modes within the source line width

of 10^9 s^{-1}. For a cavity of length 1 m, the frequency interval between resonant modes is

$$\Delta \nu_c = c \ \Delta \bar{\nu} = \frac{c}{2l} = 1.5 \times 10^8 \ \text{s}^{-1} \quad (20.10)$$

There are thus many resonant modes within the band of the spontaneous emission line (Fig. 20.3). The laser will begin to operate at a frequency close to that of the cavity mode nearest to the maximum of the spontaneous emission curve. If the power level of the laser is sufficiently high, then laser emission will occur on several lines at once. The precise resonant frequency will be between that of the cavity mode and the spontaneous emission line centre. The line will be 'pulled' away from the cavity mode frequency towards that of the line centre.

20.5   The field distribution in the plane mirror cavity

In the simple argument of Section 20.1, it was assumed that the amplitude of the wave in the cavity was constant over the cavity cross-section. In fact the effect of confining a wave to a finite area inevitably leads to diffraction. A wave launched from one end of the cavity will have spread by diffraction by the time it reaches the other end so that only part of the wave power is reflected. There will be some diffraction loss. How much? In answering this, we have a problem. Do we use the expressions for Fresnel diffraction

(Section 6.15) or Fraunhofer diffraction (Section 6.6)? The typical laser has dimensions such that *neither* treatment is appropriate. We need to go back to basics. Before so doing, let us digress and consider the general features of the problem of determining radiation distributions in cavities.

20.6   The closed cavity

If we wish to explore the resonant radiation patterns in a closed cavity with perfectly reflecting walls, then we do so by imposing the boundary condition that the electric field must vanish at all points on the walls. For a cavity of simple geometry—e.g. a rectangular box—the field distributions are easily obtained. What happens when we remove the side walls, to leave only two parallel mirrors? We no longer have simple, well-defined boundary conditions, and the problem cannot easily be dealt with.

We have in fact met an optical device consisting of two parallel, partially reflecting mirrors—the Fabry–Perot etalon (Chapter 4). We calculated the transmittance of such a system by summing the multiply-reflected and transmitted waves. We blithely ignored the fact that the plates were of finite size by taking the sums *to infinity*. We could perhaps justify this procedure by appealing to the fact that the radiance distributions so calculated agreed well with experimental results. We could explain this agreement by saying that since the reflectances of the two plates were less than unity, the amplitudes of the multiply-reflected beams decreased and eventually became negligible. Thus by (incorrectly) summing to infinity, we were including an absent contribution which was small in any case. Under what conditions will this cavalier treatment fail? If the amplitudes of the beams being combined do *not* decay, then summing to infinity will produce a nonsense. However, this is precisely the situation in the laser. Although the reflectances of the mirrors are less than unity, amplification of the beams on each transit leads to *increasing* amplitudes of successive beams and hence invalidates the Fabry–Perot approach.

The problem is therefore tackled as follows, by a method first introduced by Fox

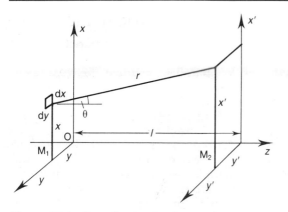

Figure 20.4 Coordinates in plane mirror system

and Li [1]. A uniform plane wave is assumed to leave one of the mirrors in the direction of the other mirror. The distribution of radiation from a small area $dx\,dy$ at (x, y) on mirror M_1 (Fig. 20.4) is calculated at a point (x', y') on mirror M_2 by use of the Kirchhoff diffraction theory (Chapter 6). The distribution for the whole wave at M_1 is obtained by integration. The distribution on M_2 will not be identical with that on M_1 because of the effects of diffraction, whereby part of the wave from M_1 will miss M_2. It is assumed that the part of the wave striking M_2 is reflected without change of amplitude. The distribution over M_2 is used as the starting wave and a similar calculation serves to determine the distribution over M_1. When this process is repeated a large number of times it is found that the form of the amplitude at the end of successive transits approaches a steady state. The amplitude decreases slightly on successive reflections (due to the presence of diffraction losses) but the *relative* values of the amplitudes at different points of the mirror remain unchanged. This is the resonant mode of the two-mirror system.

The amplitude $a_2(x', y')$ on M_2 from M_1 after a single transit is given by

$$a_2(x', y')$$
$$= \frac{i\varkappa}{4\pi} \iint a_1(x, y)\, \frac{e^{i\varkappa r}}{r}(1 + \cos\theta)dx\,dy \quad (20.11)$$

where $a_1(x, y)$ represents the amplitude distribution over M_1, assumed initially to be constant.

In the integral the value of r in the denominator may be put equal to l, the mirror separation, without significant error. In the exponential term, however, the value is highly sensitive to small changes in r, since a change of r from $n\lambda$ to $(n + \frac{1}{2})\lambda$ changes the value of the exponent from $+1$ to -1. For $l \approx 0.1$ m and $\lambda \approx 5 \times 10^{-7}$ mm, n is of order 2×10^5; thus the fractional change in r needed to change the value of the exponent from $+1$ to -1 is only $(\lambda/2)/l \approx 5 \times 10^{-6}$. Thus the approximation $r \approx l$ is inadmissible in this instance.

Since

$$r^2 = (x - x')^2 + (y - y')^2 + l^2 \quad (20.12)$$

and $x - x'$ and $y - y'$ are $\ll l$ in practice, we may write

$$r \doteq l \left[1 + \frac{(x - x')^2}{2l^2} + \frac{(y - y')^2}{2l^2} \right] \quad (20.13)$$

and eqn (20.11) becomes

$$a_2(x', y') = \frac{i\varkappa e^{i\varkappa l}}{2\pi l} \iint a_1(x, y)$$
$$\times \exp\left\{ i\varkappa \left[\frac{(x - x')^2 + (y - y')^2}{2l} \right] \right\} dx\,dy$$
$$(20.14)$$

where we have put $(1 + \cos\theta)$ equal to 2, since the maximum values of θ in cases of interest rarely exceed a degree or so.

After a large number of transits, we reach the stage where we may write

$$a_2(x, y) = \gamma a_1(x, y) \quad (20.15)$$

where γ takes account of the diminution in amplitude, but may be complex since there may be a variation of phase over the mirror surfaces. Equation (20.11) thus becomes

$$\gamma a_1(x', y') = \frac{i\varkappa e^{i\varkappa l}}{2\pi l} \iint a_1(x, y)$$
$$\times \exp\left\{ i\varkappa \left[\frac{(x - x')^2 + (y - y')^2}{2l} \right] \right\} dx\,dy$$
$$(20.16)$$

an integro-differential equation which, for a pair of plane reflecting surfaces, needs to be solved by numerical methods. For a pair of infinite strip mirrors, of width a and separation l, the form of the amplitude distribution across the width of the mirrors is

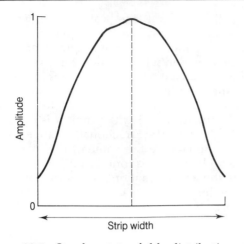

Figure 20.5 Steady state field distribution in plane mirror cavity

as shown in Fig. 20.5. These results are for a system for which $a^2/l\lambda$ (the 'Fresnel number') is 6.25. We note that the amplitude of the wave at the sides of the strip mirror is much less than that at the centre, suggesting that the power loss per transit is small. In fact, for a typical laser system, the Fresnel number is enormously higher than 6.25 and the loss by diffraction is extremely small. The dependence of the loss on Fresnel number is given in Fig. 20.6. It is clear that for a practical laser system the effects of diffraction loss are completely negligible.

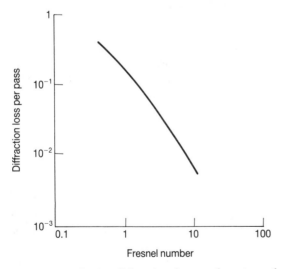

Figure 20.6 Cavity diffraction loss as function of Fresnel number

20.7 The confocal laser

The simplicity of the plane-parallel cavity conceals an experimental difficulty. The degree of parallelism must be extremely high: for even a small angle between the mirrors, multiply-reflected beams will not build up a standing-wave field, but will 'walk off' the sides of the mirrors.

Consider the system shown in Fig. 20.7. In terms of a ray picture (Fig. 20.7a), a ray initially parallel to the axis will return to the same point after four transits and will follow the same path on subsequent journeys. In wave-front terms (Fig. 20.7b) a wave-front with radius $R/2$ is reflected as a plane wave which is then reflected as a wave of radius $R/2$. This description merely serves to illustrate that a cavity of this type—the 'confocal' cavity—offers the possibility of a standing-wave system and hence of a high radiation density. The actual field distribution is discussed in Section 20.8.

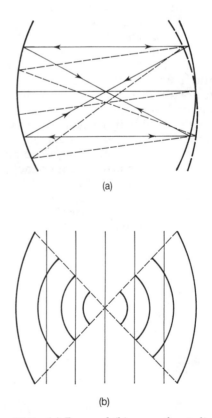

(a)

(b)

Figure 20.7 (a) Ray and (b) wave-front diagrams for confocal cavity

The confocal cavity has an immense advantage over that employing plane mirrors. Consider the effect of tilting one of the mirrors from the position indicated in Fig. 20.7(a). The axis of the system is the line through the centres of curvature of the two mirrors. Tilting one mirror shifts its centre of curvature, but since this lies on the surface of the other mirror, a new axis is created and the round-trip feature shown in Fig. 20.7(a) now applies to the new axis. Thus in contrast to the plane mirror system, small angular misalignments are unimportant.

20.8 The field distribution in the confocal cavity

This calculation follows the method given in Section 20.6, with the appropriate value for r, which needs to take account of the curvature of the mirrors. From Fig. 20.8 we have

$$r^2 = (R - t - t')^2 + (x' - x)^2 + (y' - y)^2$$

together with $2Rt = x^2 + y^2$ and $2Rt' =$

$(x')^2 + (y')^2$, leading to

$$r \doteq l \left(1 - \frac{xx' + yy'}{l^2}\right) \tag{20.17}$$

Thus for the confocal system, eqn (20.16) is replaced by

$$\gamma a_1(x, y) = \frac{i\varkappa e^{i\varkappa l}}{2\pi l} \int\int a_1(x, y)$$
$$\times \exp\left[\frac{i\varkappa(xx' + yy')}{l}\right]\bigg\} dx\, dy \tag{20.18}$$

In contrast to the plane mirror case, this equation possesses an analytical solution from which the electric field strength $E(x, y, z)$ may be written (with the origin shifted to the point midway between the mirrors):

$$E_{mn}(x, y, z)$$
$$= \frac{w_0}{w(z)} H_m\left(\frac{\sqrt{2}x}{w(z)}\right) H_n\left(\frac{\sqrt{2}y}{w(z)}\right)\exp\left(-\frac{x^2 + y^2}{w^2(z)}\right)$$
$$\times \exp\left\{-i\left[\frac{\varkappa(x^2 + y^2)}{2\rho(z)} + \varkappa z - (1 + m + n)\phi(z)\right]\right\} \tag{20.19}$$

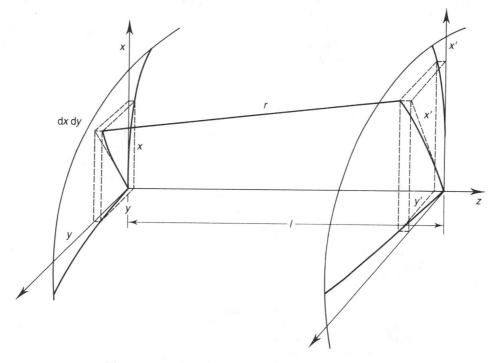

Figure 20.8 Coordiantes in spherical mirror system

where

$$w(z) = w_0 \left(1 + \frac{2z}{l}\right)^{1/2}$$

$$w_0 = \left(\frac{l\lambda}{2\pi}\right)^{1/2}$$

$$\rho(z) = z \left[1 + \left(\frac{l}{2z}\right)^{1/2}\right]$$

and

$$\phi(z) = \arctan\left(\frac{2z}{l}\right)$$

H_m and H_n are Hermite polynomials of order m, n. We note that all the possible modes contain a Gaussian factor, indicating a very rapid fall of intensity from the axis. For the lowest order $(0, 0)$ mode, we have

$$E_{00}(x, y, z)$$

$$= \frac{w_0}{w(z)} e^{-r^2/w^2(z)} \exp\left\{-i\left[\frac{\varkappa r^2}{\rho(z)} + \varkappa z\right]\right\} \quad (20.20)$$

a rotationally symmetrical distribution for which the energy density at the centre of the cavity ($z = 0$) falls to $1/e$ from the centre at a distance $w_0/\sqrt{2}$. For a cavity of length 200 mm and with $\lambda = 500$ nm, the beam radius is $w_0/\sqrt{2} = 0.126$ mm. At the cavity mirror ($z = l/2$), the beam radius is equal to w_0. This distribution contrasts with that for the plane mirror cavity (Fig. 20.5).

20.9   Unstable resonators

The design of the cavities described in the foregoing sections is conditioned by the need to build up a high standing-wave field in the (usual) situation where the amplification of the wave on a single transit is small—e.g. the order of a per cent or so. In some cases, very high single-pass gains can be obtained and in this circumstance a resonator of the type shown in Fig. 20.9 may be used. This system produces a large diameter annular beam. An advantage over the confocal cavity is that power is drawn from the whole of the active material filling the resonator. By contrast, the confocal system draws power only from the region where the standing-wave field strength is high, which is the region very close to the axis.

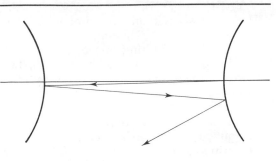

Figure 20.9   Unstable cavity

20.10   Practical laser systems

20.10.1   Doped crystal systems

The ruby laser described in Chapter 12, the first laser to operate, is in a sense atypical of doped crystal lasers in general. The disadvantage of the ruby laser is that the lower laser level is the ground state of the system. In order to create a population inversion, more than half of the Cr^{3+} ions must be pumped into the upper laser level. However, they reach there via the absorption bands (Fig. 12.6) and in making the downward transition to state F they dump a considerable amount of energy (heat) into the lattice. This is the reason that the ruby laser operates in a pulsed mode.

What is needed is a material in which the lower laser level is *not* the ground state, but a level sufficiently far above the ground state for its population, through the Boltzmann factor, eqn (12.5), to be extremely small. In this case *any* ions pumped into the upper laser level would create a population inversion, and the problem of undue lattice heating by the lasing process would be avoided.

There are many such 'four-level' systems. The doping atoms are drawn from the transition metal, lanthanide or actinide series, all of which are characterised by possessing unfilled inner electron shells. As a consequence, transitions involving the inner shells are shielded by the outermost electrons with the result that fluorescent lines are sharp. This is a desirable condition for a laser. If the fluorescent line is sharp, then a considerable amount of power can be generated in a

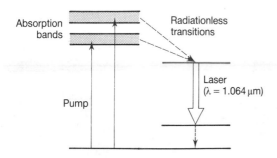

Figure 20.10 Energy levels for Nd-YAG system

narrow bandwidth, and hence a large standing-wave field in a cavity can be obtained at the frequency in question.

Figure 20.10 shows the energy level diagrams for a common doped-crystal four-level laser. In a radiationless transition the energy released appears as lattice vibrations and not as emitted electromagnetic radiation. Nd–YAG lasers are available giving CW output powers up to the order of one kilowatt at a wavelength of 1.06 μm.

20.10.2 Gas laser systems

The helium–neon laser was the first of what is now a vast collection of gas lasers, producing wavelengths that range from the far ultraviolet to the submillimetre wavelength regions. An indication of the more commonly available lasers and their wavelengths is given in Table 20.1. The ranges given merely indicate the regions where the laser emissions lie: they do *not* imply tunability (see Section 20.10.4).

The nature of the operation of the helium–neon laser makes it difficult to achieve high powers (see Problem 20.9). CW

Table 20.1 Gas lasers

Active material	Wavelength range
Excimer lasers: ArF, KrF, XeCl, XeF	193–351 nm
Rare gas ion: Ar$^+$, Kr$^+$	330–680 nm
Carbon dioxide (with He and N$_2$)	9–11 μm
Helium–cadmium	325, 442 nm
Helium–neon	633 nm, 1.153, 3.39 μm
Water	28, 78, 118 μm
Methyl fluoride	190–1222 μm

outputs in the region of milliwatts to watts are possible. In the case of other systems, scaling to large dimensions is straightforward. The CO$_2$ system, operating at a wavelength of 10.6 μm, has been operated CW at powers of the order of megawatts.

20.10.3 Chemical lasers

In the lasers described thus far, the establishment of the necessary population inversion is achieved either by excitation by photons or by electron collisions. In the chemical laser, atomic constituents of a molecule are reacted to produce the molecule, which will in many cases be in an excited vibrational state. Thus in principle a laser may be made that does not require an external source of power, but derives the energy required from the chemical energy of a reaction.

As an example, consider the HF laser, which produces an output on several vibrational transitions in the range 2.7–3.3 μm and on several rotational lines associated with each vibrational transition. The chemical reaction in question is

$$H_2 + F \rightarrow HF^* + H \qquad (20.21)$$

with an associated heat of reaction of 31.6 kcal mol^{-1}. This is sufficient to populate vibrational levels up to the third above the ground state. The transition rates between levels are such that in the steady state population inversion is achieved between three pairs of levels so that the conditions for laser action are fulfilled. Following reaction (20.21), which produces atomic hydrogen, the reaction

$$H + F_2 \rightarrow HF^* + F \qquad (20.22)$$

occurs, which serves both to produce a further supply of HF* molecules and to provide atomic fluorine atoms for reaction (20.21).

Chemical lasers using other halides of hydrogen, and of deuterium, have been built, producing very high CW powers in the 3–5 μm band. The Earth's atmosphere is highly transparent in this band: such lasers are therefore of considerable military interest.

20.10.4 Dye lasers

From the foregoing, it is clear that the wavelengths at which lasers operate are determined by the electronic structure of the lasing material used. By and large, this cannot be significantly changed. The wavelength of the emission from a ruby laser can be varied by changing the temperature, but only by about 0.5 nm. A laser whose output wavelength could be continuously tuned over a broad range of wavelengths would be of great interest, e.g. for spectroscopy or for inducing chemical reactions in specifically one isotope of an atom. How could a tunable laser be built?

The key lies in the fact that certain organic dyes fluoresce in bands some tens of nanometres wide—broad compared with an atomic fluorescent line, but not so very broad for it to be difficult to obtain a reasonably useful power at any wavelength in the band. The simple resonator system used for earlier lasers will not be suitable. However, if a dispersing system is inserted in the cavity, then resonance will occur only at one particular wavelength within the fluorescent band (Fig. 20.11). That wavelength may be chosen by adjustment of the prism or grating. With a collection of a dozen or so different dyes, continuous tuning may be achieved over the whole of the visible band.

Pumping of a dye laser system may be achieved either by exposure to an intense white-light source or, more usually and effectively, by a CW laser (e.g. Ar^+).

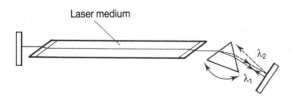

Figure 20.11 Form of cavity for dye laser system

20.10.5 The semiconductor laser

The mode of achieving the necessary population inversion in the semiconductor laser is quite different from that for the lasers described in Section 12.12. This laser consists of a *pn* junction for which the energy-band diagram, in the absence of an applied poten-

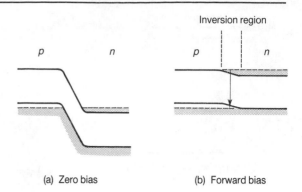

Figure 20.12 Energy level diagram for semiconductor laser

tial difference, is as shown in Fig. 20.12(a). When a forward potential is applied to the junction, the energy diagram is that of Fig. 20.12(b). Electrons flow from the *n*-type material into the *p*-type, occupying the conduction band for the material in the immediate vicinity of the junction. In this region a higher concentration of electron states in the conduction band is occupied compared with that of the hole states in the valence band. On making the downward transition between these states, the system emits recombination radiation which serves as the spontaneous emission for initiating laser action. For this type of laser, no external mirrors are needed. The lasing material is a cubic crystal with a high refractive index (Fig. 20.13). The Fresnel coefficient for normal incidence at a surface of a material with refractive index n is $[(n-1)/(n+1)]^2$. For a gallium arsenide laser, which operates at a wavelength of 0.84 μm, the value of n is

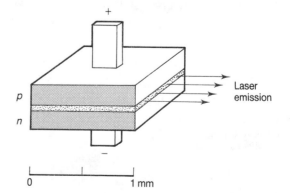

Figure 20.13 Schematic of semiconductor laser

3.6, yielding a reflectance of 0.32 at the crystal surface. The gain per pass of the laser is sufficiently high for the system to work with this value of reflectance.

20.10.6 The free electron laser

The lasers described in the foregoing sections are readily understood in terms of the quantum description of radiation. The free electron laser can be plausibly described in terms of a classical model.

Suppose that we cause an electron to oscillate about an equilibrium position with a certain angular frequency. Now expose the oscillating electron to an electromagnetic wave travelling in the direction perpendicular to the oscillation axis and with the electric vector along this axis. What happens will depend on the phase of the wave with respect to that of the oscillating electron. Consider the two extremes:

1. The electric force is in the same direction as that of the moving electron.
2. The electric force is in the opposite direction to that of the moving electron.

In the first case, the amplitude of oscillation of the electron, and hence its total energy, will have been increased. The amplitude of the wave transmitted will therefore be reduced, a condition representing absorption of the incident wave by the system. In case (2), the amplitude of the oscillating charge is reduced, the difference in energy appearing as an *increase* in the amplitude of the transmitted wave. This process entails the flow of energy from the system to the wave which is caused by the wave-stimulated emission. If we can contrive to ensure that the phase of the wave relative to that of the oscillating charge is such as to correspond to case (2), we shall obtain a resultant amplification. If we build on a cavity, we shall have a laser.

There is a snag. We cannot oscillate electrons at optical frequencies. However, all that is necessary is for the oscillation of the electron to *appear* to be at the same frequency as the wave—to someone riding on the wave. If therefore we propel the electron with a velocity approaching c in the direction of the wave, we can cause it to wiggle in syn-

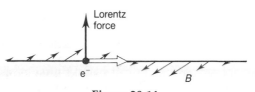

Figure 20.14

chronism with the wave frequency by, for example, arranging a row of magnets producing a spatially alternating field along the electron's path. The oscillatory displacement will arise through the Lorentz force (Fig. 20.14).

Where does the radiation that is to interact with the oscillating electrons come from? Recall that in the case of the lasers described in the preceding sections, we had no need actually to feed in radiation to be amplified. This was provided by the decaying atoms. In the case of the free electron laser, the source radiation is from the electrons themselves—an oscillating charge radiates.

The detailed story is somewhat more involved than is suggested here. The electrons are subject to longitudinal forces in addition to the transverse Lorentz force. These forces give rise to bunching in the electron beam, in a fashion reminiscent of the klystron. A further complication in the real-life FEL is that the electron current is itself not steady, but in the form of picosecond bursts from an accelerator. Nevertheless, the FEL represents a high-power tunable system, tuning being effected by varying the energy of the electron beam injected into the system. The bulk of the work done to date on these systems is at wavelengths ranging from the near infrared (few micrometres) to the millimetre wave region.

20.11 Q-switching and mode-locking

20.11.1 Q-switching

If we start to excite the atoms in a laser cavity in order to produce a population inversion, then laser action will begin as soon as the effects of gain exceed those of the losses. When laser action occurs, the population of the upper laser level is rapidly reduced and may fall below the threshold level until pumping again achieves the inversion needed. In this case, the laser output consists of a series of pulses. At no time do we have a *large* excited state population: emission occurs as soon as the threshold inversion is reached. If we could hold up the onset of laser action until a massive concentration of excited atoms was reached, then when laser action did occur we should expect a single, very intense pulse. This process—known as 'Q-spoiling'—is regularly used to produce giant pulses. A shutter (originally mechanical but subsequently electro-optic—cf. the early velocity of light experiments in Chapter 22) serves to prevent feedback from one of the mirrors until a large excitation is achieved. In this fashion, pulses of duration of the order of tens of nanoseconds, and with powers up to the gigawatt region, may be obtained.

20.11.2 Mode-locking

In Section 20.4 we noted that at high power levels a laser can operate simultaneously in many different modes, separated by a constant frequency. In crystal lasers, the residual imperfections cause some scattering of the radiation in the medium, with the result that the modes possess an approximately common phase. What will be the resultant effect of coupling N modes (assumed for simplicity to be of equal amplitudes) whose frequencies differ by a constant amount $\Delta\nu_c$?

Consider an odd number $\mathcal{N} = 2N+1$ modes centred on a frequency ν_0. The frequency of the nth mode from the centre is $\nu_0 + n\,\Delta\nu_c$. At any given point, the resultant of the $\mathcal{N}$ waves is

$$E(t) = \sum_{n=-N}^{N} \exp 2\pi i(\nu_0 + n\,\Delta\nu_c)t \quad (20.23)$$

from which

$$|E(t)|^2 = \frac{\sin^2(\mathcal{N}\pi\,\Delta\nu_c t)}{\sin^2(\pi\,\Delta\nu_c t)} \quad (20.24)$$

an expression reminiscent of that obtained in studying the diffraction grating in Chapter 6. $|E(t)|^2$ consists of maxima at intervals of $\Delta t = 1/\Delta\nu_c$, the width of the maxima being given by $\Delta\tau = 1/\mathcal{N}\Delta\nu_c$. From eqn (20.10) we see that $\Delta t = 2l/c$ and is the round-trip time for light in the cavity. The output from a laser in which $\mathcal{N}$ modes are locked together

is thus a succession of spikes of duration $1/\mathcal{N}\Delta\nu_c$, arriving at intervals of $2l/c$. For a cavity with $l = 0.15$ m, in which 51 modes combine, the output would consist of spikes of duration 20 ps, arriving at 1 ns intervals. Such pulses can be further shortened: although their spectral line width is small, it is sufficient for it to be possible, with the help of diffraction gratings, to arrange that the different wavelengths present travel different distances. By such methods, pulses of duration 12×10^{-15} s (12 fs) have been produced. These contain only about 7 wavelengths and occupy only 3.6 μm in space.

20.12 Laser characteristics

Two features of laser radiation make these devices invaluable—their high degree of coherence and the enormous radiance attainable. The importance of their coherence is illustrated by the story of holography. Although invented in 1948, this powerful technique (Chapter 13) lay dormant and unused from then until the arrival of the laser—unusable in any practical sense simply because of the absence of coherent sources of other than miniscule power.

A feature of the temporal coherence (i.e. very narrow line width) of well-stabilised lasers is the capability it provides for very precise frequency, and frequency-shift, measurements. The narrowest line width attained so far is ~ 1 Hz.

The very high radiances available from lasers enable a variety of scattering phenomena to be studied whose cross-sections were far too low for measurements to have been possible with non-laser sources. Other techniques, such as Raman spectroscopy, became vastly easier to use. A selection of such applications is briefly referred to in the following sections.

20.13 Laser applications

In the remainder of this chapter, we consider a selection of applications of lasers which illustrate the remarkable potential of the device.

The advantages of laser over incoherent sources which give rise to numerous applications are:

1. Very high spectral power density. The flux from many lasers exceeds that from incoherent sources by many orders of magnitude.
2. A high degree of coherence, making possible interference experiments with very large path differences.
3. Very small spectral line width, making high-resolution spectroscopy possible.
4. Tunability. The single-mode, tunable laser combines high radiance with the possibilities of very high spectral resolution.
5. Feasibility of producing ultra-short duration pulses, by mode-locking, making the study of extremely fast phenomena possible.

The high-power, narrow line width aspect of lasers has led to extensive development of a whole range of spectroscopies. In some instances, existing methods become more convenient, more accurate and have greatly improved resolution. In other cases new approaches are possible which could not be achieved without lasers.

20.14   Absorption spectroscopy with dye lasers

In conventional absorption spectroscopy light from a broad-band source is passed through the specimen and then goes to a spectrograph. In laser spectroscopy, the dispersing system is part of the dye laser so that all that is needed is a detector. A long Fabry–Perot etalon is generally included (Fig. 20.15). Since the radiation is highly collimated, the etalon will possess a high transmission only at wave numbers $\bar{\nu}$ given by

$\bar{\nu} = n/2d$, where d is the etalon spacing and n an integer. This serves to provide calibration marks in the spectrum. Since the wavelength of the laser may be stabilised to within one part in 10^8, absorption spectra of very high resolution are attainable. Furthermore, the high power of the laser means that the wavelength region of interest can be scanned extremely rapidly, so that short-lived radicals may be studied.

A novel and powerful method of absorption spectroscopy is photoacoustic spectroscopy, in which the heating of a gas sample produces sound waves that are detected microphonically. The signal will be a maximum when the (tunable) laser emission lies at the centre of the absorption line. This method is capable of detection at the level of parts in 10^9.

20.15   Laser Raman spectroscopy

Raman spectroscopy is an old-established method of studying molecular vibrations and rotations. The laser has revolutionised this field, as well as providing new and powerful variants of the method.

Exposure of a collection of molecules to monochromatic radiation of frequency ω_L leads to scattered radiation at frequency ω_L (Rayleigh scattering) together with sharp emission lines at frequencies lower (ω_S—'Stokes') and higher (ω_A—'anti-Stokes') than that of the incident beam. The differences in quantum energies $\hbar(\omega_L - \omega_S)$ and $\hbar(\omega_A - \omega_L)$ correspond to differences in energies of vibrational/rotational levels of the molecule. The Raman process may be described in terms of excitation to a virtual

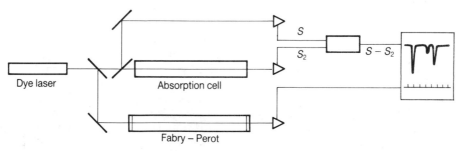

Figure 20.15 Absorption spectroscopy with dye laser

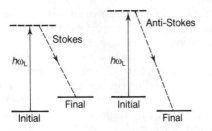

Figure 20.16 Energy levels: Stokes and anti-stokes radiation

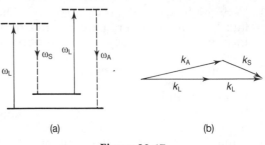

(a)   (b)

Figure 20.17

level (Fig. 20.16) followed by relaxation to a level that may be above (Stokes) or below (anti-Stokes) the initial level. Measurement of the frequency shifts $\omega_L - \omega_S$ and $\omega_A - \omega_L$ thus enables information about the molecular structure to be obtained. An important feature of Raman spectroscopy is that the selection rules governing Raman transitions are different from those governing the direct transitions observed by absorption spectroscopy. Thus the 'breathing mode' of the (tetrahedral) ammonia molecule (in which all the hydrogen atoms move outwards and inwards together) gives rise to no absorption lines, but can be studied through the Raman lines that occur.

The virtual level of Fig. 20.16 will generally *not* correspond to a 'real' energy level of the molecule. If, however, this correspondence does occur then the intensity of the Raman emission is very large. The phenomenon is known as resonance Raman scattering.

The fraction of a beam of incident photons which produces Stokes Raman emission is extremely small. Raman cross-sections are of the order of 10^{-34} m^2. The 'geometrical' cross-section of a molecule 0.5 nm across is $\sim 2 \times 10^{-19}$ m^2. Thus the Raman effect is weak. In prelaser days, the only available 'monochromatic' sources of high intensities had emission lines of substantial breadth, so that Raman lines with small displacements were swamped.

20.16   Stimulated Raman scattering

If the flux of incident radiation in a Raman scattering experiment is increased, then a threshold is reached at which the intensities of Raman emission lines abruptly increase *in*

directions lying on a cone around that of the incident radiation. Far from being at the feeble levels characteristic of the effect described in Section 20.15, the radiance in this 'stimulated' Raman scattering can be comparable with that of the exciting radiation.

The process can be understood from Fig. 20.17(a). At sufficiently high pump levels, the conditions for coherent amplification can be established. However, the radiation from different parts of the specimen will combine to give a large resultant only if their phases are matched. The emission occurs only in directions for which the resultant momentum $\hbar\mathbf{k}_S + \hbar\mathbf{k}_A$ matches that of the two input quanta $2\hbar\mathbf{k}_L$ required to produce the excitations. Thus

$$2\mathbf{k}_L = \mathbf{k}_S + \mathbf{k}_A \qquad (20.25)$$

(see Fig. 20.17b).

20.17   Brillouin scattering

In Sections 20.15 and 20.16 we dealt with the situation where the frequency of scattered radiation differed from that of the incident by an amount corresponding to a vibrational or rotational excitation of a molecule. In the solid state quantised lattice vibrations occur, arising from the collective motion of the arrays of atoms forming the solid. It would seem possible therefore that scattered radiation might occur with a frequency difference that corresponds to a quantum of lattice vibration, or phonon. This does occur, the process being known as Brillouin scattering. Figure 20.18 shows the presence of additional rings in a Fabry–Perot interferogram arising from Brillouin scattering. Since the energies of lattice vibrational quanta are

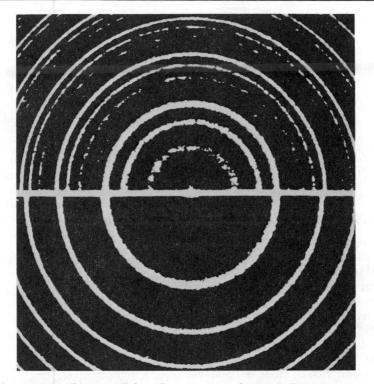

Figure 20.18   Rings in Fabry–Perot pattern due to Brillouin scattering

of the order of small fractions of an electron-volt, then the high spectral resolution of the Fabry–Perot interferometer is needed for such studies.

In a manner analogous to that for stimulated Raman scattering, stimulated Brillouin scattering also occurs at high intensities of the incident radiation.

20.18   Doppler-free spectroscopy

If a collection of atoms is exposed to radiation whose line width is very small compared with the Doppler width corresponding to the excited level, then only a very small fraction of the atoms present will be excited, so that any scattering effects will appear very weak. How can we retain the narrow line width associated with the scattering processes discussed above and at the same time ensure that *all* the atoms present can participate? This can be done by exposing the specimen to *two* (or more) exciting beams simulta-

neously. If the frequencies are ω_1 and ω_2, a transition from level i to level f will occur provided

$$E_f - E_i = \hbar(\omega_1 + \omega_2) \qquad (20.26)$$

If the atom or molecule is moving with velocity $\mathbf{v}$, then in the moving reference frame the molecule 'sees' waves with frequencies $\omega_1' = \omega_1 - \mathbf{k}_1 \cdot \mathbf{v}$ and $\omega_2' = \omega_2 - \mathbf{k}_2 \cdot \mathbf{v}$, where the $\mathbf{k}$'s are wave vectors of the two beams. Excitation of the moving molecule will occur if

$$E_f - E_i = \hbar(\omega_1' + \omega_2')$$
$$= \hbar(\omega_1 + \omega_2) - (\mathbf{k}_1 + \mathbf{k}_2) \cdot \mathbf{v} \qquad (20.27)$$

If the beams are of the same frequency and illuminate the sample from opposite directions, then $\mathbf{k}_2 = -\mathbf{k}_1$ and condition (20.27) becomes

$$E_f - E_i = h(\omega_1 + \omega_2) = 2h\omega_1 \qquad (20.28)$$

independently of the velocity. Thus all the molecules present are excited and the goal of Doppler-free spectroscopy is achieved.

20.19 The laser and fusion

The harnessing of nuclear fusion power entails the creation of extremely high temperatures (10^8 K) in order that the Coulomb repulsion of, for example, two deuterons, or D and T, can be overcome. Although the main thrust of much research in this area uses magnetically confined plasmas, a second route employs the enormous photon fluxes that can be produced by high-power laser-oscillators followed by laser amplifiers to produce pulses of enormous power (but see Section 11.5.2.) The nuclear fuel is contained in thin-walled glass spheres which are exposed to multikilojoule radiation pulses simultaneously from many directions. The glass envelope is ablated and moves away from the centre of the system, producing a compressive force on the plasma created from the fuel. The high temperatures necessary arise from the adiabatic compression that results. In order for more energy to be released through fusion than is used in achieving the fusion condition, the product of the number density n (per cubic centimetre) and the time τ for which the ignition temperature is held must exceed 10^{15}. It appears likely that this condition will be met and that fusion energy will eventually become a practicable energy source, although it is still too early to predict a likely starting date.

20.20 The laser gyroscope

In a quite different application, in which the coherence of the laser rather than its high intensity is important, the results of the Sagnac experiment (Section 22.20) are pressed into service. The source of radiation is a laser, which is placed in one arm of a three- or four-sided arrangement of mirrors that form the triangular or square laser cavity. Such a laser will operate in two independent modes, corresponding to light traversing the cavity in opposite directions. If the ring is stationary with respect to the local inertial frame of reference then the light beams traversing the cavity in opposite directions will be in phase at the end of a complete circuit. If the system rotates about an axis perpendicular to the plane of the cavity with angular velocity Ω then the phase difference $2\Delta\phi$ between the counterrotating beams arriving at one of the mirrors (partly transmitting, so serving as an output coupler) is given by

$$2\Delta\phi = \frac{4\pi\Omega A}{c\lambda} \qquad (20.29)$$

where A is the area enclosed by the optical path round the ring. Thus measurement of $\Delta\phi$ enables Ω to be found.

The sensitivity of the laser gyroscope can be greatly increased by using a flat, multiturn coil of optical fibre as the light path. If the coil has N turns of area a, then the value of A in eqn (20.29) is given by $A = Na$. For a fibre length of 500 m wound in a coil of radius 0.1 m and with radiation of wavelength 1 μm, a phase difference of 10^{-6} radian (which is easily measurable) corresponds to a rotation rate of 0.1 degrees per hour.

The laser gyroscope has many advantages over the traditional mechanical version. The latter are extremely difficult to build on account of the very high mechanical precision needed in manufacture. Since they involve high-speed rotation the problems of wear in the bearings are inevitable. In contrast the laser gyroscope has *no* moving parts. Three such lasers with mutually perpendicular axes constitute an inertial navigation system capable of locating position on the Earth's surface to within about a metre.

Reference

1. A. G. Fox and T. Li, *Bell System Technical Journal*, **40**, 453–8 (1961); *Proc. IEEE*, **51**, 80–9 (1963).

Problems

20.1 Assuming no diffraction loss, derive the expression for τ_c (eqn 20.5) for a plane-parallel cavity with mirrors of reflectance R.

20.2 In an operating helium–neon laser tube ~ 400 mm long, the gas temperature is 700 K. The output corresponds to a neon transition (atomic mass 22). What is the largest difference between the resonant frequency of the plane cavity and the centre of the emission line? How many modes could occur between the frequencies at which the spontaneous emission power drops to $1/e$ of the maximum value?

20.3 Justify the approximation for $1 + \cos\theta$ in Section 20.6 for a cavity with mirrors 5 mm square separated by 200 mm.

20.4 Justify the use of eqn (20.13) for the cavity of Problem 20.3.

20.5 Derive the approximation given in eqn (20.17).

20.6 Sketch the curves of $|E_{mn}(x, 0, 0)|^2$ and $|E_{mn}(0, y, 0)|^2$ against x and y for (a) the $(0, 0)$ mode, (b) the $(0, 2)$ mode and (c) the $(3, 1)$ mode.

20.7 Calculate the spot radius for a laser operating in the $(0, 0)$ mode for a confocal cavity 300 mm long and with $\lambda = 633$ nm, (a) at the mirror surface and (b) on the wall 5 m away from the end of the laser.

20.8 In a three-level laser system, with energy levels and Einstein coefficients as indicated, atoms are pumped from state 1 to state 3 at the rate of $N\,s^{-1}$. Establish the relationships between A_{21}, A_{31} and A_{32} if a steady population inversion is to be obtained between level 3 and level 2 *and* 90% of the transitions take place via level 2.

20.9 In Section 20.10.2 it is stated that the helium–neon laser cannot operate at high powers (of the argon laser). A helium–neon laser with a tube of diameter 1 mm works well: one that has a 10 mm diameter and one hundred times the current does not work at all. Speculate on why this should be.

20.10 Problem 20.2 shows that it is not necessary to adjust the mirror separation of the helium–neon laser to ensure that resonance will occur. Consider the case of a gallium–arsenic laser, emitting at a wavelength of 0.84 μm. Such a laser is typically ~1 mm long. In practice, adjustment of the crystal ($\equiv$ cavity) length is unnecessary. What do you conclude from this?

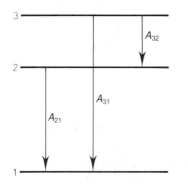

Temporal Analysis—Photon Correlation

21.1 Fluctuation with time of a light signal

Imagine a very weak beam of quasi-monochromatic light incident on a photo-emissive surface. Owing to the quantum nature of a light field, the result would be photoelectrons emitted at infrequent intervals, leading to a sequence of isolated pulses, at irregular intervals. The size of the pulses would vary since the emitted photoelectrons could have come from anywhere in the conduction band of the emitter (Fig. 21.1a). Thus the *size* of the pulse tells us nothing of interest about the *light*. The emission of a photoelectron merely signals the exchange of a photon of energy $\hbar\omega$ between the light field and the electron. We may easily arrange an electronic circuit to produce pulses of *equal* intensity and so eliminate the variation due to conduction band effects. The timing of the arrival of the pulses is then entirely determined by the characteristics of the light field.

What happens as the intensity of the beam

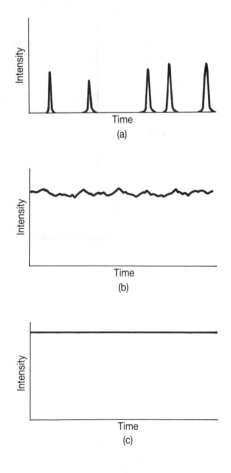

(a)

(b)

(c)

Figure 21.1 (a) Very weak light source: photo-electron pulses. (b) Stronger source: fluctuations still observable. (c) Very strong source: essentially constant signal

is increased? The pulses arrive with shorter intervals between them. Now there will be a minimum time interval over which our measuring system will be able to respond. If many photoelectric emissions occur in this interval, the system will be unable to register the appropriate number of counts, but will give an output signal proportional to the number of arriving photons (Fig. 21.1b). If the number of photons is n, the fluctuation in the signal for different intervals will be $n^{1/2}$ and the fractional variation will be $n^{1/2}/n$. Thus for a very intense beam, an essentially constant signal will result (Fig. 21.1c), since $n^{-1/2} \to 0$ as $n \to \infty$. It is 'essentially' constant because any detection system will have associated noise, which cannot be avoided.

We may readily study either the statistics of the arrival times of photons in the pulse-counting case or the fluctuations of the measured light intensity in the situation of higher intensity. What can this tell us?

21.2   What are *real* light-sources like?

In *thermal* sources, such as discharge lamps, or stars, light emission occurs by spontaneous processes from very large numbers of regions. The atoms are rugged individualists, emitting as the spirit moves them quite independently of what their neighbours are doing. They are in motion, so the frequencies of the emissions received at a detector will be Doppler-shifted. As a result the light field at a point P_1 (Fig. 21.2) will fluctuate with time in a complicated fashion. The output current from the detector will show a

similar variation, accompanied by (or, for a weak light field, swamped by) shot noise fluctuations (Section 16.7). A minute shift of the detector to P_2 should have little effect, since the times of arrival of the radiation at P_2 from different parts of the source will have not appreciably changed. If, however, P_2 is moved sufficiently far from P_1 to P_2', then the intensity fluctuations at P_2' will no longer be *correlated* with those at P_1, since the signals arriving from the source will have originated at different times from those arriving at P_1. Since the emission from each part of the source varies irregularly with time, any differences in arrival times will destroy the correlation that exists at two adjacent points.

In the case of a star, the differences in arrival times of light from different parts of the star will depend on the star's angular diameter. It would seem therefore that the measurement of the correlation between the intensity fluctuations at two detectors could give information about the angular diameter of a star and this is indeed the case.

What about the *spectral* energy distribution of the source? The light fields of different frequency will interfere and this will give rise to a fluctuation of intensity at a given point. Thus we may expect to be able to relate intensity fluctuations to the spectral composition of the light. Is this useful?

21.3   Determination of spectral composition

We have already seen how this can be done. We pass the light into a suitable dispersing system (e.g. a prism or grating spectrograph) so that different frequencies are spatially dispersed. Measurements of the power at different points yields the spectrum of the source. For higher resolution we use interferometers such as the Fabry–Perot (Section 4.23). In practice the limit achieved by the highest resolution devices corresponded to an interval of the order of 1 MHz, a resolving power of the order 10^8. Since, in prelaser days, the sharpest line sources available had widths vastly larger than 1 MHz, there was little need for higher spectral resolution. With the arrival of the laser, the associated line widths are many orders of magnitude

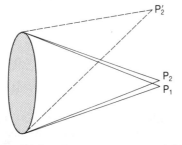

Figure 21.2 Light fluctuations at neighbouring points P_1, P_2 are highly correlated. Those between P_1 and P_2' are not

smaller than those of non-laser sources. Existing methods were incapable of being extended to attain the resolution needed and a completely new approach was needed. The relation between spectral composition and intensity fluctuations provided just such an approach, enabling the resolving power to be extended to $\sim 10^{14}$.

21.4 Temporal analysis of wave trains

Much of this book so far has been concerned with the *spatial* variation of light intensity. Usually the variation is in one plane as, for example, in a photograph of interference or diffraction fringes. We shall now consider the *temporal* variation in light intensity which we denote by the term *intensity fluctuation spectrum* (IFS).

If the light intensity being measured is sufficiently high to give a continuous output from the detector then we can examine the correlation between parts of the IFS at different times. This can yield information about the spectral composition of the light emitted by the source. This is the *analogue* method. If we are in the regime where photons can be counted, then we perform corresponding calculations on the *digital* information obtained. This is the *photon correlation* method. We find that in general the digital method produces information more efficiently than the analogue method. Nevertheless, it is convenient to consider first what can be done with the analogue approach before proceeding to the (theoretically more complicated) digital method. The experimental limitation to these methods is set by the speed with which the detector system can follow the intensity fluctuations.

21.5 Temporal correlation and spectral density

If a light field at a given point is represented by an amplitude $f(t)$ then the energy flow in an interval dt is proportional to $|f(t)|^2 \, dt$. If the light were dispersed into a spectrum $F(\omega)$ then the energy in a spectral band of width $d\omega$ will be proportional to

$|F(\omega)|^2 \, d\omega$). $|F(\omega)|^2$ and $|f(t)|^2$ must be related, since both measure energy.

In section 5.2, we introduced the Fourier transform. If the Fourier transforms of functions $f(t)$ and $h(t)$ are respectively $F(\omega)$ and $H(\omega)$, then

$$\int_{-\infty}^{\infty} f(t)h^*(t)dt$$

$$= \frac{1}{2\pi} \iint_{-\infty}^{\infty} f(t)H^*(\omega)e^{-i\omega t} \, dw \, dt \quad (21.1)$$

where $H^*(\omega)$ is the Fourier transform of $h^*(t)$. Using eqn (5.2) we obtain Parseval's theorem:

$$\int_{-\infty}^{\infty} f(t)h^*(t) \, dt = \int_{-\infty}^{\infty} F(\omega)H^*(\omega) \, d\omega \quad (21.2)$$

in its most general form. If we put $h(t) = f(t)$, this becomes

$$\int_{-\infty}^{\infty} |f(t)|^2 \, dt = \int_{-\infty}^{\infty} |F(\omega)|^2 \, d\omega \quad (21.3)$$

thus establishing the relationship between $|f(t)|$ and $|F(\omega)|$. (The functions will include any multiplying constants. For example, for an electric field amplitude $E(t)$, the magnitude of the Poynting vector is $c\varepsilon_0 |E(t)|^2$ so that $f(t) = \sqrt{c\varepsilon_0}E(t)$).

How do we determine $F(\omega)$? If we form the product of $f(t)$ and $f^*(t+\tau)$, then since the Fourier transform of $f^*(t+\tau)$ is $e^{i\omega\tau}F^*(\omega)$, we may define a function $\sigma(\tau)$ of $f(t)$ as

$$\sigma(\tau) = \int_{-\infty}^{\infty} f(t)f^*(t+\tau)dt$$

$$= \int_{-\infty}^{\infty} F(\omega)F^*(\omega)e^{i\omega\tau} \, d\omega \quad (21.4)$$

$G(\omega) \equiv |F(\omega)|^2$ represents the spectral density, or *power spectrum* (Section 5.2). Thus

$$\sigma(\tau) = \int_{-\infty}^{\infty} G(\omega)e^{i\omega\tau} \, d\omega \quad (21.5)$$

so that, on Fourier transformation,

$$G(\omega) = \frac{1}{2\pi} \int_{-\infty}^{\infty} \sigma(\tau)e^{-i\omega\tau} \, d\tau \quad (21.6)$$

$\sigma(\tau)$ is called the *autocorrelation* function. Thus from a measurement of $\sigma(\tau)$, $G(\omega)$ can be determined.

Although formally the evaluation of $G(\omega)$ from (21.6) entails shifting $f(t)$ by displacements τ for all values to infinity, in a real situation, the integral in (21.6) falls effectively to zero for a finite value of τ. Thus if $f(t)$ is a limited wave train of duration T, $f(t)f^*(t+\tau)$ will vanish when there is no overlap—i.e. for $\tau > T$.

If one wished to examine correlations between the light from two sources (e.g. two lasers) then in place of (21.4) we should need the *cross-correlation* function.

$$\sigma_{fg}(\tau) = \int_{-\infty}^{\infty} f(t)g^*(t+\tau) \, dt \quad (21.7)$$

21.6   Spatial correlations

It is of course possible to consider spatial functions $f(x)$ and $h(x)$ and results similar to those given in eqns (21.2) to (21.6) are obtained, where the *spatial* frequency replaces ω. Spatial correlations are often used in the theory of diffraction (especially X-ray diffraction). We have not followed this route in Chapter 6 but it may be stated that the autocorrelation function of a near-field (Fresnel) diffraction pattern is the same as the autocorrelation of the aperture when the incident light is a plane wave incident normally on the aperture. The spatial auto-correlation function may also be used in computations on the interferometer–spectrometer described in Section 17.14.

21.7   Shot noise

If light from a perfectly stabilised laser, which emits a simple wave train of constant amplitude and frequency, falls on a detector electrons are not emitted at a uniform rate. If the observation interval is short enough for there to be no photoelectrons in some intervals and small numbers in others the distribution is a Poisson-statistical distribution:

$$p(n, T) = \frac{\mu^n e^{-\mu}}{n!} \quad (21.8)$$

where $p(n, T)$ is the probability of n electrons being emitted in a given interval of length T and

$$\mu = \beta \int_0^T I(t) \, dt \quad (21.9)$$

where β is a constant called the quantum efficiency of the detector and I is the intensity of the incident beam in photons per second. Thus even with an idealised source of constant mean power, the received signal is subject to statistical fluctuations. In a real system there will be other sources of shot noise, e.g. due to statistical fluctuations in a photomultiplier dark current.

This noise enters into the output both for the analogue and digital case. If the interval of observation is long to make n (the average number of electrons in one interval) large then the Poisson distribution approximates to a Gaussian distribution with a standard deviation $n^{1/2}$.

21.8   Light-beating spectroscopy

In 1955 Forrester, Gudmundsen and Johnson found that when a magnetically split line from a low-pressure mercury lamp fell on a detector a beat wave of 10 GHz (1 GHz = 1 gigahertz = 10^9 hertz) was present in the output from a photoelectric detector. The detector is a square-law detector: its output at any given time is proportional to the *intensity* of the light falling on it. Thus if $E_1 \exp i\omega_1 t$ and $E_2 \exp i\omega_2 t$ represent the waves corresponding to the two components, $[E_1 \exp(i\omega_1 t) + E_2 \exp(i\omega_2 t)]$ $[E_1 \exp(-i\omega_1 t) + E_2 \exp(-i\omega_2 t)]$ represents the intensity that is proportional to

$$E_1^2 + E_2^2 + 2E_1E_2 \cos(\omega_1 - \omega_2)t \quad (21.10)$$

Thus the intensity includes a term whose frequency is the difference of the frequencies of the two components—a beat wave. The closer the frequencies of the two waves the lower the beat frequency. If sharp enough lines can be produced by lasers, frequency differences down to a few hertz (i.e. a few parts in 10^{15} of the frequency of either laser) can be measured.

21.9   Velocimeter

A simple example of light-beating spectroscopy is shown in Fig. 21.3. A stream of gas flowing through a tube is loaded with fine particles to scatter some of the light from the helium–neon laser and this light is mixed at the detector with some light from the source.

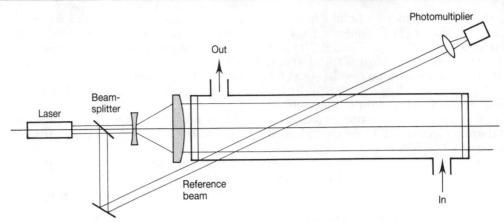

Figure 21.3 Schematic of velocimeter

The laser must be operating in a single stabilised mode. The reference signal may be obtained by using a beam-splitter, as shown in the diagram, or light scattered from a few white patches painted on the wall of the tube may be used. The density of the particles must be sufficient to give an adequate amount of scattered light but not so high as to modify the flow. Measurement of the beat frequency gives the Doppler shift due to the movement and hence the velocity.

21.10 Molecular weight determination

A more elaborate experiment of the same kind may be used to determine the molecular weight of large molecules suspended in a stationary liquid. The molecules undergo Brownian motion and each molecule scatters light with a Doppler shift proportional to a component of its velocity in the line of sight. Instead of a single beat frequency we now have a range of frequencies—a beat-frequency spectrum. This may be scanned by variable-frequency filters in the output. The distribution is Gaussian and elaborate methods are not needed to determine the molecular weight—provided only one kind of molecule is present. In this kind of experiment it is fairly easy to attain an accuracy in the region of 3%.

If a range of sizes of molecules is present, the problem becomes much more difficult. It is not easy, by means of the electrical filters, to obtain a sufficiently accurate beat spectrum over a wide range of frequencies to justify the elaborate calculation needed to interpret the results. Here the digital method has advantages.

21.11 Photon correlations

In the photon-correlation method, the information is received in the form of pulses, each of which represents the absorption of a single photon by the detector. The rate of acquisition of information is very high: in a typical experiment 10^8 photons may be registered in less than 10 seconds. The success of this method depends on (1) the development of lasers whose frequency may, if necessary, be stabilised to 1 Hz (usually a lower standard of stabilisation is sufficient); (2) the development of detectors able to record the arrival of pulses at rates of 10^7 s^{-1} or more; (3) the development of computing circuits and methods that are fast enough to deal with this information without being very expensive.

There are both experimental and computational difficulties in the photon-correlation method and the ways in which they have been overcome are among the most ingenious developments of modern physics.

21.12 Experimental difficulties

There are three important difficulties in relation to the photomultipliers. Firstly, the

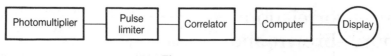

Figure 21.4

pulses given by single photons are not all of the same size (Section 21.1). This is overcome by inserting a pulse size limiter circuit at an early stage of the electronics (Fig. 21.4). Secondly, the current from the later dynodes is large enough to cause a brief fall in the voltage large enough to create a 'dead' period of about 10^{-8} s. A second photon arriving during this period is not counted. This factor limits the counting rate attainable. When the count rate is far below the maximum rate (as it often is) the dead period has no significant effect. When the counter is being pushed to its limit, allowance for the dead period must be made in the calculation. Thirdly, it is possible that the electrons on one of the later dynodes may cause the emission of soft X-rays. If these fall on an early dynode or on the photocathode they may cause the emission of spurious electron pulses. These are usually separated by a constant time interval from genuine pulses and can be allowed for in the computation. The real cure, however, is to improve the design of photon multipliers and eliminate this effect.

21.13 Computational difficulties

In many experiments it is necessary to vary one of the parameters (e.g. the angle of scattering) continuously and to calculate the autocorrelation function many times. To do this in real time by the most simple method would require an enormous number of multiplications to be carried out in a short time (perhaps a few nanoseconds) and no computer could meet this demand. On the other hand, if arrangements are made to store the information with a view to subsequent processing, a considerable amount of time on a fast computer would be needed. It has been shown that the whole information content of certain functions can be extracted by observing only the times at which zero is

crossed. This cannot be applied directly to our problem because the number of electrons in a constant time interval is never negative, but if a suitable number q is subtracted from the number n in a time interval the number $n - q$ has numerous zero crossings and by working with these the computing time is greatly reduced. This process is known as 'clipping'. Computing units that calculate the correlation function are known as correlators. A further computation after the correlator is needed but this is usually within the capacity of a good desk computer. The outline set-up of the computing system is shown in Fig. 21.4.

Some hundreds of experiments have been carried out with photon-correlation spectroscopy. As an example we describe one that uses the great speed of the method. Certain types of large molecules whose length is greater than their breadth normally have their lengths randomly arranged when in solution. A fairly small electric field aligns them so that they become liquid crystals. The photon-correlation method is used to detect the birefringence and so determines the rate at which they line up and also the rate at which they relax when the field is removed. From Fig. 21.5 it may be seen that very rapid alignment and relaxation processes can be followed [1, 2].

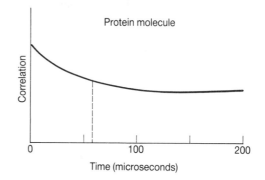

Figure 21.5 Relaxation of orientation of protein molecule on removal of electric field

21.14 Cross-correlation: the Hanbury–Brown and Twiss experiment

Suppose that images of a star are focused onto two photomultipliers separated by a distance D. If the angular size of the star is so small that it can be regarded as a point source there will be a high cross-correlation between the times of emission of pulses from the two detectors, limited only by the noise effect described in Section 21.7. If the angular size of the star is not quite so small, wave trains from points off-centre will have a time difference of arrival at the detectors and there will be a reduction in the cross-correlation which will depend on the size of the star as well as on D. By varying D it is in principle possible to measure the distribution of light across stars that are too small to be resolved in an ordinary telescope.

21.15 Interference and photon correlation

An interferometer measures light spatially. Given two beams it determines the ratio of the real amplitudes (E) and the phase difference δ. It also measures the coherence in the sense in which it was defined in Section 5.9. This we shall now call the *first-order coherence*. The photon-correlation experiments explore the variation of intensity as light reaches the detectors. When they deal with coherence as in the Hanbury–Brown and Twiss experiments, we measure coherence between two intensities ($I \propto E^2$), and this is called *second-order coherence*. For a Gaussian distribution there is a simple relation between first- and second-order coherence so that either can be obtained when the other is known. In general, however, the two orders of coherence are independent.

Interferometry was developed to use and to measure thermal sources. The half-value width of these sources was never less than about 10^{-4} nm in wavelength or 10^7 Hz in frequency. A Fabry–Perot etalon 10 cm long had a limit of resolution of about 6×10^{-5} nm or 5×10^5 Hz for the wavelength (644 nm) of the red cadmium line. Large gratings (used in fifth order) had equally good limits of resolution and these instruments were therefore capable of dealing with information concerned with or derived from these lines. It was logical for interferometry to stop here.

Lasers give lines of width 10^6 Hz (when not stabilised), 10^4 Hz (commonly when in a single mode) and can be stabilised to about 1 Hz when extreme precautions are taken. Interferometry could not make full use of these narrow lines but photon-correlation and beat-wave spectroscopy can do so. They operate well in the range 1–10 Hz and thus appear to take over just at the limit of interferometry.

It is necessary to state this conclusion with a certain degree of caution. It has been found possible to locate interference fringes to within 10^{-6} fringe and this suggests a theoretical limit for interferometry of 1 Hz. If any problems arise which photon correlation finds difficult or impossible to solve, considerable advances in interferometry may emerge in response to the demand.

References

1. H. Z. Cummins and E. R. Pike, *Photon Correlation and Light-beating Spectroscopy*, Plenum Press, New York, 1974.
2. E. R. Pike, The Malvern correlator. *Phys. Techn.*, **10**, 104–109 (1979).

The Velocity of Light and Relativistic Optics

22.1 Preamble

The progress of the measurement of the velocity of light from the recognition that it is finite to the latest accurate determinations, using lasers, went in parallel with a gradual understanding that this velocity (c) is a fundamental constant of physics. We can only outline this development. For detailed accounts of measurements up to 1972 see ref. [1] or [2]. The laser methods are described in ref. [3] or in refs. [4] and [5].

22.2 Early measurements

In 1676, Römer showed that irregularities in the times of occultation of the satellites of Jupiter were correlated with variation of the distance between the Earth and Jupiter. He deduced that light had a finite velocity near to $3 \times 10^8 \, \mathrm{m\,s^{-1}}$. Fizeau (1849) and Cornu (1874) each used a rapidly rotating toothed wheel to produce a succession of brief flashes of light. These were reflected from a distant mirror and the speed of rotation of the wheel was varied until a returning pulse was obstructed by the succeeding tooth of the wheel. Measurement of the rate of rotation gave a transit time of 5.5×10^{-5} s for a double path of 17.3 km, leading to $c = 3.15 \times 10^8 \, \mathrm{m\,s^{-1}}$.

22.3 Michelson's measurements

The American physicist, A. A. Michelson, made determinations in 1879, 1882, 1926 and 1934. In the later work he made ingenious improvements in the apparatus but did not greatly improve the accuracy. His first measurements are the most important. These showed that the velocity was equal to the ratio of the electromagnetic units then used, within the accuracy to which this ratio could be measured. This provided an experimental verification of Maxwell's electromagnetic theory of light and gave an impetus to further advances (Chapters 8 and 9).

Michelson used the rotating mirror method. Although this method was suggested by Wheatstone (1834) and used by Foucault (1860) it was Michelson who developed it to an accuracy of about 0.1%. Figure 22.1 is a plan view of the apparatus used in 1827. Light from L is successively reflected by M_1, M_2, M_3, M_4, M_5, M_4, M_3', M_2' and M_1 to a detector at E. M_1 is a precisely made octagonal mirror which rotates at a high speed, controlled by an electrically driven tuning fork. The double path was about 70 km. If the transit time is $N/8$ times the period of

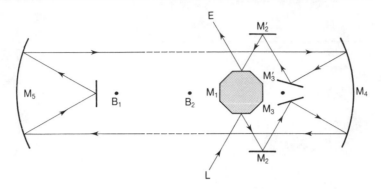

Figure 22.1 Michelson's (1927) measurement of the velocity of light. L is a slit source of light. B_1 and B_2 are surveyor's bench marks. The distance between these was accurately measured and found to be 35.3855 km. Relay lenses to keep the beam narrow enough for the small mirrors M_1, M_2, etc., have been omitted from the diagram

rotation (where N is a small integer) the direction of the final beam is the same as it is when the mirror is stationary. The speed was made to drift slowly through the desired value and the times when the returning beam was in the correct position were noted. A value of $c = 299\,798 \times 10^3\,\mathrm{m\,s^{-1}}$ was obtained. A later experiment by Michelson, Pearson and Pease gave $299\,774 \times 10^3\,\mathrm{m\,s^{-1}}$.

22.4 The electro-optical shutter

The accuracy of the experiments of Cornu and Fizeau was limited by the difficulty of making a mechanical shutter that would operate at a very high speed. By 1925, an electro-optical shutter (Section 11.4) had been invented which was capable of switching in $10^{-8}\,\mathrm{s}$. A number of measurements, using electro-optical shutters, were made in the next 25 years. The best were those of Bergstrand (1948, 1949, 1950), who obtained the value $c = 299\,792.8 \times 10^3\,\mathrm{m\,s^{-1}}$. A number of measurements of the transit times of radar pulses agreed with this value within an accuracy of about 5 in the last figure [1–3, 6, 7]. At this accuracy measurements in air had to be corrected to give the velocity in a vacuum.

22.5 Measurements of λ and ν

A new method was introduced by Essen and collaborators (1950). They measured the resonance frequency for a cavity of accurately determined dimensions and so obtained λ and ν whose product gives c. Froome (1958) used a microwave interferometer to measure the wavelength (in air) corresponding to a measured frequency. Both experiments gave a value near to $c = 299\,792.5 \times 10^3\,\mathrm{m\,s^{-1}}$ and Froome estimated an accuracy of 1 in the last figure. At this point it seemed that c had been measured with the highest obtainable accuracy but laser techniques gave a new impetus and required a reconsideration of the definition of standards of length.

22.6 Standards of length and time

The development of atomic clocks [3] made it possible to define the second as the duration of 9 162 631 770 periods of radiation corresponding to the transition between two hyperfine levels of the caesium atom. Quartz-crystal-controlled clocks can be calibrated by reference to this standard which is reproducible to within a few parts in 10^{13}. The meter was similarly defined as 1 650 763.73 wavelengths of a certain line in the spectrum of krypton. Since the line has a certain width (Section 5.5) this standard is reproducible only to about 3 in 10^8. If the frequency and wavelength are ν_1 and λ_1 for the krypton line and ν_2 and λ_2 for the caesium transition, then

$$c = \lambda_1 \nu_1 = \lambda_2 \nu_2 = \lambda_1 \nu_2 \left(\frac{\nu_1}{\nu_2} \right) \quad (22.1)$$

Since λ_1 and ν_2 are fixed by definition c can be determined if ν_1/ν_2 can be measured. In one determination this was done by generating a harmonic of ν_2 and measuring the beat frequency obtained when this was compared with the radiation from an HCN laser emitting in the far infrared. This in turn was compared with an H_2O laser, which was compared with a CO_2 laser, which was compared with a stabilized He–Ne laser, a higher frequency being obtained at each stage. Finally the wavelength of the He–Ne laser was compared by optical interferometry with that of the krypton line and it was assumed that this measurement gave the ratio of the frequencies. Thus ν_1/ν_2 was measured and a value of $c = 299\,792\,458$ m s^{-1} was obtained. A later experiment, with a different chain of lasers, gave 9 for the last figure, with an estimated error of 0.1. The accuracy is limited by the accuracy with which the krypton wavelength can be realised.

Since the standard of time can now be realised with an accuracy that is about 10^5 better than that of the krypton standard of length the Conference Generale des Poids et Mesures (CGPM) decided in 1983 to adopt the following definition: 'the metre is the length of the path travelled by light in a vacuum during a time interval of $1/299\,792\,458$ of a second'. This effectively defines the velocity of light to be $c = 299\,792\,458$ m s^{-1}. In practice length measurements have to be referred to a stabilized laser wavelength. If in the future the stability of a laser should provide a length standard of accuracy equal to or better than that of the caesium clock, this procedure will have to be reconsidered.

22.7 Variation of velocity with refractive index

Foucault showed that light travels more slowly in water than in air, as required by the wave theory. Later work by Michelson and others showed the velocity in water was that calculated from the group-velocity formula (eqn 5.15).

22.8 Relativity in relation to optics

Einstein's special theory of relativity may be stated in the following way: *the laws of physics are the same referred to any one of a set of frames of reference which are in uniform relative motion.* Since the laws of electromagnetism are the same in all these frames, the velocity of light is the same for all observers in uniform relative motion. Einstein later (about 1916) extended the theory, with modification, to frames of reference that had relative acceleration.

The theory of relativity is now supported by a large number of experiments on high-energy particles, including the conversion of mass into energy in an atomic reactor or, more dramatically, in an atomic explosion. Nevertheless, the direct proof that the velocity of light is the same for all observers together with the results of other associated optical experiments remain the basic experimental foundation of the theory. Moreover, as optics contributes to relativity so relativity contributes to optics by the understanding it gives of the photon and the relation between its mass and momentum (see Section 22.21).

22.9 Constancy of the velocity of light

Suppose the velocity of sound relative to the ground is V in the absence of a wind. Then the velocity of sound relative to the earth in the presence of a wind whose component is $v \cos \theta$ in the direction of the sound is $V + v \cos \theta$. In principle an observer could determine the wind speed and direction by measuring the velocity of sound in various directions. A maximum velocity $(V + v)$ would be found in a certain direction and a minimum $(V - v)$ in the opposite direction. The difference of speed would be $2v$ and the direction of the maximum velocity of sound would give the direction of the wind.

If the theory of relativity were not correct and light was propagated in an ether which formed an absolute frame of reference a similar experiment should determine the velocity of the earth relative to the ether. There is, however, a difficulty. We can measure the transit time for light to go to a certain point and to return. The velocity for a one-way path cannot be measured in any simple terrestrial experiment (though Römer's value is for such a path). We shall now show that, in classical theory, there is a difference between the two-way transit times for directions parallel and perpendicular to the direction of movement of the earth through an assumed ether.

For a length l in the direction parallel to the motion through the ether the two-way transit time in the classical theory is

$$T_1 = \frac{l}{c-v} + \frac{l}{c+v} = \frac{2lc}{c^2 - v^2} \qquad (22.2)$$

and, for a perpendicular direction, the velocity relative to the Earth is $(c^2 - v^2)^{1/2}$ so that the time T_2 is

$$T_2 = \frac{2l}{(c^2 - v^2)^{1/2}} \qquad (22.3)$$

If v is small compared with c,

$$
\begin{aligned}
T_2 - T_1 \\
= 2lc \left[\left(1 - \frac{v^2}{c^2}\right)^{-1/2} - \left(1 - \frac{v^2}{c^2}\right)^{-1} \right] = \frac{1}{2} T_0 \frac{v^2}{c^2}
\end{aligned}
$$
$$(22.4)$$

where $T_0 = 2l/c$ is the transit time when the Earth is stationary with respect to the ether.

The velocity of the Earth relative to the Sun is about $3 \times 10^4 \,\mathrm{m\,s^{-1}}$ and for this velocity $\frac{1}{2}v^2/c^2 = 5 \times 10^{-9}$. Optical techniques which will now be described are capable of detecting a much smaller change than this.

22.10 The Michelson–Morley experiment

In 1887, Michelson and Morley used a modification of the Michelson interferometer in an attempt to measure the velocity of the Earth through the ether. To obtain a long path in a small space, where temperature could be accurately controlled, they reflected the beams several times (Fig. 22.2). The experimenter adjusted the reference plane so as to obtain straight line fringes (Section 4.15) and set a cross-wire on one fringe seen in an eyepiece. The apparatus was floating

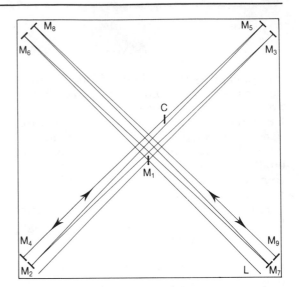

Figure 22.2 Michelson–Morley experiment. The arms of the Michelson interferometer have been 'folded'. As shown in the diagram, each arm is nearly $3\frac{1}{2}$ times the diagonal of the square. In the actual experiment twice as many transits were used. Note that M_1 is a half-silvered mirror at which the beam from the source L is divided into one beam which goes via M_2, M_3, M_4, M_5, M_4, M_3, M_2 back to M_1, where it is recombined with the other beam which goes via M_6, etc. C is the compensating plate (Section 4.15)

on a pool of mercury so that it could be rotated without mechanical disturbance. Of course, the experimenter could not, at the beginning of the experiment, know the direction of the Earth's velocity relative to the ether but, if the theory outlined above were correct, an oscillation of the fringes should have been observed as the interferometer was rotated.

No oscillation was observed. A velocity of $10 \,\mathrm{km\,s^{-1}}$ could have been detected. The experiment was repeated six months later when the Earth's velocity relative to the Sun had changed by $60 \,\mathrm{km\,s^{-1}}$ and a null result was obtained again.

22.11 Further such experiments

The experiment was repeated by Kennedy and Thorndike (1926), Illingworth (1928) and Joos (1930). The technique was refined so that a velocity of $1.5\,\mathrm{km\,s^{-1}}$ could have been detected and in each case a null result was obtained. Miller announced a small positive result but subsequent analysis of his data also gave a null result. It was later suggested that if a block of glass were inserted in one arm the ether would affect the velocity in the glass. An experiment (see ref. [3] or [6]) was carried out with a triangular interferometer in which the two beams traversed the same path in opposite directions. In this experiment an extremely low velocity of $0.5\,\mathrm{mm\,s^{-1}}$ could have been detected but again there was a null result.

22.12 The laser method

The last experiment went to the limit of conventional interferometry but with the advent of the laser a still more sensitive test became possible. The frequency of a helium–neon laser (pointing in a horizontal direction) was locked to the resonant frequency of a Fabry–Perot etalon. The light was reflected along a vertical axis about which the laser was rotated and was compared with light from another stabilized helium–neon laser which was stationary. No difference in the frequency of the rotating laser was observed though a frequency difference of 2.5 parts in 10^{15} (corresponding to an Earth ether relative velocity of about $10^{-6}\,\mathrm{m\,s^{-1}}$) could have been detected. The result of optical experiments is confirmed by experiments with microwaves, with a laser and with the Mössbauer effect, though none of these has the sensitivity of the laser experiment done by Brillet and Hall [8].

22.13 Transfer between axes in relative motion

Consider two observers who are moving with respect to one another so that A considers A′ to be moving with a velocity $-v$ and A′ considers A to be moving with a velocity $+v$ relative to himself in each case. Choose axes for A and A′ so that OX and OX' are in the same line and O coincides with O′ when $t = t' = 0$. In Newtonian mechanics, equations that are valid in the axes of A may be transferred to the system of

A′ using the following equations for the transformation of axes:

$$x' = x + vt, \quad y = y', \quad z = z' \quad \text{and} \quad t = t' \tag{22.5}$$

Also, any body that is moving with velocity u along the OX axis in the A system is moving with u' in the A′ system, where

$$u' = u + v \quad \text{and} \quad c' = c + v \tag{22.6}$$

which is not consistent with the constant velocity of light.

The following transformation equations (originally derived by Lorentz in relation to electricity) are consistent with Einstein's principle of relativity and with the constant velocity of light [7]:

$$x' = \beta(x + vt), \qquad x = \beta(x' - vt) \tag{22.7a}$$

$$t' = \beta\left(t + \frac{vx}{c^2}\right), \qquad t = \beta\left(t' - \frac{vx}{c}\right) \tag{22.7b}$$

where

$$\beta = \left(1 - \frac{v^2}{c^2}\right)^{-1/2} \tag{22.8}$$

The equations for x in terms of x' are similar if allowance is made for the positive sign of v in the A′ system and the negative sign in the A system.

22.14 Einstein's theory—addition of velocities

Suppose that a particle is moving in the common direction OX and OX' and that its velocity measured by A is u and by A′ is u'. Then

$$u = \frac{\mathrm{d}x}{\mathrm{d}t} \quad \text{and} \quad u' = \frac{\mathrm{d}x'}{\mathrm{d}t'} \tag{22.9}$$

and using the transformation equations (22.7) we obtain, remembering that v is a constant,

$$u' = \frac{u + v}{1 + vu/c^2} \tag{22.10}$$

If $u = c$, we have

$$c' = \frac{c + v}{1 + vc/c^2} = c \tag{22.11}$$

Thus the addition of any velocity, positive or negative to c, gives c again. The velocity of light emitted by a body in a system in which A is at rest but moving with velocity v relative to an observer A′ is c no matter whether it is measured by A or by A′. An experiment by Kantor gave a result not consistent with this statement but a subsequent more accurate terrestrial experiment [8] with γ-rays confirmed it. There is also astronomical evidence supporting it.

22.15 Velocity in a moving medium

In a medium of index n light travels with a velocity c/n when the medium is at rest with respect to the observer. When it is moving with velocity v in the direction of the light, eqn (22.10) gives for the resultant velocity:

$$V = \frac{c/n + v}{1 + v/cn} \qquad (22.12)$$

and when v is small compared with c we may write

$$V = \left(\frac{c}{n} + v\right)\left(1 - \frac{v}{cn}\right) = \frac{c}{n} + v\left(1 - \frac{1}{n^2}\right) \qquad (22.13)$$

The relevant experiment was made by Fizeau (Fig. 22.3). He observed interference between the two beams when water in the tube T was stationary and when it was moving as shown in the diagram. The fringe shift was in accord with eqn (22.13). Much later the experiment was repeated by Michelson and Morley and by Zeeman. After a small correction (due to dispersion in the

medium) had been taken into account eqn (22.13) was verified.

Fresnel had previously deduced eqn (22.13) from a theory in which the ether was dragged along with a moving medium but did not attain the full velocity of the medium. The experiments described in Sections 22.10 to 22.12 show that there is no ether drag and the result is a simple verification of the relativistic law of addition of velocities.

22.16 Light received when source and observer are in relative motion

Although the speed of light is unchanged by movement of the observer relative to the source, there are changes in both the observed wavelength (Doppler effect) and the observed direction (aberration effect) of the light. Let $\bar{\omega}$ and $\varkappa$ be the temporal and spatial frequencies of light received from a star S when the observer A is moving with speed $-v$ (relative to the star) along the OX axis as shown in Fig. 22.4. Let $\bar{\omega}'$ and $\varkappa'$ be the observed values when the observer A′ is at rest with respect to the star. Then

$$\frac{\omega}{\varkappa} = \frac{\omega'}{\varkappa'} = c \qquad (22.14)$$

The observer A′ is at rest in system S′ and observes that the ray from A′ makes an angle θ' with his OX' axis. He represents the waves emitted from A_1 by the expression

$$E' = \frac{E_0'}{r'} \exp i(\omega' t' - \varkappa' r' + \varepsilon') \qquad (22.15)$$

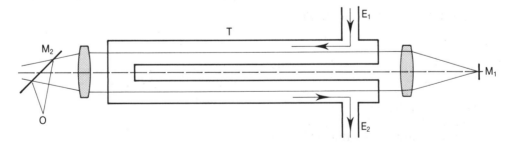

Figure 22.3 Velocity in a moving medium. L is the slit source of light. M_2 is a half-silvered mirror. Water enters the tube T at E_1 and leaves at E_2. The observer at O sees fringes due to interference between light which has gone clockwise through the system (i.e. in a direction opposite to the water flow) and a beam that has gone anticlockwise (in the same direction as the water)

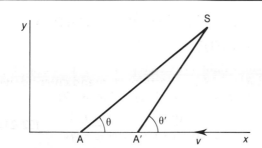

Figure 22.4 Doppler effect and aberration

where ε' is a phase constant. However, $r' = x' \cos \theta' + y' \sin \theta'$ so that

$$E' = \frac{E'_0}{r'} \exp i(\omega't' - \varkappa'x' \cos \theta' - \varkappa'y' \sin \theta' + \varepsilon') \tag{22.16}$$

The observer A represents this light by

$$E = \frac{E_0}{r} \exp i(\omega t - \varkappa x \cos \theta - \varkappa y \sin \theta + \varepsilon) \tag{22.17}$$

We must be able to obtain (22.17) by applying the transformation equations to (22.16) and the expression obtained has to be identical with (22.16). Therefore,

$$\omega t - \varkappa x \cos \theta - \varkappa y \sin \theta + \varepsilon \equiv$$

$$\beta\omega'\left(t - \frac{vx}{c^2}\right) - \beta\varkappa'(x - vt)\cos \theta' - \varkappa'y' \sin \theta' + \varepsilon' \tag{22.18}$$

remembering that v is negative for A. Since this is an identity we equate the coefficients of x and t and, using (22.14), we have

$$\omega = \beta(\omega' + v\varkappa' \cos \theta') = \beta\omega'\left(1 + \frac{v}{c} \cos \theta'\right) \tag{22.19}$$

and, similarly,

$$\varkappa \cos \theta = \beta\omega'\left(\frac{\cos \theta'}{c} + \frac{v}{c^2}\right) = \frac{\omega}{c} \cos \theta \tag{22.20a}$$

$$\cos \theta = \beta\left(\cos \theta' + \frac{v}{c}\right)\frac{\omega'}{\omega} \tag{22.20b}$$

Also using eqn (22.19),

$$\cos \theta = \frac{\cos \theta' + v/c}{1 + (v/c)(\cos \theta')} \tag{22.21a}$$

Putting $\tan^2\theta = \sec^2\theta - 1$, we obtain the alternative expression:

$$\tan \theta = \frac{\sin \theta'}{\beta(\cos \theta' + v/c)} \tag{22.21b}$$

It is easily shown that eqn (22.21b) is consistent with the inverse relation:

$$\tan \theta' = \frac{\sin \theta}{\beta(\cos \theta - v/c)} \tag{22.21c}$$

22.17 Aberration

From eqns (22.21) we may calculate $\Delta\theta = \theta' - \theta$ which is the difference in the direction of a distant source between the observations of two observers whose relative velocity is v. Now suppose that the 'two observers' are (1) an astronomer who observes the direction of a star when the Earth is at a certain point in its orbit and (2) the same astronomer who makes another observation about six months later. If v is twice the speed of the Earth (relative to the Sun), then eqn (22.21b) gives the relation between the two observed directions. This difference is called *aberration*. Suppose that one observation gives $\theta' = \pi/2$, then $\cot \theta = \beta v/c = \tan(\pi/2 - \theta) = \Delta\theta$ approximately, since $\Delta\theta \ll 1$. Taking $v = 6 \times 10^4 \ \text{m s}^{-1}$ we obtain $\Delta\theta = 2 \times 10^{-4}$ rad or $40''$. The non-relativistic formula differs from (22.21b) by the omission of the factor β, which is 1.000 002. The difference is too small to be observed.

It was thought that the aberration might be altered by filling the tube of the telescope with water, but when Airy (and also Hoek) tried this experiment no difference was observed. The relativisitic explanation of this result is simple. With the tube empty the observer sees the star in the centre of the field when the axis of the telescope is set normal to the nearly plane waves from the star. Filling the tube with water does not alter this condition.

22.18 The Doppler effect

When the source and observer are approaching one another directly with relative velocity v, $\theta = \theta' = 0$ and from eqn (22.19)

$$\omega = \beta\omega'\left(1 + \frac{v}{c}\right) \tag{22.22a}$$

or

$$\lambda' = \beta\lambda\left(1 + \frac{v}{c}\right) \qquad (22.22b)$$

A non-relativistic theory gives (22.22b) without the factor β so that the mean wavelength for two sources with velocities $+v$ and $-v$ is λ' (i.e. the same as when $v = 0$) for the non-relativistic theory and $\beta\lambda'$ in relativity theory. In two experiments [10, 11] the spectrum of light from a fast-moving beam of atoms and also of the light reflected from a mirror (so as to reverse v) has been photographed. Two lines are obtained and the spectrum of light from stationary atoms gives a third which, on non-relativistic theory, should be exactly in the centre of the other two. It is found that the mean of the other two is displaced by the amount predicted by relativity. The above is known as the *radial Doppler effect*. If the source is moving in a direction that, according to the observer, is perpendicular to the line of sight, then $\cos\theta' = 0$ and $\cos\theta = v/c$ so that (from eqn 22.19)

$$\beta\omega' = \omega \quad \text{and} \quad \lambda' = \beta\lambda \qquad (22.23)$$

This is called the *transverse Doppler effect*. According to non-relativistic theory $\lambda = \lambda'$ in this case. Owing to the difficulty of setting $\theta' = \pi/2$ to the precision required to eliminate the radial effect, the transverse effect has not so far been observed optically. It has been observed with γ-rays using the Mössbauer method [13, 14]. This depends on the emission of very narrow γ-ray lines from certain radioactive atoms which are firmly attached to a crystal lattice, of which they form part, so that there is no loss of energy in recoil. A very small change in wavelength may be detected by measuring absorption in a material containing atoms of the substance that follows the radioactive decay. A source is placed at r_1 from the centre of a rapidly rotating disc and an absorber at r_2 and the relative change of wavelength is proportional to $r_2^2 - r_1^2$. The change of absorption is measured for different speeds of the disc. The changes in wavelength obtained from these measurements agree accurately with eqns (22.23).

22.19   Reflection from a moving mirror

When light is reflected from a plane mirror moving in a direction perpendicular to its surface, the law of reflection is [7]

$$\frac{\sin i_1}{\cos i_1 + v/c} = \frac{\sin i_2}{\cos i_2 - v/c} \qquad (22.24)$$

where i_1 and i_2 are the angles of incidence and reflection. There is also a change of wavelength since the (virtual) source seen in the mirror moves with respect to the observer.

When a mirror is rotated about its own normal, so that there is no movement of the virtual source relative to the observer, there is no change in wavelength. This has been accurately verified [15].

22.20   Rotating interferometers

There are a number of experiments in which a beam of light is divided into two parts so as to produce two mutually coherent beams which go in opposite directions round a closed loop and then interfere to give a fringe pattern. Michelson proposed the experiment in 1905 but it was first carried out by Sagnac in 1913. His arrangement is shown in Fig. 22.5. When the apparatus is rotated a fringe shift proportional to the area of the

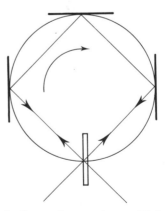

Figure 22.5   Sagnac's experiment. Light from L is split at P into two parts which traverse the interferometer in opposite directions and form interference fringes at O

loop and to the angular velocity is observed. The following theory is due. to Langevin [16]. Suppose x, y, z and t are axes for the rotating interferometer and x', y', z' and t' for the laboratory frame of reference. We confine our attention to a flat loop in the plane for which $z = z' = 0$ and the other transformation equations are

$$x' = x \cos \Omega t - y \sin \Omega t, \qquad t = t'$$

and   (22.25)

$$y' = y \sin \Omega t + y \cos \Omega t$$

where Ω is angular velocity of rotation.

In general relativity the interval s between two events is the same for all observers and is defined, in the laboratory frame, by

$$ds^2 = c^2 \, dt^2 - dl^2 = c^2 \, dt^2 - dx^2 - dy^2 \quad (22.26)$$

when both events are in the plane $z = z' = 0$. When the values of x' and y' are substituted from (22.25) we have to put

$$dx' = dx \sin \Omega t + dy \cos \Omega t$$

$$- (x \cos \Omega t - y \sin \Omega t)\Omega \, dt \quad (22.27)$$

and a similar expression for dy'. Completely written out there are 33 terms in the expression for ds^2. However, the expression is easily reduced to

$$ds^2 = c^2 \, dt^2 - 2\Omega(x \, dy - y \, dx)dt - dl^2 \quad (22.28)$$

when terms involving Ω^2 are neglected. This is justified because such terms would eventually lead to terms involving $(\Omega r/c)^2$, where r is the distance from the original and $\Omega r = v$, which is the linear velocity of a point on the rotating path. We shall see below that there is an effect proportional to v/c and we therefore neglect terms proportional to v^2/c^2. This contrasts with the situation in the Michelson–Morley experiment where there were no terms in v/c and then the v^2/c^2 terms became dominant.

Equation (22.26) implies that the interval for any two points on a ray of light is zero and then we may write

$$c^2 \, dt^2 - 4\Omega \, dA - dl^2 = 0 \quad (22.29)$$

where $dA = \frac{1}{2}(x \, dy - y \, dx)$ is the area of a triangle whose apex is at the origin and whose base is the element of the loop. Integrating round the loop we have, to the same order of approximation,

$$t_1 = \frac{dl}{c} + \frac{2\Omega A}{c^2} \quad (22.30a)$$

For the beam that traverses in the opposite direction,

$$t_2 = \frac{dl}{c} - \frac{2\Omega A}{c^2} \quad (22.30b)$$

and hence

$$t_1 - t_2 = \frac{4\Omega A}{c^2} \quad (22.31)$$

In some rotating interferometer experiments a medium fills the whole (or part) of the path. Then three cases arise: (1) the medium moves but the interferometer mirrors do not, (2) the interferometer moves in a stationary medium, (3) the interferometer and medium move together. The last case is important because a rotating fibre-loop interferometer is a possible means of inertial guidance (Section 20.20). In the coordinate system of the rotating body the velocity of the medium is zero and the velocity of light is c/n. This last affects both t_1 and t_2 equally so that eqn (22.31) for their difference remains valid.

22.21 Momentum and energy of the photon

According to relativistic theory [7] a particle whose mass is m_0 when at rest has an effective mass $m = \beta m_0$ when moving with velocity v relative to the observer. The momentum is $P = mv$ and the energy $E = mc^2$. Inserting the value of β from eqn (22.8) we find

$$E^2 = m_0^2 c^4 + c^2 P^2 \quad (22.32)$$

Since β approaches infinity as v approaches c, no material particle with a finite rest mass can have a velocity c. The photon must be assumed to have zero rest mass so that we have, for a photon,

$$E = h\nu, \qquad P = h\frac{\nu}{c} \quad (22.33)$$

and

$$E = cP \quad (22.34)$$

The mass equivalent of the energy emitted

as photons from the Sun per year is 1.4×10^{17} kg, a very small fraction of the Sun's mass (2×10^{30} kg).

22.22   Light pressure

Suppose that a parallel beam of light is incident normally on a surface at which it is absorbed. If the *energy density* in the beam is E (J m^{-3} kg m^{-1} s^{-2}), then the energy incident per unit area per second is cE and the momentum P transferred to the surface is given by $cP = E$. This is the pressure (in newtons per square metre). For light totally reflected the momentum change, and therefore the pressure, is doubled.

In prelaser experiments the highest value of E was obtained from sunlight for which the pressure is about 5×10^{-6} N m^{-2}. The area used in these experiments was only a few square centimetres and the force on an area of 2 cm^2 is only 10^{-9} N (or 10^{-4} dyne in the c.g.s. units). There was also a difficulty in avoiding (or compensating for) a pressure due to thermal action when the surface is heated. These difficulties were overcome and eqn (22.33) was verified. With lasers the limitation on the value of E is set only by the necessity of not melting the surface, and very short pulses may be used.

On an atomic scale the momentum of the photon is demonstrated by collisions between X-ray photons and electrons (Fig. 22.6). Using eqns (22.32) and (22.33) and applying the laws of conservation of energy and momentum an increase of wavelength $\Delta\lambda = (h/m_0c)(1 - \cos\theta)$, where m_e is the rest mass of the electron and θ is the angle of deflection for the photon is calculated. This change is observed with X-rays and is called the Compton effect. When an X-ray photon of energy about 10^4 electron-volts collides with one of the outer electrons of an atom (whose binding energy is a few electronvolts) it effectively collides with a free electron. When an optical photon of energy 1.5–3 volts encounters one of these electrons it does not separate the electron from the atom so the mass of the whole atom has to be used (instead of m_e) in calculating $\Delta\lambda$. The change is too small to be observed.

22.23   Interaction of gravitation and light

When a ball is thrown upwards, its velocity decreases as it rises since kinetic energy is lost to compensate for the change in gravitational potential energy. A photon that is travelling upwards must lose energy and, since its velocity does not change, the frequency must fall according to the relation:

$$h\nu_0 - h\nu = \frac{h\nu_0}{c^2} gd \qquad (22.35)$$

where ν_0 is the frequency at base level and ν at height d; g is the acceleration of gravity. This relation has been verified using the Mössbauer effect [17, 18]. The change in frequency and wavelength (for a height of about 30 m) is too small to measure by ordinary optical methods but a laser experiment may be possible.

The change in energy when a photon leaves the Sun should cause a wavelength change at the Earth of about one part in 10^6 due to the difference in gravitational potential between the Sun's surface and the Earth's surface. Such a change could be detected but there are other effects on the wavelength due to pressure, local magnetic fields, etc., which prevent unambiguous measurements being made. Some stars give a much larger effect but for them the Doppler effect is unknown.

Another interaction between light and gravitation is demonstrated by the deflection of light rays which pass near the Sun. To calculate the deflection the equations of general relativity must be used and, with them, a deflection of 1.7″ for a certain distance from the Sun is obtained. Measured values are 1.98″, 1.61″, 1.7″ and 1.72″ with estimated errors of 0.1″–0.3″. Thus the theory is fairly well verified.

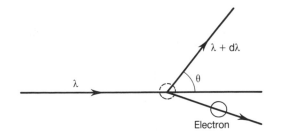

Figure 22.6   Compton effect

22.24 Nebular red shift

The American astronomer E. G. Hubble (1889–1953) found the empirical law (named after him) that lines in the spectrum of distant nebulae are displaced to the red by an amount $\Delta\lambda$ where

$$\frac{\Delta\lambda}{\lambda} = 1.7 \times 10^{-9}d \qquad (22.36)$$

where d is the distance of the nebulae in parsecs (1 parsec $= 3 \times 10^{16}$ m). This is generally believed to be due to expansion of the universe, the displacement being attributed to the Doppler effect. If this interpretation is correct the law is of great importance to cosmologists but gives nothing new about optics.

22.25 Synchrotron radiation

When an electron is moving in a circular orbit it emits radiation. Electrons can be accelerated to an energy of a few gigaelectronvolts (1 GeV $= 10^9$ eV) in a synchrotron (originally designed for high-energy physics). They are then injected into, and allowed to accumulate in, a 'storage ring' where constant speed is maintained. At 5 GeV, the factor β is about 10^4 and the power radiated, per electron, is

$$\frac{1}{6\pi\varepsilon_0} \frac{e^2 c \beta^8}{R^2} \qquad (22.37)$$

where R is the radius of the orbit. Pulses corresponding to currents of about 1 A can be obtained and the energy of the radiation emitted in a second is large so that megawatts of power are required to maintain the speed of the electrons. In a frame of reference in which the electron is at rest, the radiation is emitted over a wide range of the angle θ' between the direction of the radiation and the direction of the electrons. If θ is the angle in the laboratory frame, then from (22.21b), since $v/c \sim 1$,

$$\tan\theta = \frac{\sin\theta'}{\beta(1+\cos\theta')} \qquad (22.38)$$

Thus all the radiation is confined within a narrow cone of semi-angle β^{-1} or 10^{-4} radians for 5 GeV electrons. Within this

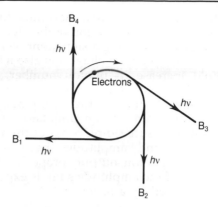

Figure 22.7 Synchrotron radiation. $B_1, \ldots, B_4$ are beam lines

angle the radiation is very intense. It may be used for absorption spectrophotometry, for biological applications using X-rays and for many other purposes. Several 'beam lines' can be taken off one ring (Fig. 22.7) and more than one experiment can be carried out on one line. Thus although the overall cost of the source is high, the cost per experiment is not excessive. The distribution of energy with wavelength for different voltages is shown in Fig. 22.8.

Two devices for increasing the output have been added. The first is to insert magnets which pull the beam sideways and then back

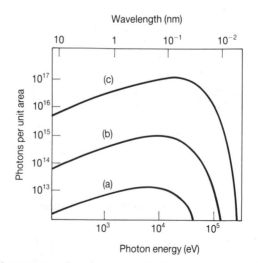

Figure 22.8 Synchrotron radiation output: (a) 3.7 GeV, (b) 5 GeV, (c) 5 GeV with 'wiggler'. Photon intensity is number per second in a wavelength interval $d\lambda = 10^{-3}\lambda$

into the circular path. For a short time during this 'wiggle' the acceleration of the electrons is increased, producing a significant increase in the highest photon voltage and also a hundredfold increase in photon number. The second device involves inserting a straight section into the ring and then using magnets to force the beam into an undulatory path. The light from succeeding peaks of the wave is coherent and amplitudes have to be added. The photon output (proportional to the square of the amplitude sum) is expected to be increased by a factor of 10^4.

22.26 Cherenkǒv radiation

It is possible to inject into a medium electrons moving with a velocity less than c but greater than the velocity of light (c/n) in the medium. A light pulse, similar to the acoustic pulse obtained at supersonic speeds, is obtained (Fig. 22.9). Huygen's construction shows that the wave-fronts form a cone of semi-angle c/nv. This angle is modified when the medium is dispersive [19]. The radiation emitted by a high-energy electron can be registered by a photomultiplier. It can thus be used for the detection of high-energy electrons or γ-rays (which release high-energy electrons) produced by nuclear reactions.

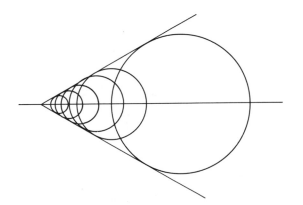

Figure 22.9 Cherenkǒv radiation

References

1. K. D. Froome and L. Essen, *Velocity of Light and Radio Waves*. Academic Press, 1969.
2. J. H. Sanders, *The Velocity of Light*, Pergamon Press, 1965.
3. B. W. Petley, *The Fundamental Constants*, Adam Hilger Ltd, 1985.
4. W. S. Trimmer, R. F. Baierlein, J. Faller and H. A. Hill, *Phys. Rev. D*, **8**, 3321 (1973).
5. W. H. McCrea, *Relativity Physics*, Methuen, 1935.
6. A. Brillet and J. A. Hall, *Phys. Rev. Lett.*, **42**, 549 (1979).
7. T. Alväger, J. M. Bailey, F. J. M. Farley, J. Kjellman and I. Wallin, *Ark. Phys.*, **31**, 145 (1979).
8. K. Brecher, *Phys. Rev. Lett.*, **39**, 1051 (1977).
9. H. J. Hay, J. P. Schiffer, T. E. Cranshaw and P. A. Egelstaff, *Phys. Rev. Lett.*, **4**, 165 (1960).
10. P. A. Davies and J. Jennison, *J. Phys. E.*, **10**, 245 (1977).
11. P. Langevin, *C.R. Acad. Sci.*, **173**, 831 (1921).
12. R. V. Pound and G. A. Rebka, *Phys. Rev. Lett.*, **3**, 439 (1959).
13. R. E. Cranshaw and J. P. Schiffer, *Nature*, **185**, 653 (1960).
14. H. Motz and L. I. Schiff, *Amer. J. Phys.*, **21**, 258 (1953).
15. T. G. Blaney, *Proc. Roy. Soc.*, **61**, 89 (1977).
16. K. M. Baird, D. S. Smith and B. G. Witford, *Opt. Comm.*, **31**, 367 (1979).

Problems

22.1 Discuss the factors limiting the accuracy of Michelson's method of determining c.

22.2 One type of electro-optical shutter used in measurements of c is the Kerr cell, making use of the electro-optical behaviour of nitrobenzene. Although this can switch a light beam very rapidly, the change *does* take a finite time. Why?

22.3 In the apparatus shown in Fig. 22.2, the distances between opposing mirrors is 2 m. If one arm is in the direction of the Earth round the Sun and if the ether existed, what would be the fringe shift observed for $\lambda = 546$ nm, assuming the separations of the two sets of mirrors are equal?

22.4 In the laser method (Section 22.12) the beat frequency of the combined laser signals is measured. Show that if the light frequency is ν, the beat frequency for a velocity relative to the ether is $v^2\nu/2c^2$. Confirm the figure quoted in Section 22.12.

22.5 Verify the statement at the end of Section 22.16 that the result of eqn (22.21c) is 'easily shown'.

22.6 Would you expect to be able to observe the aberration discussed in Section 22.17 with a telescope of diameter 1 m? If so, what problems would you expect to meet?

22.7 If you were looking at the light from excited hydrogen ions accelerated by a potential difference of 10 kV, both in their direction of motion and after reflection from a mirror (Section 22.18), what resolving power would you need in order to verify the result in eqn (22.22b) to an accuracy of 1%?

22.8 An equilateral triangular interferometer with sides 0.2 m long and with a helium–neon laser source ($\lambda = 633$ nm) is set up so that the axis perpendicular to the triangle points north–south. What frequency difference would be observed in the counterrotating beams as the Earth rotates?

22.9 So far as a high-energy X-ray is concerned an outer electron in an atom is effectively free. For a visible quantum, the electron is effectively fixed to the atom. So, consider a photon of energy 1000 eV incident on a hydrogen atom. Calculate the wavelength shift for scattering through $\pi/2$ for the two cases. What resolving powers would be needed to detect the shifts? Would you expect to detect shift in the less favourable case?

22.10 Could eqn (22.35) be used to determine the change in photon energy for a transit from a geostationary satellite (35 000 km above sea-level) and the Earth's surface? If not, obtain an expression appropriate to this problem and estimate the energy change for (a) radio waves and (b) visible light.

The Quantum Theory of Light

23.1 Introduction

The modern theory of light and of its interaction with matter has emerged as the result of a long process of development. A complete treatment demands a book to itself and involves some complicated mathematics. Here we can only outline some of the steps and state conclusions without proof. For an extensive treatment the reader should consult ref. [1] or [2].

In beginning this subject it is necessary to remind the reader that physical theories in general and quantum theory in particular deal with the results of possible experiments. They accept the results of past experiments and use them to predict the results of future possible experiments—insofar as they are predictable.

23.2 Photons in relation to modes

Atoms have stationary states which we may label state 1, state 2, etc. The result of an experiment may show that an atom is in one particular state n or it may yield probabilities C_1 that it is in state 1, C_2 that it is in state 2 and so on. We shall now show that light waves may be characterised by modes and the result of an experiment may be that a

quantum of energy is in mode n or it may be that there is a series of probabilities $C_1, ..., C_m$ for different modes.

It was shown in Chapter 14 that light can be propagated in modes in metallic or dielectric waveguides. We now consider modes of oscillation in a cubical box whose side length is L and whose walls are perfectly reflecting. The electric field E must vanish on the walls of the box and, if we put the zero of coordinates at one corner and the axes along three edges, we have, for the electric field strength,

$$E_x(x, y, z, t)$$

$$= E(t)\sin\left(\frac{\pi l_1 x}{L}\right)\sin\left(\frac{\pi l_2 y}{L}\right)\sin\left(\frac{\pi l_3 z}{L}\right) \quad (23.1)$$

with two similar equations for E_y and E_z. Note that l_1, l_2, l_3 are positive integers (including zero). Equation (23.1) satisfies the boundary conditions and also is in accord with Maxwell's equations. The mode represented by (23.1) is characterised by a wave vector, where

$$\varkappa^2_{123} = \left(\frac{\pi}{L}\right)^2 (l_1^2 + l_2^2 + l_3^2) \quad (23.2)$$

If, as in Section 14.3, we consider a cavity with *dielectric* walls (assumed infinitely thick) then we need to take account of the radiation *outside* the cavity. In this case the modes given by eqn (23.1) extend from $-\infty$ to $+\infty$.

23.3 Definition of a photon

Since $E(t)$ is independent of position the wave equation reduces to

$$\frac{\partial^2 E(t)}{\partial t^2} = -\omega^2 E(t) \qquad (23.3)$$

This is the same form as the equation for a simple harmonic oscillator. It is shown in books on quantum mechanics that the simple harmonic oscillator exists in stationary states with energies $(n + \frac{1}{2})\hbar$, where n is a positive integer including zero. We now make the quantum hypothesis that this holds also for the modes of the electromagnetic field. When $n = 1$ we associate the mode with a single photon of energy $\hbar\omega$ plus the zero-point energy $\frac{1}{2}\hbar\omega$. The difference of energy between a mode associated with one photon and one with no photon ($n = 0$) is $\hbar\omega$ so we say that this is the energy of one photon. However, a photon can be located experimentally only when it interacts with an atom or molecule so before any interaction we cannot define the position of the photon. The interaction can be (1) by absorption and release of a photoelectron, (2) by scattering, sometimes with the displacement of an electron (Compton effect), or (3) by absorption and modification of a grain on a photographic plate so that it can be developed.

23.4 The uncertainty principle

This principle of quantum mechanics can be applied to photons. When an ideal laser continuously operating in one mode emits a perfectly parallel beam of light with a precisely defined frequency,† an interaction registering the arrival of a photon can occur anywhere, so that we cannot locate the 'position' of the photon in the beam at all. If the laser operates for a time t it will emit a wave train of length ct and the photons may be detectable anywhere within this wave train. However, the finite length of the wave train

---

† The laser would have to be infinitely wide to avoid diffraction and the beam infinitely long to have an exact frequency.

implies an uncertainty in frequency of about $1/t$ (Section 5.5). Since the momentum associated with the photon is $\hbar\omega/c$ (Section 12.4) the momentum uncertainty Δp is $\hbar/ct$ and we have

$$\Delta p\, \Delta x \approx \hbar \qquad (23.4)$$

where $\Delta x = \Delta(ct)$ is the uncertainty in the x coordinate of the position where the photon is manifest by absorption, etc. This is in accord with the uncertainty principle.

If the laser is used to produce a very short pulse (less than 10^{-12} s) the frequency spread is very large. The shortest pulses that have been produced to date are of the order of femtoseconds in duration: they are effectively white light.

23.5 Planck's law

From eqn (23.2) we have

$$(\omega_{123})^2 = \left(\frac{c\pi}{L}\right)^2 (l_1^2 + l_2^2 + l_3^2) \qquad (23.5)$$

We plot $c\pi l_{1,2,3}/L$ along the x, y, z directions and so construct a cubic array of points as shown in Fig. 23.1, the length of the side of each cube being $c\pi/L$. Thus the volume associated with one mode is $(c\pi/L)^3$ and the number of points per unit volume is $(L/c\pi)^3$. A radius vector from the origin to each point gives a value of ω so each point represents a possible value of ω. The number of modes satisfying eqn (23.5) with values of ω

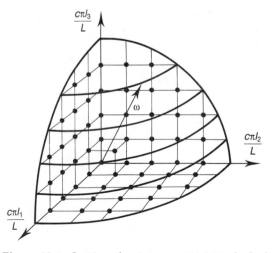

Figure 23.1 Lattice of points representing l_1, l_2, l_3

between ω and $\omega + d\omega$ is

$$\frac{4\pi\omega^2}{8} \cdot \frac{L^3}{c^3\pi^3} 2d\omega \qquad (23.6)$$

The factor $\frac{1}{8}$ is due to the fact that l_1, l_2, l_3 are always positive so that only an octant of a sphere is involved and the factor 2 because there are two states of polarization associated with each value of $\bar{\omega}$. Thus the number of modes per unit volume of real space is $\omega^2\, d\omega/\pi^2 c^3$.

According to Boltzmann's law the probability of a system being in a state of energy W_n divided by the probability for a state of energy W_0 at absolute temperature T is $\exp[-(W_n - W_0)/k_0 T]$, where k_0 is Boltzmann's constant. Then the probability of a mode of frequency ω being associated with n photons relative to the probability for zero photons is

$$P_n = \exp\left(-\frac{n\hbar\omega}{k_0 T}\right) = \exp(-na) \qquad (23.7)$$

where $a = \hbar\omega/k_0 T$.

The mean energy of the photons associated with this mode is

$$\bar{n}\hbar\omega = \frac{\sum_{n=0}^{n=\infty} n\hbar\omega \exp(-an)}{\sum_{n=0}^{n=\infty} \exp(-an)} \qquad (23.8)$$

$$= \frac{\hbar\omega(\partial/\partial a)\Sigma \exp(-an)}{\Sigma \exp(-an)} \qquad (23.9)$$

Since $\Sigma \exp(-an)$ is a geometric progression of common factor e^{-a} the sum is $n/(1 - e^{-a})$ and

$$\bar{n}\hbar\omega = \frac{\hbar\omega}{e^{-a} - 1} \qquad (23.10)$$

Using eqn (23.7) the energy per unit volume of the radiation with frequency between ω and $\omega + d\omega$ is

$$\rho(\omega)d\omega = \frac{\hbar\omega^3 \, d\omega}{c^3\pi^2 [\exp(-\hbar\omega/kT) - 1]} \qquad (23.11)$$

This is Planck's law (see Fig. 12.4). In deriving it we have made two assumptions: (1) the quantum law that the energy in a mode can change only by a multiple of $\hbar\omega$;

(2) any number of photons can be associated with one mode.

23.6   Photon statistics

The statistical distribution of photons among modes has the same underlying rules for light from thermal sources and from lasers but the forms of the distribution curves are very different. The distribution for thermal sources is a Poisson distribution which has been described in Section 21.6. The photons are thinly distributed over a number of modes much larger than the number of photons, and the most probable number for any one mode at a given instant is zero, as shown in Fig. 23.2, curves (a) and (b). When the proportion of photons grows larger the Poisson distribution takes the forms shown in curves (c) and (d), and when a very large number of photons is distributed over a much smaller number of modes the Poisson distribution becomes identical with a normal (Gaussian) distribution. Although the laser which puts all the photons in one mode is an unrealisable ideal, commonly lasers put very large numbers of photons into a much smaller number of modes and an approximately Gaussian distribution is obtained.

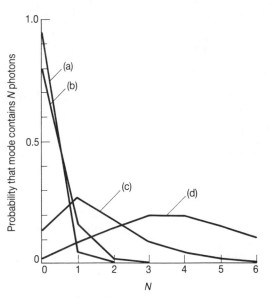

Figure 23.2 Poisson's statistical function: average number of photons per mode: (a) 0.05, (b) 0.2, (c) 2, (d) 4

23.7 Interaction with atoms: perturbation theory

In early studies of atoms it was observed that the exposure of atoms to magnetic or electric fields resulted in splitting of the spectral lines, respectively referred to as the Zeeman and Stark effect. Since a light wave is electromagnetic in character then it is clear that there will be an interaction between a light wave and an atom. If the interaction energy between the electric and magnetic fields of the light wave is small compared with the energy of the relevant electron in the atom then the effect of the disturbance can be calculated using perturbation theory. When this is done it emerges that an atom initially in its lowest energy stationary state W_0 would after time t have a probability proportional to It (where I is the intensity) of being observed in a higher state W_1 provided that the frequency of the light was near to the value derived from the equation

$$\hbar\omega = W_1 - W_0 \qquad (23.12)$$

This applied only so long as the probability is fairly small. For longer times the probability of the atom being promoted to the higher state increased asymptotically to unity, as shown in Fig. 23.3. The number of atoms in the higher state did not increase in the way shown by the curve because, all the time, atoms would be returning to the ground state by spontaneous emission. Thus the number of atoms in the upper state was determined by a competition between the processes of absorption and emission.

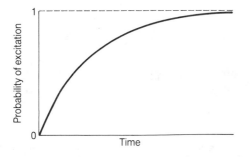

Figure 23.3   Probability of an atom being excited

23.8 Field quantisation

In the theory described in the last paragraph, the classical field equations of Clark Maxwell are assumed to be valid. Quantum theory enters only in the interaction between atoms and light, as in the original theory of Planck. With atoms, regarded as mechanical systems, the transition from classical to quantum theory involves replacing the dynamical variables by operators. By analogy it would seem that **E** and **H** should be replaced by operators to quantise the electromagnetic field. It is more convenient to start by describing the electromagnetic field by a vector potential **A** and a scalar potential ϕ. By a suitable adjustment ϕ can be made zero in classical theory (later it was found that this is not possible in quantum electrodynamics).

When the field has been described in terms of the **A** potential we may consider an assembly of atoms and photons as a single dynamical system. This is controlled by operators which create and destroy photons with simultaneous adjustments of the energy of the atoms so that, in the system as a whole, energy is conserved. Dispersion theory and scattering can be described and results very similar to those found by the semiclassical theory are obtained. It is possible to calculate the Einstein coefficients A and B independently and hence to derive Planck's law.

The quantum theory of the field is conceptually difficult, partly because the photon has no position operator. It cannot be said to be at a definite point though it is absorbed at a definite point when its effects on an atom are manifest. Partly because of this conceptual difficulty a number of workers have sought to explain all the experimental results in terms of a semiclassical theory or a modified theory (called the neoclassical theory) in which **E** and **H** are still variables whose values at any given point and time can, in principle, be measured. They have had considerable success in explaining some effects that were thought to require quantization of the field. However, there remain some crucial experiments whose results are clearly predicted by the theory of quantised fields and not by the classical or neoclassical field theories. We shall give two examples. For a detailed discussion of this matter ref. [2] or [3] should be consulted.

23.9 When do photoelectrons appear?

When a detector is suddenly exposed to a weak beam of light semiclassical theory demands an interval before any electron can be emitted in order that sufficient energy can be accumulated to balance the work function of the electron. In an experiment by Davis and Mandel [4] light from a laser was switched on and off by means of a mechanical shutter. The rise time when the light was switched on was 2–3 μs. The time required to accumulate an amount of energy equal to the work function was 20 μs. Thus semiclassical theory predicts no electrons when the on period is less than 20 μs and only one when it is less than 40 μs, etc. It was found, in accordance with the predictions of quantum field theory, that emissions were distributed at random during the periods when the detector was exposed to light (Fig. 23.4). The neoclassical theory can explain the emission of electrons in less than

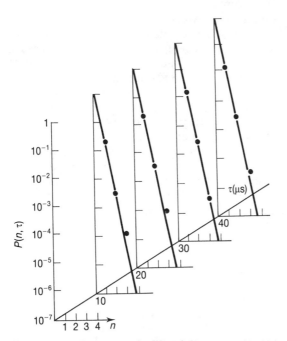

Figure 23.4 Davis and Mandel's experiment. Classically insufficient energy arrives in 10 μs for any photoelectrons to be emitted. Experimentally, 1, 2 or 3 photoelectrons may be observed in this period

20 μs as due to field fluctuations (which should not occur with a well-stabilised laser) but cannot simultaneously give detailed conservation of energy.

23.10 Taylor's experiment

In 1909, G. I. Taylor carried out an experiment in which he photographed a diffraction pattern. At first a fairly strong source of light was used so that a short exposure was sufficient (Fig. 23.5). Then the light was reduced by interposing dark filters until many days of exposure were necessary.† The average energy between the darkened source and the screen was so low that the total radiation energy in the system was much less than that of one photon. It was therefore extremely improbable that more than one photon could 'be' in the apparatus at any one time. The diffraction patterns obtained with weak light were the same as those obtained with strong light. This appears, at first sight, to favour a semiclassical theory. The quantum field theory predicts that interference will be obtained in this situation but has the conceptual difficulty of requiring a photon to interfere with itself [5]. That this is so has been said repeatedly since the time of Dirac. It is not clear to all that this has made the conceptual difficulty disappear.

In 1967, Pfleegor and Mandel [6] carried out an experiment to explore whether interference occurred between two beams of light from two independent sources. In this experiment light from two separate lasers was heavily attenuated and the two light waves combined with their directions inclined at a small angle. Detectors were arranged to

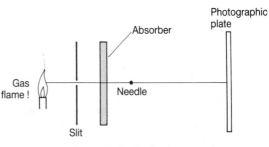

Figure 23.5 G. I. Taylor's experiment

---

† The longest exposure was 2000 hours (~3 months)!

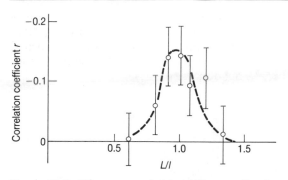

Figure 23.6 Pfleegor and Mandel's results for interference of light from two separate sources

record at spacings (L) which could be made to correspond to the interference fringe spacing (l) if interference were to occur. Thus the signals recorded by adjacent detectors in the field should alternate high/low/high/low, ..., i.e. the signals in adjacent channels should be *anticorrelated* (correlation coefficient negative).

The intensities of the intersecting beams were reduced to the level where, from a classical view, the time for the flow of energy corresponding to that of one photon was large compared with the time a photon would take to pass through the system. Crudely one could say that, with a high probability, the photon from one source was absorbed before that from the other source was emitted.

If interference occurs then a negative correlation coefficient r should be observed when the detector separation L is equal to l. The results are shown in Fig. 23.6, showing that interference effects are indeed observed under these conditions. This type of experiment has been repeated many times by different experimenters, always with the same result. Any semiclassical picture of photons fails but quantum electrodynamics correctly predicts the experimental results. The ability of a photon to appear to interfere with itself has to be accepted, together with its ability to correlate with other photons.

23.11   Conclusion

Quantum electrodynamics correctly predicts the results of all optical experiments carried out up to the present time and is also suc-

cessful in the field of high-energy physics. The conceptual difficulties make it difficult to express the results in words but involve no conflict with logic. However, there are unsolved problems. The zero-point energy is $\frac{1}{2}\hbar\omega$ for each mode and there is an infinite number of modes. If mass is associated with energy then this indicates that the universe is loaded with infinite mass. There are other places in the theory (outside optics) where expressions diverge and infinity comes on the scene. Some modification or a deeper understanding is needed. For the present we simply accept that the theory gives a correct estimation of the optical phenomena observed.

References

1. R. Loudon, *The Quantum Theory of Light*, Clarendon Press, Oxford, 1973.
2. L. Mandel, *Progress in Optics*, **XIII**, 29–68 (1976).
3. L. Mandel, 'Quantum optics', in *Open Questions in Quantum Physics* (Eds. G Terrozzi and A. A. Van der Merve), Reidel, Dordrecht, 1985.
4. W. Davis and L. Mandel, in *Coherence and Quantum Optics* (Eds. L. Mandel and E. Wolf), Plenum Press, New York, 1973.
5. G. I. Taylor, *Proc. Camb. Phil. Soc.*, **15**, 114 (1909).
6. R. L. Pfleegor and L. Mandel, *Phys. Rev.*, **159**, 1084 (1967).

Problems

23.1   Consider a long thin cavity capable only of supporting longitudinal modes ($n_2 = n_3 = 0$ in Section 23.5). Derive the one-dimensional form of expression (23.6) and determine the number of modes expected for a laser 500 mm long where the source is the Doppler-broadened line of a helium–neon discharge, the emission recurring from neon at a wavelength of 633 nm. Assume a gas temperature of 600 K.

23.2   There appears a dramatic change in Fig. 23.2 between the case of 0.2 and 2 photons per mode. Explore the region between these two values.

23.3   In Pfleegor and Mandel's experiment (Section 23.10) imagine that the power of the two independent laser sources (HeNe) to be 0.1 mW. The sources are 20 mm apart and the beams converge at a distance of 5 m. What detector separation is needed and what attenuation is needed in the beams of a detector of a few detections in time slots of 10 μs are to be recorded on each detector? Make any plausible assumptions needed.

24

The Limitations of Optical Experiments

24.1 Optics in relation to information theory

The purpose of an optical instrument is to obtain more information than is available to the naked eye. To compare the information capacity of two instruments of the same kind (e.g. two telescopes) we need a quantitative definition of information. Such a definition was developed by Shannon [1, 2] in connection with the rate of transmission of information along telegraph cables. He defined a unit of information, called a *bit*, as the amount of information that enables us to decide between two equally probable possibilities. When a coin is tossed we regard 'heads' and 'tails' as two equally probable possibilities. When it has come down we have gained one bit of information. If m coins are tossed the information involved is m bits but the permutations are such that there are $n = 2^m$ possible results provided the coins are distinguishable. Each permutation has a probability $p = 1/n$. Thus, if H is the amount of information in n trials, then

$$H = m = \log_2 n = -\log_2 p \qquad (24.1)$$

Since p is less than 1, H is always positive.

More generally, consider a set of possible events with probabilities $p_1, p_2, ..., p_l, ..., p_n$. Then the information gained when the result p_l is observed is $-\log_2 p_l$, but the probability of this event occurring is p_l. Thus, in a series of N trials, the information gained from observing the result l is $-Np_l \log_2 p_l$ and the total information is

$$-N \sum_{l=1}^{n} p_l \log_2 p_l$$

The average information per trial is

$$H = -\sum_{l=1}^{n} p_l \log_2 p_l \qquad (24.2)$$

Also

$$\sum_{1}^{n} p_l = 1 \qquad (24.3)$$

When numbers are large we define $p_l\,\mathrm{d}l$ as the probability of the event occurring between l and $l + \mathrm{d}l$: in this case

$$H = \int_{1}^{n} (-p_l \log_2 p_l)\mathrm{d}l \qquad (24.4)$$

In optics we are often concerned with the total number of bits available (e.g. on a photograph or as the result of a flashlight view of a scene) but we may also need to consider the *rate* of *transmission* or *acquisition* of information. Thus consider the case of an optical system used to view a scene that is changing rapidly. The question which then arises is whether the apparatus can accept information and put it into store rapidly enough to 'follow' the changing scene. For example, in a television system, a distorted or blurred picture will be obtained if the scene changes significantly during one frame. The theory of the rate of transmission of information along an optical fibre (Section 14.6ff) is very similar to Shannon and Weaver's theory [1] of the rate of transmission along electric cables.

24.2   Sampling theorem

The results of an optical experiment are often expressed in the form of a graph, e.g. the intensity of a star as a function of time. Every physical instrument has an upper limit to the frequencies that it can reproduce: we denote this limit by Ω_L. As a consequence the reading at a given point at t' on the graph depends on the input over an interval of order $1/\Omega_L$ before t', and readings at points that are very close together are not completely independent. Indeed, if they were completely independent the graph would represent an infinite number of readings and contain an infinite amount of information. In Appendix 24.A it is shown that we have effectively independent readings at equal intervals $1/\pi\Omega_L$ so a graph extending over a time T is completely specified by $\pi T\Omega_L$ readings. This is known as the *sampling theorem*. If the accuracy of measurement at each point is such that m levels can be distinguished then each point yields $\log_2 m$ bits of information and the total information in the graph is $\pi T\Omega_L \log_2 m$. For example, if the accuracy of the reading is about 0.1% so that $1024\,(=2^{10})$ levels can be distinguished then the information content is $10\pi T\Omega_L$. Similar considerations apply to graphs of light energy versus distance and to other graphs obtained in optics.

24.3   Information in an optical picture

Extension of the sampling theorem to two dimensions shows that the information per unit area in an optical scene is very high indeed. We have shown (Section 19.16) that independent readings can be obtained under a microscope at intervals of order $\lambda/2$ which corresponds to $2/\lambda$ per unit length or 4×10^6 per metre along any line across the picture. We may regard the picture as having independent readings at the crossing points of a square grid of mesh size $\lambda/2$, so there are 1.6×10^{13} points per square metre. If each point yields 10 bits of information the total content of the picture is $1.6 \times 10^{16}\ \mathrm{m}^{-2}$. At a single view of a high-power microscope we can see only an area of order $10 \times 10\ \mu m$, i.e. $10^{-10}\ \mathrm{m}^2$, but even in this small area the information content is greater than 10^6 bits. With a very fine grain photographic plate about 10^7 independent readings per metre or 10^{14} per square metre can be obtained, which is only about 100 times less than the maximum set by the wavelength of light. For an ordinary fairly fine grain film the number per square metre is only 10^{10}, so that the information content per unit area of ordinary photographs is much less than the theoretical maximum although it is still large.

The calculation of the information capacity of the human eye is difficult because the resolving power varies across the visual field. Various methods indicate a capacity of between 10^6 and 10^7 bits for a brief view of a scene. Since changes occurring in 0.1 s can be detected the eye can acquire information at a rate of 10^7–10^8 bits per second.

24.4   Transmission of optical information

In Sections 18.9 to 18.11 we described intrascopes used for medical examinations and similar instruments with industrial applications. These constitute methods of transmitting information to the observer from a region that is not accessible to the direct view. The above discussion shows that the number of bits transmitted is very large indeed. Much of the information may be irrelevant but the user has no difficulty in selecting those critical parts of the information which are relevant to any decision that may have to be made, e.g. a surgeon pays attention to those aspects of the 'view' seen through his instrument that affect his decision of whether to operate.

We have already considered in Section 14.8 the transmission of signals along optical fibres. The rate for fibres at present in use is of the order of 10^9 bits s^{-1}. This limit is set by the rate at which the signals can be modulated and demodulated. Since pulses of less than 10^{-14} s have been produced it is almost certain that very much higher rates of transmission will be attained when they are required.

24.5 Noise

When a signal is transmitted it is often contaminated by errors due to the transmission system. When these errors are random, which they usually are, they are called *noise* and their presence limits the rate of transmission of information. For example, consider a continuous message sent along a telephone wire and reproduced as a graph of the output versus time. There will be a finite number of sampling points, as explained in Section 24.2, but if there were no error an infinite number of levels could be distinguished at each sampling point and the rate of information transmission would be infinite. Further investigation [2] shows that the rate is proportional to $W \log_2 (1 + P/NW)$, where P is the power in the signal and N is the equivalent noise power. $2W$ is the number of points sampled per second. (This formula does not, of course, apply to a pulse code modulated system where, at each sampling point, there are just two possibilities: 'pulse' and 'no pulse'. In the pulse code method of transmission the number of sampling points is large and the probability of an error can be made very small.)

There are a number of methods of 'cleaning' a noisy signal so as to extract the 'true' information which comes from the source, from the mixture of 'true' information and from noise that is received. All of these methods reduce the rate at which the true information is received. One method is that of redundancy in which a message is, effectively, repeated. Those parts of the message that agree in the two transmissions are accepted and the missing points where they do not agree are filled in by interpolation. This, of course, leaves a possibility that some detail will be missed, but the probability that this will happen can be made very

small. Gross errors can usually be recognised; e.g. small holes in the picture on a photographic negative can be recognised and painted out.

24.6 Fidelity

A hologram may contain all the optical information about an object and yet convey no information to a person who examines it without the appropriate coherent illumination. To take a less extreme case, a picture in which the image is clear but badly distorted may be unrecognisable even though it contains all the optical information about the original object. It is as though a book had been passed through a system which produced a Greek translation in large clear type. All the information is there but it is not available. Thus the output of an optical system must not only contain as much information as possible about the object but it must also make it available. This quality is called *fidelity* and the objective of an optical designer must be both to maximise information and to give fidelity. Of course if the output is to be analysed by a computer, it has to be in the form of a series of pulses and gaps. That is 'fidelity' in the computer world.

24.7 Ultimate limits

In practice the performance of optical systems is limited by the size, the quality of the components and the residual aberrations. These affect the cost of the instrument. There are, however, certain limitations which are derived from natural laws. These limitations cannot be avoided however great the skill of the instrument maker or the money available.

The main sources of these limits are:

1. Thermal noise determined by the laws of thermodynamics
2. The wave properties of light, especially diffraction
3. The particle properties of light

24.8 Thermal noise

In the infrared region of the spectrum, the walls of any source or detector of the radiation are emitting considerable numbers of photons of the same frequency as those being measured, if the apparatus is working at room temperature. For detecting weak sources of radiation it is essential to cool the receiver, usually to liquid helium temperature. In the visible region, this direct effect is unimportant because the number of visible photons emitted by bodies at room temperature is extremely small. For many types of receiver, however, noise resulting from the random motion of electrons (Johnson noise) in the appropriate resistor sets a limit to the attainable sensitivity. Photomultipliers are nearly immune from this limitation because the current does not flow through a resistance until it has been amplified some 10^6 times in the photomultiplier tube.

24.9   Wave and particle properties of light

An ultimate limit to measurements that can be made on light is set by the uncertainty principle, which may be written in the form

$$\Delta E \; \Delta t \geqslant \frac{h}{2\pi} \qquad (24.5)$$

or

$$\Delta \nu \; \Delta t \; \geqslant \frac{1}{2\pi} \qquad (24.6)$$

Equation (24.6) is of importance in relation to the properties of light pulses of very short duration. For a pulse of 10^{-14} seconds (10 femtoseconds) $\Delta \nu$ is of the order of 10^{13} Hz. (Recall that the frequency of visible light is of the order of 10^{14} Hz.) Thus the spectral content of a pulse of very short duration with frequency ν_0 extends over a considerable region of the spectrum around ν_0.

The uncertainty principle arises in another context in which classical and quantum concepts are linked. We have seen that radiation—e.g. in an oscillating mode in a cavity—may be characterised in terms of the energy levels of a simple harmonic oscillator. A measurement of the amplitude of the wave may therefore be expressed in terms of the number N of associated quanta. Now a classical wave is characterised not only by an amplitude but also by a phase ϕ. Simultaneous measurements of N and ϕ runs into exactly the same kind of difficulty that the

measurement of position and momentum of an electron suffers. The uncertainty in the measured number of field quanta ΔN is related to the uncertainty $\Delta \phi$ in phase by the relation

$$\Delta N \; \Delta \phi \sim 1 \qquad (24.7)$$

We must be careful, however, in the way in which we interpret ΔN. In a single measurement—e.g. in a photon correlation experiment—N is an integer and ΔN would appear to be zero, implying an infinite uncertainty in ϕ. The appropriate ΔN so far as eqn (24.7) is concerned corresponds to the uncertainty in N resulting from repeated measurements of N under identical conditions. We note that if $\Delta N \sim 1/2\pi$ then $\Delta \phi \sim 2\pi$. Since phase is periodic with period 2π, an uncertainty of the order of 2π implies that all phases are equally probable or that no information on phase can be deduced.

Equations (24.5) and (24.6) taken by themselves imply that by sacrificing all knowledge of one variable we may obtain any desired accuracy in the other. It is true that with stabilised lasers the fluctuation ΔN can be made very small indeed, but in diffraction problems there is a limit to the accuracy obtainable owing to the finite wavelength of light. This uncertainty does not fall as rapidly as the uncertainty of the momentum of the photons increases.

24.10   Squeezed light

If we use pulses of light for communication there would appear to be a serious conflict between speed and accuracy. If the photons in the beams are uncorrelated then the number N in a given interval will show Poisson fluctuations. The standard deviation $\Delta N = (N^2 - N_0^2)^{1/2}$, where N_0 is the mean number of photons in the pulse, is proportional to $N_0^{1/2}$. Thus the signal-to-noise ratio for the pulse will be $N_0/\Delta N$ which is proportional to $N_0^{1/2}$. For high accuracy—to ensure that 'pulse' and 'no pulse' are unambiguously resolved—N_0 needs to be made large. However, this implies a pulse of longer duration, for a source of given power, which therefore limits the rate at which pulses can be sent, i.e. the communication rate.

This would appear to be quite a fundamental limitation which we cannot expect to overcome. In fact we *can* do better. The key lies in the fact that the above arguments apply to *uncorrelated* photons, such as emerge from any normal source. If we were able to impose some correlation between the photons in the signal, the $N_0^{1/2}$ dependence would not apply. How might this be done?

A semiconductor laser offers one possible route. The photons emitted by such a laser are a consequence of electron transitions from the conduction band to the valence band (Section 20.10.5), so the photon emission is correlated with the electron current through the laser. If the device is fed from a high-impedance source, the current noise will be mainly that of the thermal noise which will vary inversely as the source resistance. If the laser gain is sufficiently high, the current noise can be made much less than the shot noise of a random flow of electrons. Provided the quantum efficiency of the laser is sufficiently high, a significant reduction in the photon noise below the Poisson value can be achieved.

Optical parametric amplification offers another possibility. An incident 'pump' photon beam on a suitable non-linear medium yields two *correlated* signal photons in large numbers when phase-matching conditions are satisfied. Thus in the radiation field, created by the photons arising from the parametric process, the noise in the phase of the field will be less than that for a collection of uncorrelated photons. Since a communications system may use *either* amplitude *or* phase modulation, a light source capable of displaying phase noise at below the Poisson level will provide a more superior communications system than that for 'ordinary' light. Light in which correlation between photons is introduced is known as 'squeezed' light.

24.11   Interferometry

We have seen (Chapter 17) that the accuracy of spatial interferometry depends on the accuracy with which phase differences can be measured as well as on the order of interference. Given a line that is narrow enough (from a stabilized laser), a path difference of order $10^{-9}\lambda$ can be obtained and a phase difference of 10^{-6} radians can be measured so that, in principle, an accuracy of $10^{-15}\lambda$ (equivalent to 1 Hz) can be obtained, although so far experiments have only yielded a considerably lower accuracy. In temporal interferometry eqn (24.5) implies a limiting accuracy of 10^{-1} Hz for an observation lasting 1 s and 10^{-2} Hz for one lasting 10 s. We have seen in Section 20.12 that about 1 Hz is the limiting accuracy obtained with stabilised lasers.

24.12   Storage of optical information

Optical information may be stored (1) as a photograph and (2), after conversion to a pulse-code form, in a computer. The amount that can be stored photographically is very large (Section 24.3) but there is no provision for deleting items and replacing them by others, as is required in some computer operations. Efforts to develop a 'reversible' photographic process have been made but so far complete success has not been obtained.

24.13   Limitations in laser optics

The special properties of laser systems are subject to limits, both theoretical and practical. We shall now consider some of these limitations.

24.13.1   Divergence of laser beam

More than 80% of the energy of a laser operating in the TEM_0 mode lies within a cone of semi-angle approximately λ/d, where d is the diameter of the mode. This limit is set by diffraction and practical devices approach this limit very closely. A helium–neon laser with $d = 0.5$ mm would have a semi-angle of 1.3×10^{-3} radian. Expanding the beam with a pair of lenses to 50 mm yields a beam of divergence 1.3×10^{-5} radian. If such a beam is used to control the direction of a tunnel, the width of the beam at 20 km from the laser would be about 26 cm. The illuminance will have fallen by a factor of 10^{-4}. The light would, with a moderately powerful laser, still be bright enough for the centre of the spot to be located to within a few millimetres.

24.13.2 Monochromaticity

There is probably a theoretical limit to the accuracy with which the frequency of a laser can be held constant by any one method of stabilisation, but no reason why more accurate methods of stabilisation should not be achieved. The present limit is about 1 Hz in 10^{15}/Hz.

24.13.3 Shortness of pulse

Pulses as short as 6×10^{-15} s have been produced. Since the period of the light wave in question is about 10^{-15} s, it is a moot point whether pulses shorter than 2×10^{-5} s could be regarded as light.

24.13.4 Power in a continuous beam

The limit here is determined by the ability of the laser material to withstand the heat associated with generating the laser beam. If the maximum power that can be applied is P_{max}, the maximum laser power will be εP_{max}, where ε is the efficiency of laser power generation. This is generally less than 1%. Outputs in excess of a kilowatt have been obtained with the Nd–YAG solid state laser. Several kilowatts have been obtained with a CO_2 laser in which the gas is blown transversely across the path of the beam.

24.13.5 Energy in laser pulses

A pulsed laser has five important parameters connected with the energy. These are (1) the total energy (W) in one pulse, (2) the peak power (P), (3) the duration of the pulse (T), (4) the repetition rate (R) (i.e. the number of pulses per second) and (5) the average power over a period long enough to contain many pulses (P_a). The parameters W and P are ultimately limited by the number of atoms that can be excited. In a gas laser it may be possible to excite all the atoms or molecules but at the moderately low pressures normally required the number of atoms per unit

volume is small compared with the number for a solid. However, in ruby and Nd–YAG lasers only the 'impurity' atoms emit laser radiation and these are a small fraction of the atoms of the solid. In practice there is a limit to the energy that can be pumped into a solid without disintegrating it through the cracks due to uneven heating or melting it. There is one limit if the repetition rate is so low that the solid can cool between pulses and another if the pulses follow so rapidly that P_a rather than P is the relevant parameter.

24.14 High-power systems

The most powerful lasers are designed for experiments on 'inertial confinement'. In this device, pellets (about 5 mm in diameter) of a mixture of solid deuterium and tritium are bombarded by a set of pulsed laser beams from different directions. The object is to use the momentum of the photons to compress the material to a density of 10^3 gm cm^{-3} and to produce a very high temperature (in excess of 10^8 degrees) causing a small thermonuclear explosion by a fusion reaction. Lasers for these experiments have so far produced energies of tens of kilojoules at powers of tens of terawatts. The efficiency of converting the laser energy into compressed plasma is of the order of 10–20%. When fusion is achieved then at least five times as much energy must be supplied to initiate the explosion. If it occurs the explosion will release several hundred times as much energy. This device is estimated to cost about $\$5 \times 10^8$ and probably represents the limit of techniques at present available. The lasers used in industry for cutting, welding, etc., are usually about 10^4 less in power (and cost).

24.15 The future

We conclude this book by asking what has the future in store for optics? Classical optics (what was known before 1950) is so near the fundamental limits (Section 24.7) that only marginal advances can be expected. In laser optics, non-linear optics etc., the field is wide open for further developments. In spectroscopy, laser methods and also the development of very powerful sources (e.g. storage rings filled by synchrotrons) will maintain the momentum of present advance for many years.

On the practical side, the optical invasion of the field of communications and of computer hardware is likely to continue at an increasing rate. Optical methods have advantages in speed and robustness. At present they are too costly and too large.

Optics will undoubtedly contribute much to astronomy by the development of new instruments. It is also possible that astronomers may discover new properties of light by studying the emission of light from very powerful sources and its transmission over long distances.

In 1950 it seemed that optics was approaching an end. The laser was not foreseen. It may be that there is still to come another 'break through' which we cannot now predict [3].

Appendix 24.A The sampling theorem

Since the frequency content of any real signal $f(t)$ will be restricted to values not exceeding Ω_L, the Fourier transform $g(\Omega)$ of $f(t)$ will be zero for $\Omega > \Omega_L$. Over the range $0 < \Omega \leqslant \Omega_L$, $g(\Omega)$ may be represented by a Fourier series

$$g(\Omega) = \sum_{n=-\infty}^{\infty} c_n \exp\left(\frac{i\pi n\Omega}{\Omega_L}\right) \quad (24.8)$$

where $c_0, c_1, c_2,\ldots$ are constants. From Fourier's theorem

$$f(t) = \frac{1}{2\pi} \int_{-\infty}^{\infty} g(\Omega)\, e^{-i\Omega t}\, d\Omega \quad (24.9)$$

Substituting for $g(\Omega)$ and noting that since $g(\Omega)$ is negligible for values of $\Omega > \Omega_L$, we have

$$f(t) = \sum_{n=-\infty}^{\infty} \frac{1}{2\pi} \int_{-\Omega_L}^{\Omega_L} \exp\,[i\left(\frac{n\pi}{\Omega_L} - t\right)\Omega]\, d\Omega \quad (24.10)$$

$$= \frac{1}{2\pi} \sum_{n=-\infty}^{\infty} c_n \frac{\sin(\Omega_L t - n\pi)}{\Omega_L t - n\pi} \quad (24.11)$$

Each term in this series has the form shown in Fig. 24.1, which is the case for $n = 0$. Maxima of the function occur where $t_n = n\pi/\Omega_L$ for the nth curve. Thus a measure-

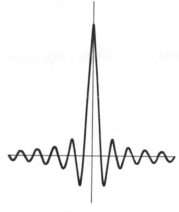

Figure 24.1

ment of $f(t)$ at each t_n gives c_n, so that n measurements made at intervals spaced by $1/\pi\Omega_L$ provide sufficient information completely to characterise the function $f(t)$. If the duration over which the function is recorded is T, then the number of sampling points required to determine the curve is $T/(1/\pi\Omega_L) = \pi\Omega_L T$.

References

1. C. E. Shannon and W. Weaver, *A Mathematical Theory of Communication*, University of Illinois Press, 1949.
2. R. G. Gallagher, *Information Theory and Reliable Communication*, John Wiley & Sons, 1968.
3. R. Loudon and P. L. Knight 'Special issue on squeezed light', *J. Mod. Opt.*, **34**, 709–1020 (1987).

Problems

24.1 If before a measurement the probability that a quantity lying in the range $x - \frac{1}{2}dx$ and $x + \frac{1}{2}dx$ is $p_0(x)\, dx$ and if after the measurement the probability is $p_1(x)\, dx$, calculate the increase in information provided by the measurement.

24.2 If the probabilities in Problem 24.1 arise from the presence of random errors in the measurements, sketch the forms of the functions $p_0(x)$ and $p_1(x)$.

24.3 A television-type screen has an array of phosphor dots 400×500 and produces pictures at the rate of 25 per second. If the intensity of each spot can be set at one of eight different levels, at what rate must information be supplied? The frequency of a typical television carrier wave is $3-4 \times 10^6$ Hz. Comment.

24.4   Design a system for use in tunnelling that would produce a reduction in spot size at the end of the tunnel by a factor of ten compared with that described in Section 24.13.1.

Answers, Hints and Solutions

Problems 2

2.1 Consider rays between points in media n_1, n_2 via points p, p' on the surface where the angle of incidence via p is θ. Show from geometry that, to first order, the difference in path is pp' $(n_1 \sin \theta_1 - n_2 \sin \theta_2)$.

2.2 Similar procedure to that for 2.1.

2.3 Using eqn (2.14),

$$n'(\sin \theta_1 + \sin \theta_4) = n(\sin \theta_2 + \sin \theta_3)$$

$$n'[\sin \tfrac{1}{2}(\theta_1 + \theta_4)\cos \tfrac{1}{2}(\theta_1 - \theta_4)]$$

$$= n \sin \tfrac{1}{2}(\theta_2 + \theta_3)\cos \tfrac{1}{2}(\theta_2 - \theta_3) \quad (2.15)$$

and, using eqn (2.7),

$$\sin \tfrac{1}{2}(\theta_1 + \theta_4) = \frac{n}{n'} Q \sin \frac{\alpha}{2} \quad (2.16)$$

where

$$Q = \frac{\cos \tfrac{1}{2}(\theta_2 - \theta_3)}{\cos \tfrac{1}{2}(\theta_1 - \theta_4)} \quad (2.17)$$

If $\theta_1 > \theta_4$ then $(\theta_1 - \theta_4) > (\theta_2 - \theta_3)$ because deviation at a single surface increases when the angle of incidence increases and $Q > 1$.

If $\theta_1 < \theta_4$ then $(\theta_1 - \theta_4) < (\theta_2 - \theta_3)$; but both these differences are negative so that the magnitude $(\theta_1 - \theta_4)$ is greater than $(\theta_2 - \theta_3)$ and Q is again greater than 1.

When $\theta_1 = \theta_4$ and $\theta_2 = \theta_3$ we have $Q = 1$ and this is the only minimum value for Q.

2.4 55.3.

2.5 $37°$.

2.6 Limits are 1.5263 to 1.5267, a range of $\pm 0.014\%$. Refractive index of air at STP $= 1.00029$, differing from unity by $\sim 0.03\%$. Correction necessary.

2.7 Numerical aperture $= 0.156$, corresponding to an entrance half-angle of $8.97°$. For beam of radius 0.2 mm, converging lens of focal length ~ 1.3 mm is needed.

Problems 3

3.1 Image 600 mm from surface on side of incidence.

3.4 89 mm; 1.81.

3.5 78.9 mm; 0.63 mm; $24.8°$.

3.6 $+300$ mm.

3.9 83 mm.

3.11 2.

Problems 4

4.1 $7.409 \times 10^{14} \text{ s}^{-1}$, $\quad 4.655 \times 10^{14} \text{ s}^{-1}$; $5.490 \times 10^{14} \text{ s}^{-1}$, $3.449 \times 10^{14} \text{ s}^{-1}$.

4.2 546 nm.

4.3 589 ± 1 nm.

4.4 4.21 mm.

4.5   4 mm.

4.6   Principal maxima occur when $\delta/2 = m\pi$, at which value $\sin(N\delta/2)$ is zero and the ratio $\sin(N\delta/2)\sin(\delta/2)$ tends to N. Adjacent zeros are at $N = \pm\pi$, separated by $4\pi/N$. Hence the ratio of line width to separation is $(2\pi/N)\pi = 2/N$ or 5×10^{-5}.

4.7   The order $m \sim 2 \times 5 \times 10^{-3}/500 \times 10^{-9}$ $= 20\,000$. Calculate wavelengths corresponding to neighbouring orders $+0.7$ at the centre.

Order	Wavelength (nm)
20 000.7	499.998
19 999.7	500.007 ± 0.001
19 998.7	500.030

The middle figure strictly is the only one falling within the stated range, but the closeness of the first figure would suggest that more measurements would be desirable, perhaps with a different etalon.

4.8   For the given limits on the separation, the order at the centre for $\lambda = 404.656$ nm lies between 49 419 and 49 429. f is known to be 0.41. Calculate thicknesses for orders 49 419.41 to 49 429.41 and from these the orders corresponding to the other two wavelengths. Look for a value that gives fractional orders 0.85 ± 0.05 and 0.19 ± 0.05.
Ans. 9.999 74 ± 0.000 02 mm.

4.9   The argument ignores multiple reflections. Oddly enough, when these are taken into account, the result still holds.

4.10   Note that if r_1 is the amplitude reflection coefficient for air/film, then that for film/air is $-r_1$ (eqn 4.39). Let the transmitted amplitude at the air/film interface be t. Then the emerging amplitudes in the reflected beam are given by r_1, $r_2 t_1^2$, $-r_1 r_2^2 t_1^2$, $r_1^2 r_2^3 t_1^2$, ..., with phase differences 0, 2δ, 4δ, 6δ, ... The terms form a geometric progression.

4.11   Minimum optical thickness = $546/2$ nm: first order for 546 nm. Will be second order for 273 nm. No peak at longer wavelength. For thicker spacer, optical thickness = 546 nm—

second order. This is first order for 1092 nm and third order for 364 nm. Not exact because of variation of refractive index of spacer with wavelength.

Problems 5

5.3   The pulse duration corresponds to three cycles.

5.4 $$U = \frac{\lambda^2(A\lambda^2 - B)}{(A\lambda^2 + B)^2}$$

5.5 2.039×10^8 ms; 0.659×10^8 ms.

5.6 $E_0(1 + \sqrt{2})$, $\pi/2$; $E_0^2(3 + 2\sqrt{2})$.

5.7 $4E_0^2/5$.

5.8   0.55.

Problems 6

6.3   12.7 mm.

6.4   0.025 mm.

6.5   From the structure factor of eqn (6.22), the width of the first-order line is $\Delta\lambda/\lambda \sim 1/24\,000 \sim 4 \times 10^{-5}$. From this calculate Doppler width, $\Delta\lambda/\lambda \sim 7 \times 10^{-7}$.

6.6   Calculate separation between adjacent zones using (6.25). If 0.002 mm is barely resolved, there is no point in drawing circles where the separation on the plate is less than a few times this limit—say 0.005 mm.

$$\rho_{101} - \rho_{100} = 0.0050 \text{ mm}.$$

Thus there is no point in drawing more than ~ 50 zones.

Problems 7

7.1 20×83 mm.

7.2   (a) Plane-polarised: $\mathbf{E}$ vector at $\theta = \arctan(E_1/E_0)$ to x direction.
      (b) RH circularly polarised.
      (c) LH elliptically-polarised: axes $\parallel Ox$, Oy.

7.4   When analyser aligned, intensities on each side $\alpha \sin 2.5°$. Rotation through α produces intensities $\sin^2(2.5 + \alpha)$ and $\sin^2(2.5 - \alpha)$. If $\alpha = 0.02°$, percentage difference ≈ 3.

7.6   (a) Nothing.
      (b) Fringes change in visibility; for four orientations, they disappear.
7.7   From eqns (7.9) and (7.11b), bright bands occur when $\delta = (2N+1)\pi$. Wavelengths λ_k where

$$\frac{1}{\lambda_k} = \frac{2(N+k)+1}{2e\,\Delta n}$$

Plot $1/\lambda_k$ versus k: slope gives $1/e\,\Delta n$. (0.035: assumes no variation with wavelength.)

7.8 $0.1258 \pm 0.0002 \text{ g ml}^{-1}$.

Problems 8

8.1 194 V m^{-1}; 0.515 A m^{-1}.
8.2   Power dP in annulus $r \to r + dr$ is given by

$$dP = 2\pi r \,|S(r)|\,dr$$

where

$$|S(r)| = |E(r) \times H(r)|$$
$$= |E(r)|^2/377$$
$$= E_0^2\, e^{-2r^2/r_0^2}/377$$

Total power

$$P = \int_0^\infty \frac{E_0^2\, e^{-2r^2/r_0^2}/2\pi r \ dr}{377}$$
$$= \frac{\pi r_0^2 E_0^2}{2 \times 377}$$
$$= 5 \times 10^{-3}$$

Whence

$$E_0 = 3650 \text{ V m}^{-1}.$$

8.3   Reflected power $= \dfrac{(1-n)^2}{(1+n)^2}$

Transmitted power $= \dfrac{2n}{(1+n)^2}$

The two add to unity. If the plate thickness is of the order of the coherence length of the light, interference effects would occur.

8.4   0.997.
8.5   400 nm; 257 nm.
8.6   If $E = E_0 \cos \omega t$ then $H = H_0 \sin \omega t$. Sign of Poynting vector reverses at intervals of a quarter period. Hence time-average value of power flow is zero.

8.7 $53°$; $127°$.

Problems 9

9.1   (a) 3.4×10^{-12}; (b) 4.4×10^{-7}.
9.2   11.8 nm.

9.3   Amplitude reflectance

$$= \frac{(n - ik - 1)}{(n - ik + 1)}$$
$$= \frac{(n^2 + k^2 - 1) - 2ik}{(n+1)^2 + k^2}$$
$$= |R|\, e^{i\phi}$$
$$\phi = \text{arc tan}\left(\frac{-2k}{n^2 + k^2 - 1}\right)$$
$$= -29.7° \text{ or } -0.165\pi.$$

9.4   Equation (9.21) gives
      $1 - R = 1.22 \times 10^{-4}$.
9.5 $149°C$.

Problems 10

10.1 At STP, 2.7×10^{25} molecules m^{-3}. In volume λ, $n = 3.3 \times 10^6$.
      $N^{-1/2} \sim 5 \times 10^{-4} \ll 1$.
      Atmospheric scale height $= 8$ km. Pressure at 100 km $\sim \exp(-100/8) \sim 3.7 \times 10^{-6}$, whence $n \sim 12$. Large fluctuations are expected.
10.2 2.4×10^7.
10.3 Acoustic wave acts as grating of spacing λ_s, the acoustic wavelength. Hence by Bragg's equation

$$2\lambda_s \sin \theta/2 = \lambda_i$$

From eqn (10.10),

$$\omega_i'' - \omega_i = \pm\, \omega_i \frac{n v_s}{c} \frac{\lambda_i}{\lambda_s}$$
$$= \pm\, \omega_s$$

since

$$\frac{v_s}{\lambda_s} = \frac{\omega_s}{2\pi} \quad \text{and} \quad \omega_i \lambda_i = 2\pi c/n$$

Problems 11

11.1 $\arctan\left[\dfrac{(n_e^2 - n_i^2 \sin^2\theta_i \, \cos^2\theta_i)^{1/2}}{n_i \sin\theta_i \, \cos^2\theta_i}\right]$

11.2 $\arctan\left(\dfrac{n_y^2 - n_x^2}{n_x^2 - n_z^2}\right)$

11.3 18.9 kV.

11.4 337 volts.

11.5 1.04 tesla.

Problems 12

12.1 Gives $\omega = 2.23 \times 10^{15}$ s which corresponds to $\lambda = 857$ nm.
Maximum of solar emission is at ~ 555 nm.
Maximum of $\rho(\omega)$ versus ω does *not* occur at the frequency corresponding to the maximum of the curve of $\rho(\lambda)$ versus λ. (See footnote to Section 12.5.)

12.2 3.9 eV.

12.3 2.95 eV.

12.4 20.8 μm.

12.5 6.32×10^7.

Problems 13

13.1 0.25 mm.

13.2 Change of $\lambda/2$ between object and reference waves causes a maximum to become a minimum. Hence there must be negligible vibration over the exposure time.

Problems 14

14.1 A phase difference of $\pi/2$ with respect to the E component.

14.2 $79.9°$.

14.3 7.49π.

14.4 $21.0°$.

14.6 That the fibre is straight.

14.7 27 ns.

14.8 (a) 60 km; (b) 24 km.

Problems 15

15.1 8×10^9 V m^{-1}.

15.3 319 K. This assumes that dn_o/dT and dn_e/dT are independent of wavelength: they probably are not.

15.4 Frequency of back-scattered wave is shifted. For example indicated it is $\sim 2 \times 10^{10}$ s—small; cf. light frequency of 3×10^{15} s.

15.5 Note $I = c\varepsilon_0 E_{rms}^2 = \frac{1}{2}c\varepsilon_0 E_0^2$ (1.98×10^7 W m^{-2}).

Problems 16

16.1 $V(633\text{ nm}) = 0.252$.
Flux $= 0.252 \times 583 \times 10^{-3}$
$= 0.1471$ m

16.2 Area of beam $= 1.59 \times 10^{-7}$ m.
Solid angle $= (2 \times 10^{-3})^2/4 = 10^{-6}$ sr.

	Radiant	Luminous
Exitance	6290 W m^{-2}	9.24×10^5 lm m^{-2}
Intensity	10^3 W sr^{-1}	1.47×10^5 lm sr^{-1}

16.3 Area of beam $= 7.07 \times 10^{-6}$ m^2.
In 1 second, 4 joules flow into cylinder of length c and cross-section 7.07×10^{-6} m. Hence radiant energy density

$$= \frac{4}{7.07 \times 10^{-6} \, c}$$
$$= 1.89 \times 10^{-3} \text{ J m}^{-3}$$

16.4 Diameter of patch on Moon $= 3.8 \times 10^5 \times 10^{-4}$ km.
Area $= 1.13 \times 10^9$ m^2.
Irradiance $= 8.82 \times 10^{-9}$ W m^{-2}.
Energy of 1 photon $= 4.07 \times 10^{-19}$ J.
Number of photons m^{-2} s $= 2.17 \times 10^{10}$.
Number of photons on 20 cm^2 detector $= 4.34 \times 10^7$.

16.5 In 0.1 s, 4.34×10^6 photons received. *If* noise is photon shot noise only, then

$$\text{SNR} = 4.34 \times 10^6 / (4.34 \times 10^6)^{1/2}$$
$$= 2.1 \times 10^3$$

16.6 For visible region, $h\nu \sim 2$ eV. At room temperature, kT 1/40 eV. Hence second term of eqn (16.4b) is $e^{-80} \sim 10^{-35}$.
This is rather small, cf. unity.

16.7 $0.30 \times (3.5)^{12} = 1.01 \times 10^6$.

16.8 If 10^4 cones occupy 25 mm, average separation ~ 0.05 mm. From Chapter 3, calculate lens–retina distance

(20 mm). Adjacent cones subtend angle $0.05/20 = 2.5 \times 10^{-3}$ radian. Eye should not be able to do better than this, but it *does*. Calculations assume *static* eye. In reality, eye makes constant 'jiggling' movements so that many receptors scan the retinal image. The brain then does a spot of processing.

Problems 17

17.1 $\pm 3 \times 10^{-5}$ of fringe width.
17.2 Most of the time atmospheric pressure keeps within $\pm 0.2\%$ of the average value. $(n-1) \alpha$ pressure and $n = 1.000\,29$. Thus neglect of pressure effects would introduce an error of $\sim \pm 6 \times 10^{-7}$. One micron in 1 metre is 1 in 10^6. No need to read the barometer.
17.3 Refractivity of 50% mixture is $\frac{1}{2}(3.5 + 28) \times 10^{-5} = 15.7 \times 10^{-5}$. Thus need to cover a range 28×10^{-5} (pure Ar) to 15.7×10^{-5}. Range of optical path is $L(28 - 15.7) \times 10^{-5} = 200\lambda$, since compensator range is 200 fringes. Hence $L = 0.81$ m. Accuracy depends on detection method. Visually, with ~ 0.1 fringe detection, would give $\sim 0.2\%$ of Ar in mixture. A phase-sensitive detection system could be 3–4 orders of magnitude better than this.
17.4 $L(t) =$

$L[1 + (0.47t + 0.000\,52t^2) \times 10^{-6}]$.
17.5 Maximum reflection if the beams reflected from opposite faces are in phase, i.e. if $2nd \cos \theta = \lambda$ where θ is the angle of refraction. $1.51 \sin \theta = \sin 45°$, so $\theta = 27.9°$. Hence $d = 3.75\ \mu$m.
From Fresnel equations, amplitude reflectances are $\tan(27.9° - 45°)/\tan(27.9° + 45°) = -0.095$ for p component and $\sin(27.9° - 45°)/\sin(27.9° + 45°) = -0.308$.
Mean reflectance $= -0.201$.
Ignoring multiple reflections, maximum occurs when waves from opposite sides in phase.

Resultant intensity $= (2 \times 0.201)^2$
$= 0.16$

After reflection at first mirror, beam makes transmission through beamsplitter: emerging intensity $= 0.16 \times 0.84 = 0.13_6$ – amplitude 0.37. Beam from other mirror will have same intensity and amplitude.

Problems 18

18.1 Exit pupil 22 mm from eyepiece—slightly larger than ideal. Diameter 2.5 mm—good match for eye in daylight.
18.2 $2.1°$.
18.3 61%.
18.4 If you start with a beam in an arbitrary direction on the face of the prism, you will already have algebraic mayhem by the time you establish the direction of the refracted ray. No need for this. You merely need to establish that the ray after refraction (*inside* the prism) returns parallel to itself. This is trivial. At each total reflection one of the three direction cosines changes sign. Thus if the incident direction is $\alpha\mathbf{i} + \beta\mathbf{j} + \gamma\mathbf{k}$ and reflection is successively at the planes $x = 0, y = 0, z = 0$, then the beam directions are $-\alpha\mathbf{i} + \beta\mathbf{j} + \gamma\mathbf{k}$, $-\alpha\mathbf{i} - \beta\mathbf{j} + \gamma\mathbf{k}$ and $-\alpha\mathbf{i} - \beta\mathbf{j} - \gamma\mathbf{k}$.
18.8 37 mm.
18.9 Seven-mirror: 77%; nineteen mirror: 76%.

Problems 19

19.4 (a) 3.0", (b) 0.7", (c) 0.056", (d) 0.023".
19.5 The angular diameter of the Airy disc is twice the angle α given by eqn (16.5) so the required limit of resolution is 0.025" and a mirror of diameter 5.5 m is needed.
19.7 1400 lines mm^{-1}. $R = 1.4 \times 10^5$.
19.8 140 lines mm^{-1}. $R = 1.4 \times 10^5$.
19.9 7×10^4.

Problems 20

20.1   Time for N transits $= Nl/v$. After each transit, power falls by factor R. After N transits $\rightarrow R^N$. If time constant is τ_c then after time τ_c, power has fallen to $1/e$ of initial value.

$$R^N = e^{-1}$$
$$N \ln R = -1$$

If $R \sim 1$, we can write $R = 1 - \varepsilon$, $\varepsilon \ll 1$ and $\ln(1 - \varepsilon) \doteq -\varepsilon$.
Hence $N = 1/(1 - R)$, so

$$\tau_c = \frac{Nl}{v} = \frac{l}{v(1 - R)}$$

20.2   Assuming Doppler broadening, line width

$$\Delta \nu = \nu \sqrt{\frac{kT}{Mc^2}}$$
$$= 8.1 \times 10^8 \text{ s}^{-1}$$

Cavity mode separation $= c/2l$
$$= 3.75 \times 10^8 \text{ s}$$
There will therefore always be a mode within 1.87×10^8 s of the centre.
There will be five modes if one is within 0.6×10^8 s of the line centre: otherwise four.

20.3   Minimum value of $\theta = \arctan(5\sqrt{2}/200)$.
$\cos \theta = 0.999\,376$.

20.4   The next term in the binomial expansion for r is

$$-\frac{1}{8l^4}\left[(x - x')^4\right.$$
$$\left. + 2(x - x')^2(y - y')^2 + (y - y')^4\right]$$

Maximum value of $x - x'$, $y - y'$ is 5. Hence this term is -1.9×10^{-7}—three orders of magnitude smaller than the terms of eqn (20.13).

20.5   Note that the products of the segments of perpendicular chords of a circle are equal.

20.6   The appropriate Hermite polynomials are: $H_0 = 1$, $H_1(x) = 2x$, $H_2(x) = 4x^2 - 2$, $H_3(x) = 8x^3 - 12x$.

20.7   (a) 0.246 mm; (b) 1.00 mm.

20.8   Inversion requires $A_{21} > A_{32}$. 90% condition requires $A_{32} = 9A_{31}$.

20.9   See Fig. 12.11. Unless there are many wall collisions the atoms cannot return to the ground state from the first excited levels. They are then kicked up to the next level by electron collisions and so destroy the population inversion.

20.10   Cavity   mode   separation $= c/2l = 1.5 \times 10^{11}$ s^{-1}.
This is very large, cf. that of the He–Ne cavity. If the spontaneous line width of the semiconductor radiation were of the order of 10^9 s^{-1}, the cavity would need tuning. The fact that no tuning is necessary indicates that the semiconductor line width is $\geqslant$ that of the He–Ne system.

Problems 22

22.1   Among the points to be considered (although some may not be limiting) are: How large is the final image of the slit? What determines its size? What speed of rotation is involved? Is it necessary to correct for varying barometric pressure? Are the flatnesses of the mirrors important? You can think of others.

22.2   With no field, the directions of the nitrobenzene molecules are random. With the field on, they are aligned. They have to rotate against viscous forces. This takes time (e.g. for nitrobenzene about 32 ps; carbon disulphide is even better—2 ps).

22.3   Time   difference   (eqn   22.4) $= 2.3 \times 10^{-16}$ s.
One   period   of   light   wave $= 1.8 \times 10^{-15}$ s $\equiv 1$ fringe.
Hence 0.13 fringe. Shift for one rotation $= 0.26$ fringe.

22.6   Diffraction-limited   resolution   of   a 1 m telescope ($\lambda = 500$ nm) is $\sim 0.1''$. However,   the   atmosphere   will degrade this, unless the telescope is in orbit. Muse on problem of keeping the system rigid over a period of 6 months and having good seeing on the nights in question.

22.7   Find $\beta \sim (1 - 1.1 \times 10^{-5})$:   need   to measure line positions to $\sim 10^{-7}\lambda$, so resolving power of 10^7 required.

22.8   13 Hz.

22.9 Free electron assumption,
$\Delta\lambda = 0.0024$ nm.
Bound to atom $\rightarrow \Delta\lambda = 1.3 \times 10^{-6}$ nm.
Resolving powers: (a) 500; (b) 10^7.
No. Although resolving powers in excess of 10^7 can be achieved in the optical region, the techniques required (e.g. Fabry–Perot etalons) are not available for the X-ray region.

Problems 23

23.1 Doppler line width $\Delta\omega_D = \omega(kT/M)^{1/2}/c \sim 4.7 \times 10^9$ s^{-1}. Mode separation $= c\pi/L = 1.9 \times 10^9$ s^{-1}. 2–3 modes.

23.3 Order 10^{-11}.

Problems 24

24.1 $$\Delta H = \int_{-\infty}^{\infty} p_0(x)\ln p_0(x)\ \mathrm{d}x$$
$$- \int_{-\infty}^{\infty} p_1(x)\ln p_1(x)\ \mathrm{d}x$$

24.2

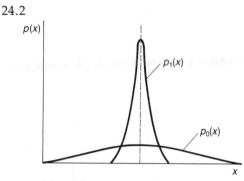

Note that
$$\int_{-\infty}^{\infty} p_0(x)\ \mathrm{d}x = \int_{-\infty}^{\infty} p_1(x)\ \mathrm{d}x = 1$$

24.3 On the face of it, $400 \times 500 \times 25 \times 8 = 40 \times 10^6$ s^{-1}. This supposes that the information is supplied serially. If, for example, the eight-level information were supplied along eight wires, a smaller bandwidth would suffice.

24.4 The system of Section 24.13.1 expands the beam diameter by a factor of 100. This could, for example, be done with a 4 mm focal length microscope objective followed by a lens with $f = 400$ mm and not less than 50 mm in diameter. For a thousandfold expansion, a lens becomes rather large and cumbersome and a parabolic mirror would be better. This would be 0.5 m in diameter and with focal length of 4 m.

Index